Springer Tracts in Electrical and Electronics Engineering

Series Editors

Springer Tracts in Electrical and Electronics Engineering (STEEE) publishes the latest developments in Electrical and Electronics Engineering - quickly, informally and with high quality. The intent is to cover all the main branches of electrical and electronics engineering, both theoretical and applied, including:

- Signal, Speech and Image Processing
- Speech and Audio Processing
- Image Processing
- Human-Machine Interfaces
- Digital and Analog Signal Processing
- Microwaves, RF Engineering and Optical Communications
- Electronics and Microelectronics, Instrumentation
- Electronic Circuits and Systems
- Embedded Systems
- Electronics Design and Verification
- Cyber-Physical Systems
- Electrical Power Engineering
- Power Electronics
- Photovoltaics
- Energy Grids and Networks
- Electrical Machines
- Control, Robotics, Automation
- Robotic Engineering
- Mechatronics
- Control and Systems Theory
- Automation
- Communications Engineering, Networks
- Wireless and Mobile Communication
- Internet of Things
- Computer Networks

Within the scope of the series are monographs, professional books or graduate textbooks, edited volumes as well as outstanding PhD theses and books purposely devoted to support education in electrical and electronics engineering at graduate and post-graduate levels.

Review Process

The proposal for each volume is reviewed by the main editor and/or the advisory board. The books of this series are reviewed in a single blind peer review process.

Ethics Statement for this series can be found in the Springer standard guidelines here https://www.springer.com/us/authors-editors/journal-author/journal-author-helpdesk/before-you-start/before-you-start/1330#c14214

Springer Tracts in Electrical and Electronics Engineering

Springer Tracts in Electrical and Electronics Engineering (STEEE) publishes the latest developments in Electrical and Electronics Engineering - quickly, informally and with high quality. The intent is to cover all the main branches of electrical and electronics engineering, both theoretical and applied, including:

- Signal, Speech and Image Processing
- Speech and Audio Processing
- Image Processing
- Human-Machine Interfaces
- Digital and Analog Signal Processing
- Microwaves, RF Engineering and Optical Communications
- Electronics and Microelectronics, Instrumentation
- Electronic Circuits and Systems
- Embedded Systems
- Electronics Design and Verification
- Cyber-Physical Systems
- Electrical Power Engineering
- Power Electronics
- Photovoltaics
- Energy Grids and Networks
- Electrical Machines
- Control, Robotics, Automation
- Robotic Engineering
- Mechatronics
- Control and Systems Theory
- Automation
- Communications Engineering, Networks
- Wireless and Mobile Communication
- Internet of Things
- Computer Networks

Within the scope of the series are monographs, professional books or graduate textbooks, edited volumes as well as outstanding PhD theses and books purposely devoted to support education in electrical and electronics engineering at graduate and post-graduate levels.

Review Process

The proposal for each volume is reviewed by the main editor and/or the advisory board. The books of this series are reviewed in a single blind peer review process.

Ethics Statement for this series can be found in the Springer standard guidelines here https://www.springer.com/us/authors-editors/journal-author/journal-author-helpdesk/before-you-start/before-you-start/1330#c14214

D. Nirmal · J. Ajayan
Editors

Modeling of AlGaN/GaN High Electron Mobility Transistors

Springer

Editors
D. Nirmal
Department of Electronics
and Communication Engineering
Karunya Institute of Technology
and Sciences
Coimbatore, Tamil Nadu, India

J. Ajayan
Department of Electronics
and Communication Engineering
SR University, Warangal
Hasanparthy, Telangana, India

ISSN 2731-4200 ISSN 2731-4219 (electronic)
Springer Tracts in Electrical and Electronics Engineering
ISBN 978-981-97-7505-7 ISBN 978-981-97-7506-4 (eBook)
https://doi.org/10.1007/978-981-97-7506-4

This Springer imprint is published by the registered company Springer Nature Singapore Pte Ltd.
The registered company address is: 152 Beach Road, #21-01/04 Gateway East, Singapore 189721, Singapore

Preface

Owing to wide band gap, outstanding power handling capability, extremely high on and off state breakdown voltages, high sheet charge density in the channel, high saturation velocity of electrons, high electron mobility and high temperature operation, GaN HEMT is the most promising transistor technology for RF power applications such as 5G communications, space and defense applications. Accurate small signal models are required to predict the RF power performance of RF electronic circuits. Therefore, small signal models of GaN HEMTs have been included in this book. GaN HEMTs are widely used in power amplifier designs. Precise large signal models are very much essential for power amplifier design and circuit optimization. Therefore, this book also focuses on accurate large signal modeling of GaN HEMTs. Accurate and compact physical models were also included in this book which will help the RF circuit designers to design and optimize GaN HEMT-based power amplifiers and integrated circuits. It is also important to consider the effects of self-heating for the accurate prediction of RF power performance of GaN HEMTs. Hence, this book also deals with thermal resistance modeling of GaN HEMTs. Gate leakage current is also a serious concern in high power GaN HEMTs. Charge-based compact models and surface potential-based models were included in this book to study the impact of gate leakage current on the RF power performance of GaN HEMTs. In order to model I–V and C–V characteristics of GaN HEMTs, physics-based compact models are discussed in detail. This book also deals with analytical modeling of intrinsic charges and surface potential of GaN HEMTs, physical modeling of charge trapping, neural network-based GaN HEMT models, numerical-based GaN HEMT models, modeling of short channel effects in GaN HEMTs, modeling of parasitic capacitances and resistances, modeling of current collapse and kink effects in AlGaN HEMTs, etc.

Coimbatore, India — D. Nirmal
Hasanparthy, India — J. Ajayan

Contents

Editors and Contributors

About the Editors

Dr. D. Nirmal is Full Professor in Electronics and Communication Engineering and Associate Dean Engineering and Technology, Karunya Institute of Technology and Sciences, India. He specialized in VLSI Design after his Bachelor of Engineering and received his Ph.D. in Information and Communication Engineering from Anna University. His research interests include Nanoelectronics, GaN Technology, Device and Circuit Simulation—GSL, Sensors, HEMT, Beyond 5G, Nanoscale device design and modelling. He is a founding chair of IEEE Electron Device Society Coimbatore chapter and Currently Region 10 IEEE EDS Vice chair and volunteered several committees in IEEE. He has funding project tune of 1.2 Cores from various agencies like DRDO (Defence Research and Development Organization), Ministry of Electronics and Information Technology, ISRO (Indian Space Research Organization) and AICTE (All India Council of Technical Education). He is a recipient of various awards namely IEI-Young Engineer award, IETE Smt. Manorama Rathore Memorial Award 2022 from IETE and Young Scientist Award 2019 from the Academy of Sciences.

Prof. Nirmal has made more than 150+ peer-reviewed research publications and three patents to his credit. He is also Fellow of IETE and Senior IEEE member. He has delivered many Keynote talks, and lectures in National and International Level conferences/Faculty Development programs.

Dr. J. Ajayan received his B.Tech. Degree in Electronics and Communication Engineering from Kerala University in 2009, M.Tech. and Ph.D. Degree in Electronics and Communication Engineering from Karunya University, Coimbatore, India, in 2012 and 2017 respectively. He is a Professor and Dean, Post Graduation Programs at SR University, Telangana, India. He has published more than 150 research articles (including 110 SCI articles) in various journals (IEEE Transactions on Electron Devices, Elsevier, Taylor and Francis, Springer, IOP Science, Wiley) and international conferences. He has published 6 books, more than 25 book chapters, and 3 patents. He is a reviewer of more than 30 journals in various publishers which includes IEEE, IET, Elsevier and Springer journals. He is a Guest Editor of Special Issue on P2P Computing for Beyond 5G Network (B5G) and Internet-of-Everything (IoE) by Peer-to-Peer Networking and Applications, Springer and Special Issue on Energy Harvesting Devices, Circuits and Systems for Internet of Things by Microelectronics Journal, Elsevier. He has served more than 10 conferences as technical advisory/reviewer committee. His areas of interest are microelectronics, semiconductor devices, nanotechnology, RF integrated circuits and photovoltaics.

Contributors

J. Ajayan Department of Electronics and Communication Engineering, SR University, Warangal, Telangana, India

Kalyan Biswas ECE Department, MCKV Institute of Engineering, Howrah, West Bengal, India

Matteo Buffolo Department of Information Engineering (DEI), University of Padova, Padova, Italy

Ashutosh Chakrabarty Centre for Internet of Things, Department of Electronics and Communication Engineering, ITER, Siksha O Anusandhan University, Bhubaneswar, Odisha, India

Gaurav Chatterjee Department of Electronics and Communication Engineering, The LNM Institute of Information Technology, Jaipur, India

Meenakshi Chauhan Department of Electronics and Communication Engineering, The LNM Institute of Information Technology, Jaipur, India

Alessandro Chini Department of Engineering "Enzo Ferrari", Università Di Modena E Reggio Emilia, Modena, Italy

Marcello Cioni Department of Engineering "Enzo Ferrari", Università Di Modena E Reggio Emilia, Modena, Italy;
STMicroelectronics, Catania, Italy

Sanghamitra Das Depatrment of ECE, Silicon Institute of Technology, Bhubaneswar, India

Taraprasanna Dash Department of ECE, Siksha 'O' Anusandhan (Deemed to be University), Bhubaneswar, India

Carlo De Santi Department of Information Engineering (DEI), University of Padova, Padova, Italy

Neha Garg Department of Instrumentation, Shaheed Rajguru College of Applied Sciences for Women, University of Delhi), New Delhi, India

Rachita Ghoshhajra CSE Department, MCKV Institute of Engineering, Howrah, West Bengal, India

Devika Jena Department of ECE, Siksha 'O' Anusandhan (Deemed to be University), Bhubaneswar, India;
School of Electronic Sciences, Odisha University of Technology and Research, Bhubaneswar, India

K. Jena Department of Electronics and Communication Engineering, The LNM Institute of Information Technology, Jaipur, India

Sneha Kabra Department of Instrumentation, Shaheed Rajguru College of Applied Sciences for Women, University of Delhi), New Delhi, India

Abdul Naim Khan Department of Electronics and Communication Engineering, The LNM Institute of Information Technology, Jaipur, India

Gaudenzio Meneghesso Department of Information Engineering (DEI), University of Padova, Padova, Italy

Matteo Meneghini Department of Information Engineering (DEI), University of Padova, Padova, Italy

B. Mounika Department of Electronics and Communication Engineering, SR University, Warangal, Telangana, India

D. Nirmal Department of Electronics and Communication Engineering, Karunya Institute of Technology and Sciences, Coimbatore, Tamilnadu, India

Paolo Pavan Department of Engineering "Enzo Ferrari", Università Di Modena E Reggio Emilia, Modena, Italy

J. Charles Pravin Department of ECE, Kalasalingam Academy of Research and Education, Krishnankoil, Tamil Nadu, India

Francesco Maria Puglisi Department of Engineering "Enzo Ferrari", Università Di Modena E Reggio Emilia, Modena, Italy

S. Routray Department of Electronics and Communication Engineering, SRM Institute of Science and Technology, Chennai, India

V. Sandeep Department of ECE, Mangalore Institute of Technology & Engineering, Mangalore, India

Angsuman Sarkar ECE Department, Kalyani Government Engineering College, Kalyani, West Bengal, India

Arghyadeep Sarkar Department of Electrical and Computer Engineering, McMaster University, Hamilton, Canada

Raghunandan Swain Department of Electronics and Telecommunication, Parala Maharaja Engineering College, Berhampur, Odisha, India

Aruna Tripathy School of Electronic Sciences, Odisha University of Technology and Research, Bhubaneswar, India

Giovanni Verzellesi Department of Sciences and Methods for Engineering (DISMI), EN&TECH Center, University of Modena and Reggio Emilia, Reggio Emilia, Italy

Nicolò Zagni Department of Engineering "Enzo Ferrari", Università Di Modena E Reggio Emilia, Modena, Italy

Enrico Zanoni Department of Information Engineering (DEI), University of Padova, Padova, Italy

Compact Surface Potential-Based AlGaN/GaN HEMT Models

J. Ajayan, D. Nirmal, and B. Mounika

Abstract Modeling of GaN HEMTs is a hot field of research. The convergence and accuracy of circuit simulations heavily rely on the compact models used. For technology optimization and circuit design, a precise physical model of the GaN HEMT is essential. This chapter highlights the surface potential (SP)-based compact GaN HEMT models. A good GaN HEMT model should be taken into account of trapping effects as well as self-heating effects. SP-based modeling of GaN HEMTs requires the computation of intrinsic charges and surface potentials for the modeling of drain current and gate leakage current. This chapter also covers the modeling of E-Mode p-GaN gate GaN HEMTs. Now, most of the models used are either simplified approximation-based models or semi-empirical models or numerical calculations-based models. Among various physics-based compact analytical models, surface potential (SP)-based models are gaining attention for future compact AlGaN/GaN HEMT modeling because of two reasons: (1) A single equation can describe all regions of operations of the AlGaN/GaN HEMTs and (2) inherent symmetry.

Keywords Surface potential · Drift–diffusion · Intrinsic charge · Gate current · Charge trapping effects

1 Introduction

AlGaN/GaN HEMTs (high electron mobility transistors) have recently emerged as highly potential transistors for high frequency and high power applications in solid state millimeter wave and microwave power devices due to their excellent properties such as high electron mobility/velocity, high charge density, high breakdown voltage/

J. Ajayan (✉) · B. Mounika
Department of Electronics and Communication Engineering, SR University, Warangal, Telangana, India
e-mail: email2ajayan@gmail.com

D. Nirmal
Karunya Institute of Technology and Sciences, Coimbatore, Tamil Nadu, India

D. Nirmal and J. Ajayan (eds.), *Modeling of AlGaN/GaN High Electron Mobility Transistors*, Springer Tracts in Electrical and Electronics Engineering,
https://doi.org/10.1007/978-981-97-7506-4_1

field and high thermal conductivity (Mounika et al. 2023, 2022, 2022; Mounika and Ajayan 2023b; Angen Franklin et al. 2023; Kumar et al. 2022a, 2022b; Arivazhagan et al. 2021, 2021b; Arivazhagan et al. 2019; Fletcher et al. 2022; Husna Hamza et al. 2022, 2021; Murugapandiyan et al. 2021, 2022, 2021, 2017; Ajayan et al. 2021, 2022; Augustine Fletcher et al. 2020, 2019; Nirmal et al. 2018). The AlGaN/GaN HEMTs are expected to replace traditional silicon (Si) and silicon carbide (SiC)-based power devices in near future for RF power applications. Physics-based compact analytical models of AlGaN/GaN HEMTs are required for fast and accurate simulation of AlGaN/GaN HEMTs-based circuits which is essential for completely exploiting the potential of these transistors. At present, a majority of models used are either simplified approximation-based models or semi-empirical models or numerical calculations-based models (Wu et al. n.d.). The main disadvantage of numerical calculations-based model is that it is very slow. A semi-empirical model requires the use of a large number of parameters. In order to overcome these difficulties, physics-based compact analytical models are desirable. Among various physics-based compact analytical models, surface potential (SP)-based models are gaining attention for future compact AlGaN/GaN HEMT modeling because of two reasons: (1) A single equation can describe all regions of operations of the AlGaN/GaN HEMTs and (2) inherent symmetry.

2 Sourabh Khandelwal Model

The description of the parameters used in Sourabh Khandelwal model (Khandelwal et al. 2012, 2013) of AlGaN/GaN HEMT is given in Table 1.

2.1 *Surface Potential Calculation*

To calculate ψ in GaN HEMTs, it is essential to model E_f in the quantum well (QW). E_f can be calculated from a consistent solution of Poisson's and Schrodinger's equations in the channel, which poses severe challenges. n_s can be estimated by assuming triangular potential as (Khandelwal et al. 2012)

$$n_s = DV_{th}\left[\ln\left(1 + e^{\frac{E_f - E_0}{V_{th}}}\right) + \ln\left(1 + e^{\frac{E_f - E_1}{V_{th}}}\right)\right] \tag{1}$$

$$E_{0,1=\gamma_{0,1} n_s^{2/3}} \tag{2}$$

$$n_s = \frac{\varepsilon}{qd}\left(V_{g0} - E_f - V_x\right) \tag{3}$$

$$V_{g0} = V_g - V_{off} \tag{4}$$

Table 1 The parameters and their description

Parameters	Description
n_s	2-DEG density
E_f	Fermi level position
E_0	First energy level position
E_1	Second energy level position
d	AlGaN layer thickness
D	Density of states
q	Charge of electron
ε	AlGaN permittivity
$C_g = \varepsilon/d$	Capacitance of gate/unit area
$\gamma_{0,1}$	2.12×10^{-12}, 3.73×10^{-12} V/m$^{4/3}$
ψ	Surface potential
V_g	Gate bias
V_x	Channel voltage
V_{off}	Cut off voltage
V_{th}	Thermal voltage221532
L	Gate length
W	Gate width
$\boldsymbol{\psi_d}$	Surface potential at drain
$\boldsymbol{\psi_s}$	Surface potential at source
υ_{sat}	Saturation velocity
$\boldsymbol{\lambda}$	Channel length modulation parameter
$\boldsymbol{\mu_0}$	Low field mobility of carriers

To determine E_f and n_s, in 2012, Sourabh Khandelwal et al. (2012) divided the modulation of E_f with V_g into 3 separate regions and they are.

Sub-Voff Region: $V_g < V_{off}$.

Moderate 2-DEG Region: $V_g > V_{off}$ & $E_f < E_0$.

Strong 2-DEG Region: $V_g > V_{off}$ & $E_f > E_0$.

Regional equations for $V_g > V_{off}$ valid in regions 2 and 3 can be combined to get a single equation for E_f as (Khandelwal et al. 2012)

$$E_{f,above=V_{g0}(1-H(V_{g0}))} \tag{5}$$

where

$$H\left(V_{g0}\right) = \frac{V_{g0} + V_{th}\left[1 - \ln\left(\beta V_{gon}\right)\right] - \frac{\gamma_0}{3}\left(\frac{C_g V_{g0}}{q}\right)^{2/3}}{V_{g0}\left(1 + \frac{V_{th}}{V_{god}}\right) + \frac{2\gamma_0}{3}\left(\frac{C_g V_{g0}}{q}\right)^{2/3}} \tag{6}$$

Here,

$V_{gon(d)} = V_{gon}(d)\alpha_n(d)/\sqrt{V^2_{gon(d)}+\alpha^2_n(d)}$ can be computed using interpolation function:

$$\beta = C_g/qDV_{th} \tag{7}$$

$$\alpha_n = e/\beta \tag{8}$$

In sub-V_{off} region $|E_f|$«E_1 and E_0. Hence, E_f-E_0 and E_f-E_1 $\approx E_f$

$$n_{s,sub/} - V_{off} = 2DV_{th}\exp(V_{g0}/Vth) \tag{9}$$

From (3) for $V_x = 0$, the equation for E_f (sub-V_{off}) can be obtained as

$$E_{f,sub-Voff} = V_{g0} - \frac{2qDV_{th}}{C_g}\exp\left(\frac{V_{g0}}{V_{th}}\right) \tag{10}$$

But for compact models, a single equation for E_f is preferred.

By combining (5), (6) and (10), a single equation which is valid for all regions can be obtained as

$$E_{f,unified} = V_{g0} - \frac{2V_{th}\ln\left(1 + e^{\frac{V_{g0}}{2V_{th}}}\right)}{1/H(V_{g0,p}) + (C_g/qD)e^{-V_{g0}/2V_{th}}} \tag{11}$$

Here, $V_{g0,p} = V_{g0}$ for the above V_{off} and $V_{g0,p}$ is in the order of V_{th} if $V_g < V_{off}$. Equation (11) is valid for all regions; however, the accuracy is in millivolts. To obtain accuracy in femtovolts, (11) can be modified as

$$E_{f1} = E_{f,unified} - \frac{p}{q}\left(1 + \frac{pr}{2q^2}\right) \tag{12}$$

Here, p, q and r are given by (Mounika et al. 2023)

$$p = \frac{C_g}{q}V_{gef} - \sum_{i=0}^{1} DV_{th}\ln(\xi_i + 1) \tag{13}$$

$$q = -\frac{C_g}{q} - \sum_{i=0}^{1}\frac{D}{1+\xi_i^{-1}}\left(1 + (2/3)k_i V_{gef}^{-1/3}\right) \tag{14}$$

$$r = \sum_{i=0}^{1}\frac{\frac{2}{9}V_{gef}^{-4/3}Dk_i\left(1 + \xi_i^{-1}\right) + \frac{D}{V_{th}}\left(1 + \frac{2}{3}k_i V_{gef}^{-1/3}\right)^2}{\left(1 + \xi_i^{-1}\right)^2} \tag{15}$$

Here,

$$k_{0,1} = \gamma_{0,1}\left(\frac{C_g}{q}\right)^{2/3} \tag{16}$$

$$\mathrm{V_{gef}} = \mathrm{V_g} - \mathrm{V_{off}} - \mathrm{E_f} \tag{17}$$

$$\xi_{0,1} = \exp\left(\frac{\mathrm{E_{f,unified}} - \mathrm{k_{0,1}} V_{gef}^{2/3}}{V_{th}}\right) \tag{18}$$

The $\mathrm{E_{f,unified}}$ estimated from (18) can be used to compute SP on both sides of the channel:

$$\mathrm{SP},\ \psi = \mathrm{V_x} + \mathrm{E_f} \tag{19}$$

2.2 *Intrinsic Charge Calculation*

The intrinsic source, drain and gate charges in the HEMTs have to be estimated precisely for accurate transient circuit simulations and computation of capacitances. The precise modeling of intrinsic charges needs correct allocation of the channel charge to the HEMT terminals. Usually, this is done by considering drain charge as (Khandelwal et al. 2012)

$$\mathrm{Q_d} = \mathrm{x_p Q_{ch}} \tag{20}$$

and source charge as

$$\mathrm{Q_s} = \left(1 - \mathrm{x_p}\right)\mathrm{Q_{ch}} \tag{21}$$

Here $\mathrm{Q_{ch}}$ and $\mathrm{x_p}$ represent channel charge and partitioning factor respectively. Ward et al. (1981) introduced a more rigorous way of computing $\mathrm{Q_d}$ and $\mathrm{Q_s}$ as follows:

$$Q_d = \int_0^{\mathrm{L}} (x/L) Q_{ch}\left(V_g, V_x\right) dx \tag{22}$$

$$Q_s = \int_0^{\mathrm{L}} (1-(x/L)) Q_{ch}\left(V_g, V_x\right) dx \tag{23}$$

Therefore, total gate charge ($\mathrm{Q_g}$) can be computed as (Khandelwal et al. 2012)

$$\mathrm{Q_g} = -(\mathrm{Q_s} + \mathrm{Q_d}) \tag{24}$$

$$Q_g = -\int_0^L qWn_s(V_g, V_x)d_x = -\int_0^L qWC_g(V_{g0} - \psi(\mathrm{x}))d_x \tag{25}$$

d_x can be written as (Cheng and Wang 2011)

$$d_x = \frac{\mathrm{L}[\mathrm{V}_{g0} - \psi + \mathrm{V}_{\mathrm{th}}]}{(V_{g0} - \psi_{\mathrm{m}} + \mathrm{V}_{\mathrm{th}})(\psi_d - \psi_{\mathrm{s}})} d\psi \tag{26}$$

By integrating (26), we will get (Khandelwal et al. 2012)

$$x = \frac{\mathrm{L}(\psi(\mathrm{x}) - \psi_{\mathrm{s}})}{(\mathrm{V}_{g0} - \psi_{\mathrm{m}} + \mathrm{V}_{\mathrm{th}})}\left(V_{g0} + \mathrm{V}_{\mathrm{th}} - \frac{\psi(\mathrm{x}) + \psi_{\mathrm{s}}}{2}\right) \tag{27}$$

Here,

$$\psi_m = (\psi_{\mathrm{s}} + \psi_{\mathrm{d}})/2 \tag{28}$$

Integrating (25) using (26) from source to drain, we will get (Khandelwal et al. 2012)

$$Q_g = \frac{C_G LW}{(V_{g0} - \psi_m + V_{th})}\left[V_{go}^2 + (1/3)(\psi_d^2 + \psi_s^2 + \psi_s\psi_d)\right] - V_{go}(\psi_d + \psi_s - V_{th}) - V_{th}\psi_m \tag{29}$$

$$Q_d = -\frac{C_g LW}{120(V_{g0} - \psi_m + V_{th}^2)} x \begin{bmatrix} 12\psi_d^3 + 8\psi_s^3 + \psi_s^2\left(16\psi_d - 5(V_{th} + 8V_{g0})\right) + 2\psi_s\begin{pmatrix} 12\psi_d^2 - 5\psi_d(5V_{th} + 8V_{g0}) + \\ 10(V_{th} + V_{g0})(V_{th} + 4V_{g0}) \end{pmatrix} + \\ 15\psi_d^2(3V_{th} + 4V_{g0}) - 60V_{g0}(V_{th} + V_{g0})^2 + 20\psi_d(V_{th} + V_{g0})(2V_{th} + 5V_{g0}) \end{bmatrix} \tag{30}$$

The intrinsic capacitances can be calculated as (Khandelwal et al. 2012)

$$\mathrm{C_{GG}} = \mathrm{dQ_g/dV_g} \tag{31}$$

$$\mathrm{C_{gs}} = -\mathrm{dQ_s/dV_g} \tag{32}$$

$$\mathrm{C_{gd}} = -\mathrm{dQ_d/dV_g} \tag{33}$$

The experimental validation of the model is illustrated in Fig. 1a and b respectively. This model didn't use any fitting parameters.

2.3 Drain Current Model

The drain current (I_{ds}) can be computed by considering drift–diffusion (DD) charge transport model with velocity saturation effect as (Khandelwal et al. 2013; Khandelwal and Fjeldly 2012)

$$I_{ds} = \frac{\mu_{eff} C_g}{\sqrt{1 + \theta_{sat}^2 \psi_{ds}^2}} \frac{W}{L} \left(V_{g0} - \psi_m + V_{th}\right)(\psi_{ds})(1 + \lambda V_{ds}) \tag{34}$$

$$\psi_{ds} = \psi_d - \psi_s \tag{35}$$

$$\theta_{sat} = \frac{\mu_{eff}}{\upsilon_{sat}} L \tag{36}$$

Effective mobility of carriers can be estimated as (Khandelwal et al. 2013)

$$\mu_{eff} = \frac{\mu_0}{1 + UA\left(V_{g0} - \psi_m\right) + UB\left(V_{g0} - \psi_m\right)^2} \tag{37}$$

Here, UA and UB represent the vertical electric field dependence of charge mobility, and they can be obtained from experimental results. Generally, GaN HEMTs have a separation from the source (S)/drain (D) contact to the edge of the gate, and these gaps are called access areas. Access regions function like a nonlinear resistor since quasi-saturation of the charge velocity happens in these areas. This resistive effect can be modeled as (Khandelwal et al. 2013)

$$R_{s(d)} = R_c + \frac{R_{sh} L_{sg}(dg)}{W(1 + \kappa V_g)} \left(1 + \frac{(\mu_0 V_s(d) - si(di))^{\gamma}}{\upsilon_{sat} L_{sg(dg)}}\right)^{1/\gamma} \tag{38}$$

R_s = resistance of the source access region.
R_d = resistance of the drain access region.
R_c = resistance of the S/D contact.
R_{sh} = sheet resistance of the access region.
$L_{sg(dg)}$ = length between gate edge and the source (drain).
$V_{s(d)-si(di)}$ = Voltage drop across the S and the D access areas.
γ and κ represent the model parameters.

If operating power of the HEMT increases, the temperature of the HEMT also increases well above the surrounding temperature, leading to a major change in its characteristics. This effect is known as the self-heating effect. The experimental validation of the drain current model is plotted in Fig. 2.

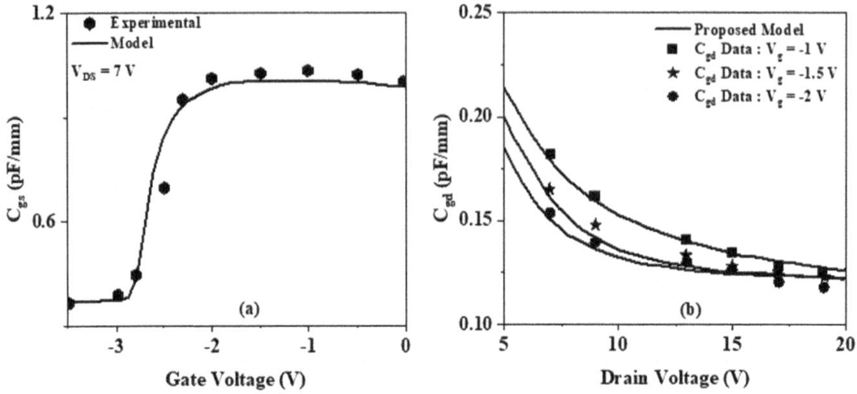

Fig. 1 The experimental validation of the model (Khandelwal et al. 2012)

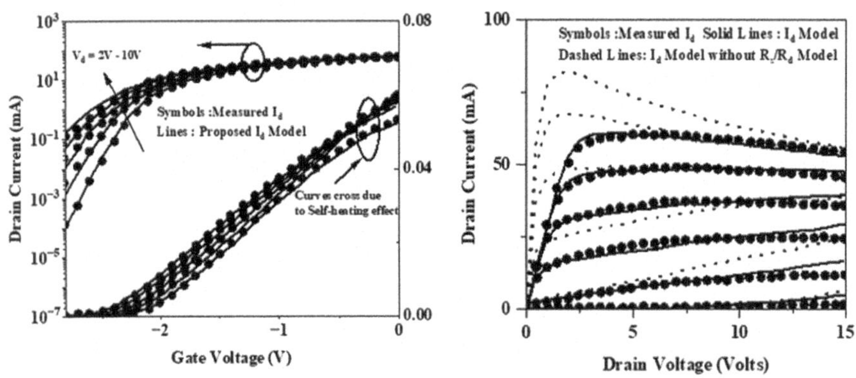

Fig. 2 The experimental validation of the drain current model (Khandelwal et al. 2013)

2.4 Gate Current Model

The GaN HEMTs have a Schottky gate contact in which there is a current flow in and out of the gate contact depending on the bias voltages. It is demonstrated that traps play a key role in the gate leakage (I_g), and Frenkel–Poole (FP) model illustrates the reverse leakage current accurately (Zhang et al. 2006; Yan et al. 2010). Based on the FP model, Ig (reverse gate current) can be estimated as (Zhang et al. 2006)

$$I_g = C_{trap} E_s \exp\left(-\frac{\varphi_d}{V_{th}} + \beta\sqrt{E_s}\right) \tag{39}$$

$$C_{trap} = C_{trap0} \ln\left(1 + \exp\left(\frac{V_{ds}}{V_{trap}}\right)\right) \tag{40}$$

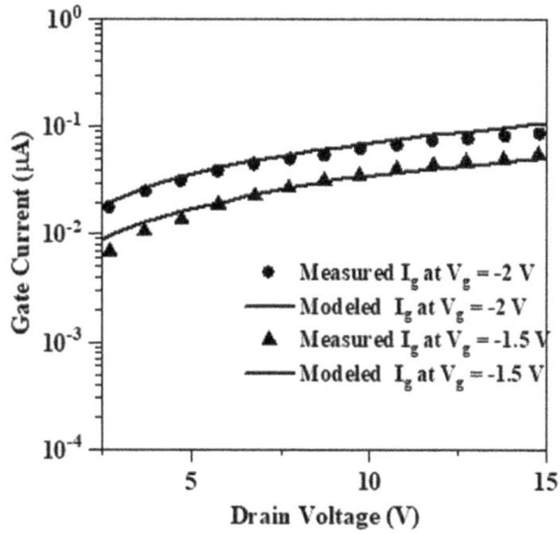

Fig. 3 The validation of gate leakage model with experimental data (Khandelwal et al. 2013)

C_{trap} = trap density parameter.
E_s = surface electric field @ AlGaN—Metal interface.
ϕ_d = height of the barrier @ AlGaN—Metal interface

$$\beta = \sqrt{q/\pi\varepsilon_0\varepsilon_s}/V_{th} \quad (41)$$

ε_s = Relative dielectric permittivity @ high frequency

$$E_s = \frac{q\sigma_p - qn_s}{\varepsilon} \quad (42)$$

σ_p = concentration of polarization charge
V_{trap} and C_{trap0} represent model parameters.
The validation of gate leakage model with experimental data is depicted in Fig. 3.

3 Sudip Ghosh Model

Poole–Frenkel emission and thermionic emission are the two key mechanisms for the I_g (gate leakage current) in the reverse- and forward-bias conditions, respectively. High gate leakage current in AlGaN/GaN HEMTs leads to the increase of noise figure (NF) and decrease of breakdown voltage. There are three components that contribute the gate leakage in GaN HEMTs under wide operating voltages. Under reverse-bias condition with high reverse voltage, the I_g is dominated by Poole–Frenkel (PF) emission. If the reverse voltage is low when the GaN HEMT is reverse-biased, the I_g is contributed by trap-assisted tunneling (TAT) mechanism. Finally, thermionic emission (TE) leads to the formation of I_g under forward-bias (Ghosh et al. 2015).

3.1 Poole–Frenkel (PF) Model

The analytical relationship between J_{PF} (current density) and E (electric field) for PF-model conduction is expressed as (Sze 2001)

$$J_{PF} = E.C.\exp\left(\beta\sqrt{E} + \alpha\right) \tag{43}$$

$$\alpha = -\frac{\varphi_d}{V_{th}} \tag{44}$$

$$\beta = \frac{\left(\frac{q}{\pi\varepsilon_s}\right)^{1/2}}{V_{th}} \tag{45}$$

Here, C represents the trap concentration parameter. The E at the metal–AlGaN interface can be computed as (Yan et al. 2010; Wang et al. 2014)

$$E = \frac{q\sigma_p - C_g\left(V_{g0} - \psi\right)}{\varepsilon_s} \tag{46}$$

To determine the current (I_{PF}), it is required to integrate the J_{PF} along the quantum well (QW) length from S to D terminals:

$$I_{PF} = W\int_0^L J_{PF}.dx \tag{47}$$

In (47), dx is given by (Khandelwal et al. 2012; Khandelwal and Fjeldly 2012)

$$dx = \frac{L\left(V_{th}+V_{g0}-\psi\right)}{(\psi_d-\psi_s)\left(V_{th}-\psi_m+V_{g0}\right)}d\psi \tag{48}$$

$$dx = L.K\left(V_{th} - \psi + V_{g0}\right) \tag{49}$$

$$\frac{dE}{dx} = \frac{C_g}{\varepsilon_s LK.\left(V_{th} - \psi + V_{g0}\right)} = \frac{C_g^2}{(A - B.E)L} \tag{50}$$

$$A = \varepsilon_s.K\left(q\sigma_p + V_{th}C_g\right) \tag{51}$$

$$B = K.\varepsilon_s^2 \tag{52}$$

Equation (47) can be modified by replacing dx with dE as

$$I_{PF} = CLW.e^{\alpha}\int_{E_S}^{E_D} e^{\beta\sqrt{E}}.E.\frac{(A - B.E)}{C_g^2}dE \tag{53}$$

$$I_{PF} = e^{\alpha}.\frac{CLW}{C_g^2}\left[e^{\beta\sqrt{E}}.\frac{2A}{\beta^4}\left(6\beta\sqrt{E} - 6 - 3\beta^2 E + \beta^3 E^{1.5}\right) - e^{\beta\sqrt{E}}.\frac{2B}{\beta^6}\begin{pmatrix} 20\beta^3 E^{1.5} - 5\beta^4 E^2 + \beta^5 E^{2.5} + \\ 120\beta\sqrt{E} - 120 - 60\beta^2 E \end{pmatrix}\right]_{E=E_S}^{E=E_D} \tag{54}$$

3.2 Thermionic-Emission (TE) Model

Under forward-bias condition, TE mechanism dominates in GaN HEMTs and the I-V characteristics of a Schottky contact can be expressed as (Rhoderick and Williams 1988)

$$J_{TE} = J_{TE0}\left[\exp\left(\frac{V}{\eta Vth}\right) - 1\right] \tag{55}$$

$$J_{TE0} = A^* T^2 \exp\left(-\frac{\varphi_b}{V_{th}}\right) \tag{56}$$

J_{TE0} = Reverse saturation current density.
A^* = Effective Richardson's constant.
η = Ideality factor

$$V = V_g - \psi \tag{57}$$

Therefore, I_{TE} can be computed as

$$I^{TE} = W\int_0^L J_{TE}.dx \tag{58}$$

$$I_{TE} = \mathrm{LW}.J_{TE0}.K\left[\exp\left(\frac{V_g - \psi}{\eta V_{th}}\right).V_{th}\eta.\left(\psi + \eta V_{th} - V_{g0} - V_{th}\right) - \psi\left(V_{g0} + V_{th}\right) + \frac{\psi^2}{2}\right]_{\psi=\psi_S}^{\psi=\psi_D} \tag{59}$$

3.3 Trap-Assisted Tunneling (TAT) Current Model

TAT mechanism dominates at reverse-bias condition in GaN HEMTs with low bias voltages. Current due to TAT mechanism exhibits the same temperature dependence as PF emission (Turuvekere et al. 2013). TAT current density is given by (Turuvekere et al. 2013)

$$J_{TAT} = J_{TATO}\left[\exp\left(\frac{V - V_0}{\eta_2 V_{th}}\right) - 1\right] \tag{60}$$

$$V = V_g - \psi \tag{61}$$

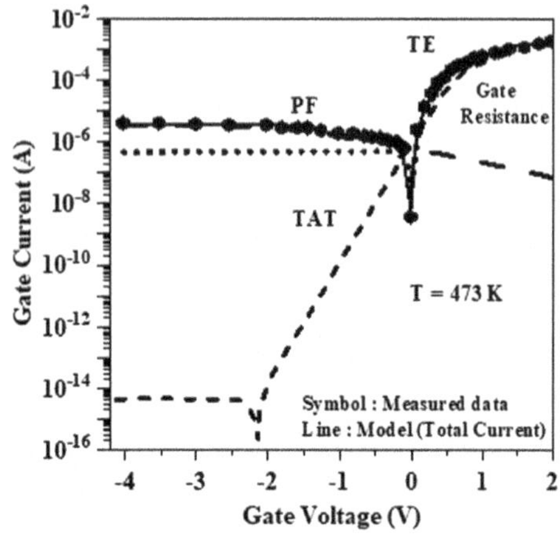

Fig. 4 The validation of Sudip Ghosh model (Ghosh et al. 2015) with experimental data

J_{TATO} = Reverse saturation current density.
η_2 = Ideality factor.
V_0 = Fitting voltage

$$I_{TAT} = \mathrm{LW}.J_{TAT}.K\left[\exp\left(\frac{V_g - \psi - \mathrm{V}_0}{\eta \mathrm{V}_{th}}\right).V_{th}\eta.(\psi - \mathrm{V}_{g0}) + \psi\left(\frac{\psi}{2} - \mathrm{V}_{g0} - \mathrm{V}_{th}\right)\right]_{\psi=\psi_S}^{\psi=\psi_D} \quad (62)$$

I_{PF}, I_{TE} and I_{TAT} depend on area (WL) of the GaN HEMT. The validation of Sudip Ghosh model (Angen Franklin et al. 2023) with experimental data is illustrated in Fig. 4.

4 Wanling Deng Model

In 2015, Wanling Deng et al. (2015) developed a compact and explicit SP-based current model for AlGaN/GaN HEMTs which was also taken into the account of the self-heating effect (SHEs) by considering the temperature-dependent free-charge mobility.

4.1 Surface Potential (SP) Calculation

Considering the self-consistent solution to Poisson's and Schrödinger's equations in the channel, the fundamental carrier control equations for 2DEG along the quantum well are calculated as (Cheng and Wang 2011; Wang et al. 2022)

$$n_s = \frac{\varepsilon}{\mathrm{qd}}\left(V_{gs} - \mathrm{E_f} - \phi_\mathrm{n} - \mathrm{V_{off}}\right) \quad (63)$$

$$n_s = \mathrm{D}\phi_{th} \ln\left[1 + \exp\left(\frac{\mathrm{E_f} - \mathrm{E_0}}{\phi_{\mathrm{th}}}\right)\right] \tag{64}$$

Here, E_0 is given in (2), ϕ_n, ϕ_{th} and D indicate the quantum well potential, thermal voltage and density of states respectively.

Like the Sourabh Khandelwal model (Khandelwal et al. 2012), to determine E_f and n_s, the modulation of E_f with V_g is divided into 3 separate regions and they are.

Sub-Voff Region: $V_{gs} < V_{off}$.

Moderate 2-DEG Region: $V_{off} < V_{gs} < V_{gth}$ & $E_f < E_0$.

Strong 2-DEG Region: $V_g > V_{gth}$ & $E_f > E_0$.

Here, V_{gth} is obtained by the condition of $E_0 = E_f$.

In strong 2DEG-region, Eq. (64) can be approximated as

$$n_s = D(\mathrm{E_f} - \mathrm{E_0}) = \mathrm{D}\left(\mathrm{E_f} - \gamma_0 n_s^{2/3}\right) \tag{65}$$

From Eqs. (63) and (65), we can write

$$A\left(V_{g0} - E_f\right)^2 = \mathrm{D}\left[\mathrm{E_f} - \gamma_0 A^{2/3}\left(V_{g0} - \mathrm{E_f}\right)^{2/3}\right] \tag{66}$$

$$A = \frac{\varepsilon}{\mathrm{qd}} \tag{67}$$

$$V_{g0} = \mathrm{V_{gs}} - \phi_{\mathrm{n}} - \mathrm{V_{off}} \tag{68}$$

Equation (66) can be rewritten as

$$A^2\gamma_0^3\left(V_{g0} - E_f\right)^2 = \left[-\frac{A}{D}V_{go} + E_f\left(1 + \frac{A}{D}\right)\right]^3 \tag{69}$$

Equation (69) can be written in the form of

$$aE_f^3 + \mathrm{bE_f^2} + \mathrm{cE_f} + \mathrm{e} = 0 \tag{70}$$

where

$$a = \left(\frac{\mathrm{A}}{\mathrm{D}} + 1\right)^3 \tag{71}$$

$$b = 3V_{g0}\left(-\frac{\mathrm{A}}{\mathrm{D}}\right)\left(\frac{\mathrm{A}}{\mathrm{D}} + 1\right)^2 - A^2\gamma_0^3 \tag{72}$$

$$c = 3\mathrm{V}_{g0}^2\left(\frac{A}{D}\right)^2\left(\frac{\mathrm{A}}{\mathrm{D}} + 1\right) + 2\mathrm{A}^2 V_{g0}\gamma_0^3 \tag{73}$$

$$e = V_{g0}^{3}\left(-\frac{A}{D}\right)^{3} - A^{2}V_{g0}^{2}\gamma_{0}^{3} \tag{74}$$

Equation (70) can be solved by using the technique reported by Fan et al. (1989) as

$$E_f = \frac{-Y_2^{1/3} - Y_1^{1/3} - b}{3a} \tag{75}$$

$$Y_{1,2} = A_0 b + 1.5a\left(-B_0 \pm \sqrt{\Delta}\right) \tag{76}$$

$$\Delta = B_0^2 - 4A_0C_0 \tag{77}$$

$$A_0 = b^2 - 3ac \tag{78}$$

$$B_0 = bc - 9ae \tag{79}$$

$$C_0 = c^2 - 3be \tag{80}$$

Equation (75) is valid only for strong 2DEG-region ($V_g > V_{gth}$ & $E_f > E_0$) (Cheng and Wang 2011; Wang et al. 2022)

$$V_{gth} = \varphi_n + V_{off} + \gamma_0(\varphi_{th} D \ln 2)^{2/3} + \frac{\varphi_{th} D}{A}\ln 2 \tag{81}$$

In sub-V_{off} region ($V_{gs} < V_{off}$), $E_f < 0$ and $|E_f| >> E_0$, then Eq. (64) can be simplified as

$$n_s = \varphi_{th} D \exp\left(\frac{E_f}{\varphi_{th}}\right) \tag{82}$$

In moderate 2DEG-region, The E_f Vs n_s exhibits a linear relationship (Kola et al. 1988). Therefore, by combining moderate and sub-Voff regions, we will get

$$n_s = \varphi_{th} D \exp\left(\frac{|E_f|}{\varphi_{th}}\right) + E_f k_3 \tag{83}$$

At $E_f = E_0$,

$$k_3 = \frac{fD}{\left(\gamma_0 D n_{s0}^{-1/3} + 1\right)} \tag{84}$$

$$n_{s0} = \varphi_{th} D \ln 2 \tag{85}$$

$$f = \frac{1}{\exp(-10^3 V_{g0})+1} \tag{86}$$

Here, f is a function which smoothly links the sub-Voff and moderate 2DEG-regions.

Substituting (83) in (63), we will get

$$\varphi_{th} D \exp\left(-\frac{|\mathrm{E_f}|}{\varphi_{\mathrm{th}}}\right) = \mathrm{AV_{g0}} - \mathrm{E_f}(k_3 + \mathrm{A}) \tag{87}$$

$$x_g - \mathrm{x_f} = \delta \exp(-|\mathrm{x_f}|) \tag{88}$$

Equation (88) represents the normalized form of (87). Equation (88) can be solved explicitly with the help of the Lambert W function ($\mathrm{W_0}$) which gives

$$E_f = \varphi_{\mathrm{th}}.x_f = \varphi_{\mathrm{th}}\left\{x_g + \mathrm{sgn}(\mathrm{V_{g0}}) W_0\left[- \mathrm{sgn}(\mathrm{V_{g0}})\delta \exp(-|\mathrm{x_g}|)\right]\right\} \tag{89}$$

$$\delta = \frac{\mathrm{D}}{(\mathrm{k_3} + \mathrm{A})} \tag{90}$$

$$x_g = \mathrm{V_{g0}} A/\varphi_{th}/(k_3 + \mathrm{A}) \tag{91}$$

From the above expressions, SP can be computed as

$$\psi_S = \mathrm{E_f} + \varphi_{\mathrm{n}} \tag{92}$$

4.2 Drain Current Model with Self-Heating Effect

To obtain a compact drain current model, Deng et al. (2015) used the typical SP-based expression as (Cheng and Wang 2011; Wang et al. 2022)

$$I_{ds0} = \frac{\mu_{\mathrm{eff0}}.\varepsilon}{d.\sqrt{\theta^2\psi^2+1}}\frac{W}{L}.\left(V_{gs} + \varphi_{\mathrm{th}} - \psi_{\mathrm{m}} - \mathrm{V_{off}}\right)\psi \tag{93}$$

θ = fitting parameter

μ_{eff0} = effective mobility @ room temperature ($\mathrm{T_0}$)

$$\mu_{eff0} = \frac{\mu_0}{1 + \mathrm{p1}(\mathrm{V_{G0}} - \psi_{\mathrm{S}}) + \mathrm{p2}(\mathrm{V_{G0}} - \psi_{\mathrm{S}})^2} \tag{94}$$

μ_0 = low field mobility.

p1 and p2 represent degradation parameters. Self-heating effect can be taken into account by adopting temperature (T)-dependent mobility in the model:

$$\mu_{eff}(T) = \mu_{\mathrm{eff0}}\left(\frac{T_0}{T}\right)^{\alpha} \approx \mu_{eff0}\left[1 + \beta\left(\frac{1}{\mathrm{T}} - \frac{1}{\mathrm{T_0}}\right)\right] \tag{95}$$

β is selected as 5 × 10² K for better approximation. Finally, the rise in T caused by SHEs in a GaN HEMT can be obtained as

$$T - \mathrm{T_0} = \mathrm{I_{ds}}.V_{ds}.R_{th} \tag{96}$$

Here, R_{th} indicates the thermal resistance of the GaN HEMT.

Substituting (95) into (93), together with (96), the drain current (I_{ds}) model including the self-heating effect (SHEs) is obtained as

$$I_{ds} = \mathrm{I_{ds0}}\left(1 - \frac{\beta}{\mathrm{T_0}} \frac{I_{ds}V_{ds}R_{th}}{I_{ds}V_{ds}R_{th} + \mathrm{T_0}}\right) \tag{97}$$

The solution of the above quadratic equation is given by

$$I_{ds} = \frac{-\,\mathrm{b_0} + \sqrt{\mathrm{b_0^2} - 4\mathrm{a_0c_0}}}{2\mathrm{a_0}} \tag{98}$$

Here,

$$a_0 = \mathrm{R_{th}}.\mathrm{V_{ds}} \tag{99}$$

$$b_0 = \mathrm{T_0} - \left(1 - \frac{\beta}{\mathrm{T_0}}\right)R_{th}.V_{ds}.I_{ds0} \tag{100}$$

$$c_0 = -\,\mathrm{T_0}.I_{ds0} \tag{101}$$

The validation of Deng et al. (2015) model with experimental data is depicted in Fig. 5.

4.3 Binit Syamal Model

In 2016, Syamal et al. (2016) developed a scalable and compact SP-based model for GaN HEMTs that took into account transient charge trapping and quasi-steady state effects. Figure 6 illustrates the locations of charge traps in GaN HEMTs. In GaN HEMTs, charge traps can be found at the interface, bulk and surface. These charge traps may lead to various effects such as drain-lag and gate-lag transients (Binari

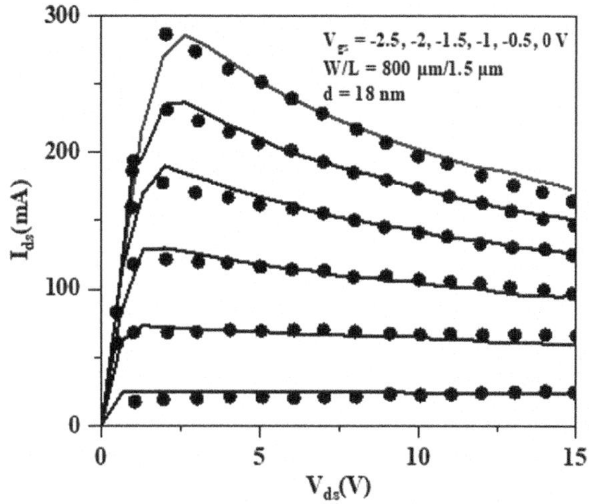

Fig. 5 The validation of Deng et al. (2015) model

et al. 1997, 1997), transconductance (g_m) frequency dispersion (Kruppa et al. 1995; Kohn et al. 1999) and current collapse (CC) (Khan et al. 1994; Klein et al. 1999). In Fig. 6, T1, T2, T3 and T4 represent surface traps, barrier traps, interface traps and substrate/buffer traps, respectively.

Interface traps can degrade the drain current, and they can also shift the threshold voltage (V_T) of the GaN HEMTs. In order to model the charge trapping effects, assume acceptor-like traps at the AlGaN/GaN interface having a trap density of N_t (cm^{-2}). Acceptor-like traps can either be neutral with a charge state of "0" or negative (ionized) with a charge state of − 1. Assuming that traps are present below the gate, Q_{nT} (the effective electron-trap density) can be calculated as (Sah 1971)

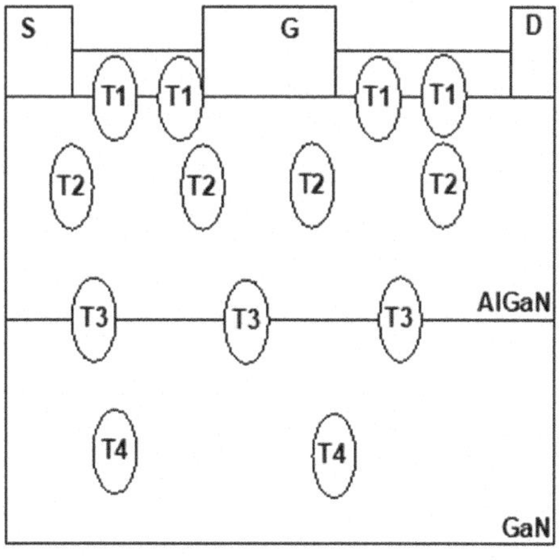

Fig. 6 Locations of charge traps in GaN HEMTs (Syamal et al. 2016)

$$Q_{nT} = qN_t \cdot \frac{e_{ps} + nc_{ns}}{e_{ps} + pc_{ps} + e_{ns} + nc_{ns}} \tag{102}$$

$$c_{ns} = v_n \sigma_n \tag{103}$$

$$c_{ps} = v_p \sigma_p \tag{104}$$

$$c_{ps} = \sigma_p v_p \tag{105}$$

$$c_{ns} = \sigma_n v_n \tag{106}$$

Here,

σ_n = electron capture cross section.
σ_p = hole capture cross section.
v_n = electron thermal velocity.
v_p = hole thermal velocity.
e_{ns} = electron emission rate.
e_{ps} = hole emission rate.
n = electron concentration.
p = hole concentration.

In the case of acceptor-like traps, the hole and electron emission rates can be expressed as

$$e_{ps} = \sigma_p v_p n_i e^{-E_T/v_{th}} \tag{107}$$

$$e_{ns} = \sigma_n v_n n_i e^{E_T/v_{th}} \tag{108}$$

Here,

E_T = Interface trap energy level.

2DEG-density (n_s) can be expressed as (Zhang et al. 2014)

$$n_s = \frac{C_d}{q}\left(V_g - E_f - V_C - V_{OFF}\right) \tag{109}$$

E_f equation which is valid from subthreshold region to strong inversion region is given by

$$E_f(V_C) = V_g - \Delta E_f - V_{OFF} \tag{110}$$

SP (ϕ_s) of GaN HEMT can be computed as

$$\varphi_s(V_c) = V_c + E_f(V_c) \tag{111}$$

The trap charge-dependent drain current is given by

$$I_{ds} = \mu_{\text{eff}} C_d \frac{W}{L_{eff}} \left(V_g + 2\text{v}_{\text{th}} - \overline{\varphi_{\text{s}}} - \text{V}_{\text{OFF}} - \frac{\overline{\text{Q}_{\text{nT}}}}{\text{C}_{\text{d}}} \right) \tag{112}$$

$$\overline{Q_{nT}} = \left(\text{Q}_{\text{nT,seff}} + \text{Q}_{\text{nT,deff}}\right)/2 \tag{113}$$

$$\overline{\varphi_s} = \left(\varphi_{\text{s,eff}} + \varphi_{\text{d,eff}}\right)/2 \tag{114}$$

Here,

$Q_{nT,seff}$ = Effective electron trap density at source.

$Q_{nT,deff}$ = Effective electron trap density at drain.

Interface traps significantly degrade the S/D access resistances. The R_c of GaN HEMTs with interface traps can be modeled as

$$R_C = \frac{\text{L}_{\text{gc}}}{\text{W.q.n}_{\text{s}}.\mu_{rc}} \sqrt{\left(1 + \xi_{\text{rc}} \left(\frac{\mu_{rc}(V_c - \text{V}_{\text{ci}})}{L_{gc}.v_{sat,c}}\right)^2\right)} \tag{115}$$

μ_{rc} = mobility in c

$v_{sat,c}$ = Saturation velocity in c with c = S (D).

$L_{gc} = L_{gs}$ (L_{gd}) and ξ_{rc} is a fitting parameter.

4.4 *Sayed Ali Albahrani Model*

The high value of the critical electric field is one of GaN's finest features. This enables GaN HEMTs to achieve a significantly higher RF power density. In other words, GaN HEMTs are able to produce the same amount of output power with a substantially smaller device size as compared to conventional III/V or silicon-based devices. GaN HEMT design still faces a number of obstacles, one of which is significant gate leakage, which has an adverse effect on the functionality of the component and circuits. High gate leakage leads to the increase in noise figure and degradation of breakdown voltage. Additionally, it affects circuits where the gate current from a DC source can substantially charge a capacitor or circuits with high load resistance at the gate terminal. In 2020, Albahrani et al. (2020) developed a compact SP-based gate current model for AlGaN/GaN Schottky HEMTs (Fig. 7).

The gate leakage current under reverse voltage is mainly attributed by the Fowler–Nordheim (FN) tunneling and Poole–Frenkel (PF) emission (Turuvekere et al. 2014) at medium-to-high reverse bias, whereas thermionic emission (TE) plays a key role in the forward-bias condition.

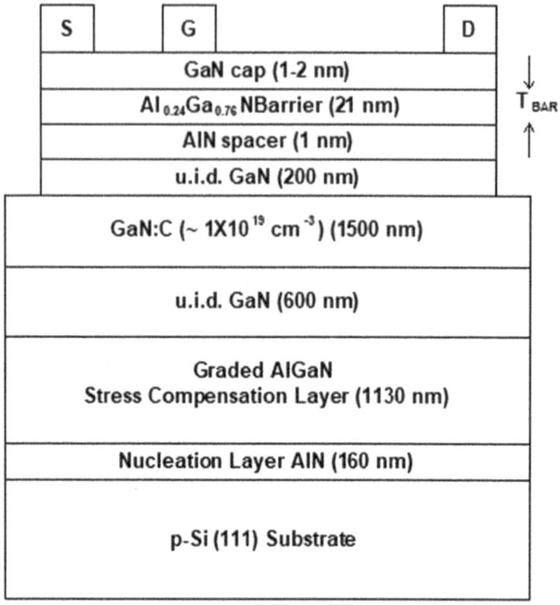

Fig. 7 AlGaN/GaN Schottky HEMT structure (Albahrani et al. 2020)

4.5 *TE Model*

The predominant conduction mechanism in a Schottky contact under forward-bias condition is TE. The J-V characteristics of a Schottky contact can be expressed as (Sze 2001)

$$J_{TE} = \mathrm{J_{TE0}}\left(\exp\left[{}^{V}/_{\eta V_{th}}\right]-1\right) \tag{116}$$

V = Applied bias.
η = Ideality factor.
V_{th} = Thermal voltage (K_BT).
K_B = Boltzmann's constant.
T = Temperature.
J_{TE0} = Reverse saturation current density

$$J_{TE0} = \mathrm{A}^{*}T^{2}\exp\left(-{}^{\mathrm{q}\varphi_{\mathrm{TE}}}/_{\mathrm{V_{th}}}\right) \tag{117}$$

Φ_{TE} = Height of the Schottky barrier.
A^* = Effective Richardson's constant

$$A^{*}= 4\pi\,\mathrm{qm}^{*}k_B^2/h^3 \tag{118}$$

H = Planck's constant.

m^* = Effective electron mass in the AlGaN barrier.

Let I_{TED} and I_{TES} represent the TE current components of gate-to-drain and gate-to-source regions, respectively:

$$I_{TE(S.D)} = \frac{1}{2} L.W.NF.J_{TE0}\left[\exp\left(\frac{V_{G_i(S_i,D_i)}}{\eta V_{th}}\right) - 1\right] \quad (119)$$

NF = Number of fingers.

V_{GiDi} = intrinsic gate-to-drain bias.

V_{GiSi} = intrinsic gate-to-source bias.

To simplify the model, V_{GiSi} and V_{GiDi} can be replaced with their respective effective voltages as given below

$$V_{GS.eff} = \beta_S V_{GiSi} + (1-\beta_S) V_{GiDi} = V_{GiDi} + \beta_S V_{DiSi} \quad (120)$$

$$V_{GD.eff} = \beta_D V_{GiSi} + (1-\beta_D) V_{GiDi} = V_{GiDi} + \beta_D V_{DiSi} \quad (121)$$

Here, β_S and β_D represent model parameters and $0 \le \beta_S \le 1$, $V_{GiDi} \le V_{GS,eff} \le V_{GiSi}$, $0 \le \beta_D \le 1$, $V_{GiDi} \le V_{GD,eff} \le V_{GiSi}$,

$$\phi_{TES} = \phi_{TE} \quad (122)$$

$$\phi_{TED} = \phi_{TE} + \Delta\phi \quad (123)$$

Here, ϕ_{TED} and ϕ_{TES} indicate the Schottky barrier heights of drain- and source-side and $\Delta\phi$ is a model parameter. For the precise modeling of I–V characteristics, it is essential to include the temperature dependence of the height of the Schottky barrier height and the ideality-factor η as

$$\varphi_{TE.T} = \varphi_{TE} + K_{\varphi_{TE}}\left({}^{T}/{}_{T_n} - 1\right) \quad (124)$$

$$\eta_T = \eta + K_\eta\left({}^{T}/{}_{T_n} - 1\right) \quad (125)$$

Here, Tn = nominal temperature

$K_{\phi TE}$ and $K\eta$ are the model parameters.

4.6 PF Model

The application of gate bias creates a vertical electric field ($E\perp$) in the AlGaN layer which leads to the flow of electrons. In addition to this current flow, the vertical electric field in the AlGaN layer also leads to the emission of the electrons from the trap states onto the continuum of states. This process is called the PF emission.

The current density due to the PF emission process is expressed as (Sze 2001)

$$J_{PF} = \mathrm{E}_{\perp} C_{PF} \exp\left(-\mathrm{q}\varphi_{\mathrm{PF}}/\mathrm{V}_{\mathrm{th}}\right) \tag{126}$$

$$\varphi_{PF} = \varphi_{\mathrm{PF0}} - \sqrt{\mathrm{qE}_{\perp}/\pi\tilde{\varepsilon}} \tag{127}$$

$$\tilde{\varepsilon} = \mathrm{k}\varepsilon \tag{128}$$

Here, k, C_{PF} are model parameters.

ϕ_{PF0} = barrier height with zero-electric-field.

$\tilde{\varepsilon}$ = dielectric permittivity of the AlGaN layer at high-frequency.

Increase in $E\perp$ leads to the decrease of potential barrier ϕ_{PF}:

$$E_{\perp} = \left[\mathrm{q}\sigma_{\mathrm{p}} - \mathrm{C_g}(V_{G0} - \psi)\right]/\varepsilon \tag{129}$$

σ_p = piezoelectric polarization charge (barrier) + (spontaneous polarization charge in barrier - spontaneous polarization charge in buffer)

$$C_g = \mathrm{T_{BAR}}/\varepsilon \tag{130}$$

T_{BAR} = AlGaN thickness

$$V_{G0} = \mathrm{V_G} - \mathrm{V_{OFF}} \tag{131}$$

The vertical electric fields at the S and D sides of the quantum well are represented by $E_{\perp,S}$ and $E_{\perp,D}$ respectively:

$$E_{\perp(S,D)} = \left[\mathrm{q}\sigma_{\mathrm{p}} - \mathrm{C_g}\left(V_{Gi(sio,Dio)} - \psi(\mathrm{S_i}, D_i)\right)\right]/\varepsilon \tag{132}$$

$$V_{G_iS_iO} = V_{G_iS_i} - \mathrm{V_{OFF}} \tag{133}$$

$$V_{G_iD_iO} = V_{D_iS_i} - \mathrm{V_{OFF}} \tag{134}$$

The PF current components are given by

$$I_{PF(S,D)} = \frac{1}{2}.L.W.NF.E_{\perp(S,D)} C_{PF} \exp\left(-\frac{\varphi_{\mathrm{PF(S,D)}}}{V_{th}}\right) \tag{135}$$

$$\varphi_{PF(S,D)} = \varphi_{\mathrm{PF0}} - \sqrt{\mathrm{q.E}_{\perp(\mathrm{S,D})}/\pi\tilde{\varepsilon}} \tag{136}$$

The effective surface potentials are expressed as

$$\psi_{Si,eff} = \beta_S \psi_{Si} + (1 - \beta_S)\psi_{Di} \tag{137}$$

$$\psi_{Di,eff} = \beta_D \psi_{Si} + (1 - \beta_D)\psi_{Di} \tag{138}$$

Here, the model parameter $0 \leq \beta_S \leq 1$ and $0 \leq \beta_D \leq 1$

The source and drain barrier heights are given by

$$\varphi_{PFS0} = \varphi_{PF0} \tag{139}$$

$$\varphi_{PFD0} = \varphi_{PF0} + \Delta\varphi \tag{140}$$

Here, $\Delta\phi$ is the model parameter.

V_{OFF} (difference in work function) can be slightly modified as

$$\tilde{V}_{OFF} = V_{OFF} + \Delta V_{OFF} \tag{141}$$

Here, ΔV_{OFF} is the model parameter.

For the precise modeling of I-V characteristics, it is essential to include the temperature dependence of the height of the Schottky barrier height, ϕ_{PF0}, as

$$\varphi_{PF0,T} = \varphi_{PF0} + K_{\varphi_{PF}}\left({}^{T}/_{T_n} - 1\right) \tag{142}$$

Here, $K_{\varphi_{PF}}$ is a model parameter.

4.7 FN Model

The electric field across the AlGaN barrier layer enhances with increase in gate bias toward more negative direction. At E_C (critical electric field), if the width of the barrier at the Fermi level of the metal becomes less than the critical barrier width (d_c), the electrons from the Fermi level of metal tunnel through the barrier (AlGaN) result in an additional gate leakage current, known as the Fowler–Nordheim (FN) current. The E-J_{FN} dependence of Fowler–Nordheim tunneling can be expressed as (Sze 2001)

$$J_{FN} = \mu.q.n_{FN}.E_{\perp} \tag{143}$$

$$n_{FN} = D\Delta E_{\perp} \exp\left(-{}^{B}/_{\Delta E_{\perp}}\right) \tag{144}$$

μ = electron drift mobility in AlGaN barrier
n_{PF} = FN induced electron density.
D = a constant

$$\Delta E_{\perp} = \mathrm{E}_{\perp} - \mathrm{E}_{\mathrm{C}} \tag{145}$$

$$B = 8\pi\sqrt{2\mathrm{m}^{*}(\varphi_{FN})^{3}/(3hq)} \tag{146}$$

$$E_{C} \approx \varphi_{FN}/d_{C} \tag{147}$$

$$J_{FN} = \mathrm{E}_{\perp} C_{FN} \Delta E_{\perp} \exp\left(-\mathrm{B}/_{\Delta \mathrm{E}_{\perp}}\right) \tag{148}$$

$$C_{FN} = \mu \mathrm{qD} \tag{149}$$

There are two sources of FN currents and they are gate-to-drain FN current source and the gate-to-source FN current source. The two FN current sources can be represented by I_{FNS} and I_{FND}. The FN current sources can be calculated as

$$I_{FN(S,D)} = \frac{1}{2} L.W.NF.E_{\perp,(S,D)} x C_{\mathrm{FN}} \Delta E_{\perp,(S,D)} \exp\left(\frac{-B}{\Delta E_{\perp,(S,D)}}\right) \tag{150}$$

Like in PF and TE models, here also, ϕ_{FN} (effective barrier height) can be modified as

$$\varphi_{FNS} = \varphi_{\mathrm{FN}} \tag{151}$$

$$\phi_{FND} = \phi_{\mathrm{FN}} + \Delta\phi \tag{152}$$

For the precise modeling of I-V characteristics, it is essential to include the temperature dependence of the height of the barrier height, ϕ_{FN} as

$$\phi_{FN,T} = \phi_{\mathrm{FN}} + K_{\varphi_{\mathrm{FN}}}\left(^{T}/_{T_{n}} - 1\right) \tag{153}$$

Here, $K_{\varphi_{FN}}$ is a model parameter.

5 Jie Wang Model

In 2020, Wang et al. (2020) modeled the gate leakage current of p-GaN Gate HEMTs. Enhancement-mode (E-mode) transistors are commonly used in circuits due to their typically OFF characteristics and simplicity of usage. Adopting a p-GaN gate that raises the conduction band at the channel surface in equilibrium is one approach of producing E-mode transistors. Due to its high threshold voltage, low ON-resistance and excellent stability, p-GaN gate HEMTs are now the first commercially available E-mode devices. The metallization layer can be a Schottky metal contact or an ohmic metal to the p-GaN on top of the p-GaN gate. The use of Schottky metal offers lower gate leakage current and higher V_{th} than ohmic metal. The Poisson equation including holes is given by

$$d^2\phi/dx^2 = q/\varepsilon_{GaN}\cdot(N_{Mg} + n_{GaN} - p_{GaN}) \tag{154}$$

E_{GaN} = GaN permittivity.
N_{Mg} = Mg doping (unintended) in the channel.
n_{GaN} = electron density in the GaN.
p_{GaN} = hole density in the GaN

$$n_{GaN} = n_{i-GaN}.\exp\left[(\varphi - \Phi_F - V_C)/V_T\right] \tag{155}$$

$$p_{GaN} = n_{i-GaN}.\exp\left[(-\varphi + \Phi_F)/V_T\right] \tag{156}$$

n_{i_GaN} = Intrinsic charge density
V_C = Voltage applied between the substrate and the channel.
n_{i_GaN} = intrinsic charge density.
V_C = Voltage applied between substrate and the channel

$$\Phi_F = V_T\left(\ln(N_0/n_i) + 2^{-1.5}(N_0/n_i)\right) \tag{157}$$

N_0 = Channel charge density.
V_T = Thermal voltage.
The SP is given as

$$V_G - V_0 - \varphi_{GaN} = \gamma\sqrt{\begin{array}{l}\frac{\varphi_{GaN}}{V_T}N_{Mg} + n_{i\_GaN}\exp\left(-\frac{-V_C - \Phi_F}{V_T}\right)\left[\exp\left(\frac{\varphi_{GaN}}{V_T}\right) - 1\right] \\ + n_{i\_GaN}\exp\left(\frac{\Phi_F}{V_T}\right)\left[\exp\left(\frac{-\varphi_{GaN}}{V_T}\right) - 1\right]\end{array}} \tag{158}$$

γ = body factor

ϕ_{GaN} = SP @ AlGaN/GaN interface.
V_0 = Fitting parameter.
The drain current can be modeled as

$$I_D = -\mu \mathrm{W}\left(\mathrm{q_i}.\frac{d\phi_{GaN}}{dy} - \mathrm{V_T}.\frac{dq_i}{dy}\right) \tag{159}$$

The gate current of p-GaN Gate HEMT is contributed by both thermionic emission (TE) and hopping transport (HT) mechanisms. The gate current components are given by.

If $V_G > 0$ V,

$$J_{TE} = \mathrm{T}^2.J_{TE0}.Area.\left[\exp\left(\frac{\varphi_{pGaN} - \varphi_{GaN}}{V_T.n_{TE}}\right) - 1\right] \tag{160}$$

n_{TE} = Ideality factor
J_{TE0} = forward saturation current density.
If $V_G < 0$ V.

$$J_{HT} = \mathrm{Peri}.\mathrm{J_{HT0}}.\left[\exp\left(\frac{\phi_{pGaN}}{V_T.n_{HT}}\right) - 1\right].\exp\left[-\mathrm{T}^{-\alpha}\right] \tag{161}$$

Here,
Peri = perimeter of the gate.
J_{HTO} = reverse saturation current density.
nHT is the ideality factor, and.
α = temperature coefficient of $\alpha = 0.25$.

6 Summary

In this chapter, we reviewed the various SP-based models of GaN HEMTs. FN tunneling, PF emission and TE emission are responsible for gate leakage current in GaN HEMTs. All the models pointed out in this chapter showed excellent agreement with experimental readings. The SP-based models included important device physics such as charge trapping effects, the 2DEG and the self-heating effects. The SP-models have great potential for RF circuit simulations as well as precise and reliable simulation of power switching and analog circuits based on GaN HEMTs. The scalable models can be employed to precisely optimize device structure for the desired performance in the design of MMIC power amplifiers, in addition to being suitable for device characterizations.

References

Ajayan J, Nirmal D, Ramesh R, Bhattacharya S, Tayal S, Leo Joseph LMI, Raju Thoutam L, Ajitha D (2021) A critical review of AlGaN/GaN-heterostructure based Schottky diode/HEMT hydrogen (H2) sensors for aerospace and industrial applications. Measurement 186:110100

Ajayan J, Nirmal D, Mohankumar P, Mounika B, Bhattacharya S, Tayal S, Augustine Fletcher AS (2022) Challenges in material processing and reliability issues in AlGaN/GaN HEMTs on silicon wafers for future RF power electronics & switching applications: A critical review. Mater Sci Semicond Process 151:106982

Albahrani SA, Heuken L, Schwantuschke D, Gneiting T, Burghartz JN, Khandelwal S (2020) Consistent surface-potential-based modeling of drain and gate currents in AlGaN/GaN HEMTs. IEEE Trans Electron Devices 67(2):455–462

Angen Franklin S, Jebalin BK, , Subhash Chander IV, Raj Kumar, Ajayan J, Nirmal D (2023) Investigation on Fe-Doped AlGaN/GaN HEMT at 148 GHz Using E-FPL Technology for High-Frequency Communication Systems 12:035006

Arivazhagan L, Nirmal D, Reddy PPK et al (2021b) A numerical investigation of heat suppression in HEMT for power electronics application. SILICON 13:3039–3046

Arivazhagan L, Nirmal D, Godfrey D, Ajayan J, Prajoon P, Augustine Fletcher AS, Amir Anton Jone A, Raj Kumar JS (2019) Improved RF and DC performance in AlGaN/GaN HEMT by P-type doping in GaN buffer for millimetre-wave applications. AEU-Int J Electron Commun 108:189–194

Arivazhagan L, Nirmal D, Jarndal A, Huq HF, Chander S, Bhagyalakshmi S, Reddy PK, Ajayan J, Varghese A (2021) Applicability of double channel technique in AlGaN/GaN HEMT for future biosensing applications. Superlattices Microstruct 160:107086

Augustine Fletcher AS, Nirmal D, Arivazhagan L, Ajayan J, Varghese A (2020) Enhancement of Johnson figure of merit in III-V HEMT combined with discrete field plate and AlGaN blocking layer. Int J RF Microw Comput Aided Eng. 30:e22040. https://doi.org/10.1002/mmce.22040

Augustine Fletcher AS, Nirmal D, Ajayan J, Arivazhagan L (2019) Analysis of AlGaN/GaN HEMT using discrete field plate technique for high power and high frequency applications. AEU—International J Electron Commun 99(2019):325–330

Binari SC, Redwing JM, Kelner G, Kruppa W (1997) AlGaN/GaN HEMTs grown on SiC substrates. Electron Lett 33(3):242–243

Binari SC, et al. (1997) GaN-based electronic devices for high-power, high speed, and high-temperature applications. In: Proc. Int. Conf. Nitride Semiconductor, pp 476–478

Cheng X, Wang Y (2011) A surface-potential-based compact model for AlGaN/GaN MODFETs. IEEE Trans Electron Devices 58(2):448–454

Deng W, Huang J, Ma X, Liou JJ (2015) An explicit surface potential calculation and compact current model for AlGaN/GaN HEMTs. IEEE Electron Device Lett 36(2):108–110

Fan S (1989) A new extracting formula and a new distinguishing means on the one variable cubic equation. Natural Sci. J. Hainan Teachers College 2(2):91–98

Fletcher ASA, Nirmal D, Arivazhagan L et al (2022) A 28-GHz Low-Loss AlGaN/GaN HEMT for T_X/R_X switches in 5G Base stations. J Electron Mater 51:1215–1225

Ghosh S, Dasgupta A, Khandelwal S, Agnihotri S, Chauhan YS (2015) Surface-potential- based compact modeling of gate current in AlGaN/GaN HEMTs. IEEE Trans Electron Devices 62(2):443–448

Husna Hamza K, Nirmal D, Augustine Fletcher AS, Arivazhagan L, Ajayan J, Natarajan R (2021) Highly scaled graded channel GaN HEMT with peak drain current of 2.48 A/mm. AEU-Int J Electron Commun 136:153774

Husna Hamza K, Nirmal D, Augustine Fletcher AS, Ajayan J, Natarajan R (2022) Enhanced drain current and cut off frequency in AlGaN/GaN HEMT with BGaN back barrier. Mater Sci Eng: B 284:115863

Khan MA, Shur MS, Chen QC, Kuznia JN (1994) Current/voltage characteristic collapse in AlGaN/GaN heterostructure insulated gate field effect transistors at high drain bias. Electron Lett 30(25):2175–2176

Khandelwal S, Fjeldly TA (2012) A physics based compact model of I-V and C–V characteristics in AlGaN/GaN HEMT devices. Solid-State Electron 76:60–66

Khandelwal S, Chauhan YS, Fjeldly TA (2012) Analytical modeling of surface-potential and intrinsic charges in AlGaN/GaN HEMT devices. IEEE Trans Electron Devices 59(10):2856–2860

Khandelwal S et al (2013) Robust surface-potential-based compact model for GaN HEMT IC Design. IEEE Trans Electron Devices 60(10):3216–3222

Khandelwal S, Fjeldly TA (2012) A surface-potential-based drain current model for study of non-linearities in AlGaAs/GaAs HEMTs. In: Proc. Compound Semicond. IC Symp pp 1–4

Klein PB, Freitas JA Jr, Binari SC, Wickenden AE (1999) Observation of deep traps responsible for current collapse in GaN metal–semiconductor field-effect transistors. Appl Phys Lett 75(25):4016

Kohn E, Daumiller I, Schmid P, Nguyen NX, Nguyen CN (1999) Large signal frequency dispersion of AlGaN/GaN heterostructure field effect transistors. Electron Lett 35(12):1022–1024

Kola S, Golio JM, Maracas GN (1988) An analytical expression for Fermi level versus sheet carrier concentration for HEMT modeling. IEEE Electron Device Lett 9(3):136–138

Kruppa W, Binari SC, Doverspike K (1995) Low-frequency dispersion characteristics of GaN HFETs. Electron Lett 31(22):1951–1952

Kumar JSR, Nirmal D, Ajayan J et al (2022a) Investigation on $L_G = 50$ nm Tapered T-Gated AlGaN/GaN HEMT on Silicon Wafer with a f_T/f_{max} of 264/312 GHz for beyond 5G (B5G) Applications. SILICON 14:11315–11322

Kumar JSR, Nirmal D, Hooda MK et al (2022b) Intensive study of field-plated AlGaN/GaN HEMT on silicon substrate for high power RF applications. SILICON 14:4277–4282

Mounika B, Ajayan J (2023b) Sandip Bhattacharya, "An intensive study on effects of lateral scaling and gate metals on the RF/DC performance of recessed T-gated Fe-doped AlN/GaN/SiC HEMTs for future RF and microwave power applications." Microelectron Eng 271–272:111948

Mounika B, Ajayan J, Bhattacharya S, Nirmal D (2022) Recent developments in materials, architectures and processing of AlGaN/GaN HEMTs for future RF and power electronic applications: A critical review. Micro Nanostructures 168:207317

Mounika B, Ajayan J, Bhattacharya S, Nirmal D (2022) Investigation of back barrier material effects on the scalability of Fe-doped recess-gated AlN/GaN HEMTs for next generation RF power electronics. Micro Nanostructures 171:207431

Mounika B, Ajayan J, Bhattacharya S (2023) Investigation on impact of AlxGa1-xN and InGaN back barriers and source-drain spacing on the DC/RF performance of Fe-doped recessed T-gated AlN/GaN HEMT on SiC wafer for future RF power applications. Micro Nanostructures 175:207504

Murugapandiyan P, Ravimaran S, William J, Ajayan J, Nirmal D (2017) DC and microwave characteristics of 20 nm T-gate InAlN/GaN high electron mobility transistor for high power RF applications. Superlattices Microstruct 109:725–734

Murugapandiyan P, Nirmal D, Tanvir Hasan Md, Arathy Varghese J, Ajayan AS, Fletcher A, Ramkumar N (2021) Influence of AlN passivation on thermal performance of AlGaN/GaN high-electron mobility transistors on sapphire substrate: A simulation study. Mater Sci Eng: B 273:115449

Murugapandiyan P, Tanvir Hasan Md, Rajya Lakshmi V, Mohd Wasim, Ajayan J, Ramkumar N, Nirmal D (2021) Breakdown voltage enhancement of gate field plate $Al_{0.295}Ga_{0.705}N$/GaN HEMTs. Int J Electron 108:1273–1287

Murugapandiyan P, Nirmal D, Ajayan J, et al. (2022) Investigation of influence of SiN and sio_2 passivation in gate field plate double heterojunction $Al_{0.3}Ga_{0.7}N/GaN/Al_{0.04}Ga_{0.96}N$ high electron mobility transistors. Silicon 14:1421–1429

Nirmal D, Arivazhagan L, Augustine Fletcher AS, Ajayan J, Prajoon P (2018) Current collapse modeling in AlGaN/GaN HEMT using small signal equivalent circuit for high power application. Superlattices Microstruct 113:810–820

Rhoderick EH, Williams RH (1988) “Current-transport mechanisms”, in Metal- Semiconductor Contacts, 2nd edn. Clarendon, Oxford, U.K., pp 98–100

Sah CT (1971) Equivalent circuit models in semiconductor transport for thermal, optical, auger-impact, and tunnelling recombination–generationtrapping processes. Phys Status Solidi A 7(2):541–559

Syamal B, Zhou X, Chiah SB, Jesudas AM, Arulkumaran S, Ng GI (2016) A comprehensive compact model for GaN HEMTs, including quasi-steady-state and transient trap-charge effects. IEEE Trans Electron Devices 63(4):1478–1485

Sze SM (2001) “MIS diode and charge-coupled devices”, in Physics of Semiconductor Devices, 2nd edn. Wiley, New York, NY, USA, pp 402–407

Turuvekere S, Karumuri N, Rahman AA, Bhattacharya A, DasGupta A, DasGupta N (2013) Gate leakage mechanisms in AlGaN/GaN and AlInN/GaN HEMTs: Comparison and modeling. IEEE Trans Electron Devices 60(10):3157–3165

Turuvekere S, Rawal DS, Dasgupta A, Dasgupta N (2014) Evidence of Fowler–Nordheim tunneling in gate leakage current of AlGaN/GaN HEMTs at room temperature. IEEE Trans Electron Devices 61(12):4291–4294

Wang M et al (2014) Investigation of surface- and buffer-induced current collapse in gan high-electron mobility transistors using a soft switched pulsed I−V measurement. IEEE Electron Device Lett 35(11):1094–1096. https://doi.org/10.1109/LED.2014.2356720

Wang J et al (2020) Surface-potential-based compact model for the gate current of p-GaN Gate HEMTs. IEEE Trans Electron Devices 67(9):3564–3567

Wang Y, Wu Q, Yan B, Xu R, Xu Y (2022) A Compact surface potential model for flexible radio frequency AlGaN/GaN high-electron-mobility transistor. IEEE Trans Microw Theory Tech 70(1):315–322. https://doi.org/10.1109/TMTT.2021.3102237

Ward DE (1981) Charge based modeling of capacitance in MOS transistors. In: Integr. Circuits Lab., Stanford Univ., Stanford, CA, Tech. Rep. G201–11

Wu Q, et al. (n.d.) A scalable multiharmonic surface-potential model of AlGaN/GaN HEMTs. IEEE Trans Microw Theory Tech 66(3):1192–1200

Yan D, Lu H, Cao D, Chen D, Zhang R, Zheng Y (2010) On the reverse gate leakage current of AlGaN/GaN high electron mobility transistors. Appl Phys Lett 97(15):153503–1–153503–3

Zhang J, Syamal B, Zhou X, Arulkumaran S, Ng GI (2014) A compact model for generic MIS-HEMTs based on the unified 2DEG density expression. IEEE Trans Electron Devices 61(2):314–323

Zhang H, Miller EJ, Yu ET(2006) Analysis of leakage current mechanisms in Schottky contact of GaN and Al0.25Ga0.75N/GaN grown by molecular beam epitaxy. J Appl Phys 99(2):023703–1–023703–6

Physical Modelling of Charge Trapping Effects

Matteo Buffolo, Carlo De Santi, Gaudenzio Meneghesso, Matteo Meneghini, Enrico Zanoni, Nicolò Zagni, Marcello Cioni, Alessandro Chini, Francesco Maria Puglisi, Paolo Pavan, and Giovanni Verzellesi

Abstract The performance, stability, and reliability of GaN transistors are affected by charge trapping effects, thus calling for dependable characterization and modelling techniques for these phenomena. In this work, after a review of the different GaN deep levels and a description of the possible compact and TCAD physical modelling approaches that can be adopted to model charge trapping effects, we present a selection of relevant cases where modelling has substantially contributed to the physical understanding of these phenomena. Device technologies taken into consideration comprehend both GaN RF and GaN power transistors. Analyzed charge trapping effects include barrier, surface, and buffer trap-related current collapse effects, dynamic R_{ON} effects, as well as threshold-voltage instabilities.

Keywords Reliability · Physical modelling · Fe-doping · Carbon doping · Buffer trapping effects · Current collapse

1 Introduction

Despite GaN transistors have already found their way to market and their share in both RF and power commercial applications is steadily increasing, charge trapping effects still represent a limit for the performance, stability, and reliability of GaN

M. Buffolo · C. De Santi · G. Meneghesso (✉) · M. Meneghini · E. Zanoni
Department of Information Engineering (DEI), University of Padova, 35131 Padova, Italy
e-mail: gauss@dei.unipd.it

N. Zagni · M. Cioni · A. Chini · F. M. Puglisi · P. Pavan
Department of Engineering "Enzo Ferrari", Università Di Modena E Reggio Emilia, 41125 Modena, Italy

M. Cioni
STMicroelectronics, 95121 Catania, Italy

G. Verzellesi
Department of Sciences and Methods for Engineering (DISMI), EN&TECH Center, University of Modena and Reggio Emilia, 42122 Reggio Emilia, Italy

D. Nirmal and J. Ajayan (eds.), *Modeling of AlGaN/GaN High Electron Mobility Transistors*, Springer Tracts in Electrical and Electronics Engineering,
https://doi.org/10.1007/978-981-97-7506-4_2

technologies, so that they require to be carefully characterized, understood, and modelled, in view of their minimization or trade-off with other electrical parameters within the process/device optimization loop.

Traps able to cause detrimental effects in the electrical behavior of GaN transistors can virtually be present in any layer and at any interface forming the device structure (Meneghini et al. 2021). Traps can either be associated to intrinsic crystallographic defects or to impurities incorporated, intentionally or not, during heterostructure growth. Traps charge up or discharge through capture or emission of mobile carriers, following fast bias changes more slowly than device capacitances that govern the "prompt" response of the device, and, by this means, they can induce dynamical effects that worsen the device figure of merits and their stability. Moreover, one of the most common effects of prolonged application of large bias voltages is the generation of new traps, which typically results in the permanent degradation of electrical performance and eventually to device failure.

Considering RF amplification, the most harmful trap-related effect is the so-called RF current collapse resulting in a reduction in the maximum drain current and simultaneous increase in the minimum drain-source voltage explorable by the operating point during the RF sweep (i.e., knee-voltage walk-out). The overall result is a compression in the RF output power and a degradation of the power added efficiency compared to the theoretical limits achievable by the given transistor that can be calculated from the DC output characteristics. When GaN transistors are used in power switching converters, instead, the most detrimental trap-related effect is the dynamic increase in the ON-state resistance (R_{ON}) compared to its DC value, resulting in an undesirable increment in power losses. Finally, in transistors with insulated or p-GaN gate structures, also the gate dielectric bulk and its interface with the underlying semiconductor or the p-GaN region can act as trapping sites. Potential detrimental effect for device performance are the threshold-voltage (V_{TH}) instabilities, adding up to other trapping effects like dynamic R_{ON}.

In this Chapter, we present an overview of charge trapping effects in GaN transistors, the comprehension of which has substantially been aided by modelling, providing physical insight into the observed phenomena, validating theoretically experimental results, and/or allowing to discriminate between different alternative interpretations. In all cases considered, modelling has played a role within the device optimization loop, by increasing the dependability of adopted experimental techniques and providing hints to the technologists about material and/or device weak points. The Chapter is organized as follows: in Sect. 2.2, GaN deep levels are reviewed and a thorough, updated database of associated, reported energy levels is provided. In Sects. 2.3 and 2.4, compact and TCAD modelling approaches are introduced. In Sect. 2.5, a selection of charge trapping effects relevant for GaN RF transistors is presented and the role of modelling in their analysis is highlighted. The same is done for GaN power transistor in Sect. 2.6. Conclusions are eventually drawn in Sect. 2.7.

2 Deep Levels in GaN

Semiconductor defects, i.e., each localized deviation from the ideal periodicity of the crystal lattice, can have detrimental effects on the performance and lifetime of semiconductor devices, as already discussed. The specific impact of the structural defect on device operation depends on its location, concentration, and on its properties as a trapping/recombination center (Buffolo et al. 2022). While the two former properties strongly depend on the growth and fabrication process, the latter are typically determined by the physical origin of the defect itself.

Generally speaking, lattice imperfections can be divided into two macro-categories: native defects and impurities-related defects. The former class includes all the types of defects that involve an improper re-arrangement of the atoms forming the host lattice (Ga or N, in the case of gallium nitride), such as vacancies, interstitials, antisites, voids, dislocations, etc. (Yam et al. 2011; Lyons et al. 2021). On the other hand, the second class of defects represents a further deviation from the ideal periodicity of the crystal, since also atoms not belonging to the ideal unit cell can contribute to its deformation. Impurities-related defects, which also include dopant atoms, often originated from residues of the precursors adopted for the growth process, or from unwanted contaminants. For GaN, these mainly include, not being limited to, Si, Mg, H, O, C, and Fe.

Evidences of the presence of semiconductor defects can either be directly found through in-depth material-level investigations carried out through sophisticated techniques—such as positron annihilation spectroscopy (PAS), SIMS, atomically-resolved TEM, etc., (Meneghini et al. 2023)—or by indirect probing through the analysis of the effects that such defects have on specific performance figures of the devices/materials which contain them. This latter approach, though being less indicative in terms of the physical origin of the defect itself, is much more versatile, and allows to relate main defect properties (concentration, activation energy, capture cross-section, capture barrier, Franck–Condon shift, etc., (Alkauskas et al. 2016)) to specific performance- and lifetime-limiting processes that can occur during device operation.

Among the aforementioned defects properties, activation energy and capture cross-section represent the most important ones, since, to a first level of approximation, they allow to the determine the emission rate of a trapped carrier, and therefore the impact of the related defects on the dynamic properties of a GaN HEMT (Santi et al. 2021). Experimentally, these parameters can be estimated by (indirectly) evaluating the variation in the emission rate with temperature and by building the related Arrhenius plot. These plots represent a characteristic signature of the emission properties of a specific defect, which can be used to identify correspondences with similar defect-related processes previously reported in literature (Bisi et al. 2013). This methodology ultimately allows to discriminate the presence of a specific defect, i.e., of a specific structural imperfection, in different devices, thus helping manufacturers in improving their growth/processing technology.

In conjunction with the graphical analysis of trap signatures, detected defects can also be directly comparted in terms of their activation energy, i.e., their energetic positioning within the forbidden gap of the semiconductor material. Referring again to the defects database presented in Bisi et al. (2013), and further extended to more than 480 records of traps in Meneghini et al. (2021), the activation energies, determined by means of different techniques, can be grouped on the basis of the assignment of the physical origin performed by the authors, as schematically reported in Table 2.1. The main properties and considerations regarding the trap levels related to the most common defects and impurities found in GaN will be discussed in the following paragraphs.

3 Native Defects

This family of crystalline defects mainly includes nitrogen antisites (N_{Ga}) nitrogen interstitial (N_I), nitrogen vacancies (V_N), gallium interstitials (Ga_I), gallium vacancies (V_{Ga}), or multi-dimensional clusters of the formers, referred to as extended defects (e.g., dislocations). The spread in activation energy of typical native defects within the bandgap of GaN is reported in Fig. 2.1.

Nitrogen antisite defects are assumed to form a mini-band of levels between E_C-0.66 eV and EC-0.5 eV (Boturchuk et al. 2018; Cho et al. 2003, 2001; Kindl et al. 2009; Umana-Membreno et al. 2008; Stuchlíková et al. 2010; Hacke et al. 1994; Hacke et al. 1994; Chung et al. 2000; Chen et al. 2012; Honda et al. 2012; Tokuda 2016; Nakano 2018). No direct correlation between the levels associated to N_{Ga} and extended defects are reported in literature, to date.

Nitrogen interstitials (N_I) were found to originate trap states between $E_C - 1.2$ eV and $E_C - 0.76$ eV (Asghar et al. 2004; Fang et al. 2000; Look et al. 2005; Belahsene et al. 2015; Lee et al. 2019; Boturchuk et al. 2018; Polyakov et al. 2016; Zhang et al. 2015), with average preferential energy positioning at $E_C - 1.02$ eV, $E_C - 0.89$ eV, and $E_C - 0.79$ eV: the first and last one of these levels have also been preliminarily associated with an extended defect in Boturchuk et al. (2018). Considering also the other species of nitrogen-related native defects, N_I exhibits the highest activation energy, which allows them to also act as efficient recombination centers, thus limiting the efficiency of GaN-based optoelectronic devices in which are eventually contained.

Nitrogen vacancies represent the most commonly observed type of native defect in GaN, appearing in more than 40 records related to 25 different reports. The related trap states are typically positioned within the $E_C - 0.27$ eV to $E_C - 0.089$ eV band (Look et al. 2005; Boturchuk et al. 2018; Polyakov et al. 2016; Cho et al. 2003, 2008, 2001; Chung et al. 2000; Chen et al. 2012; Honda et al. 2012; Umana-Membreno et al. 2002; Soh et al. 2004; Park et al. 2010; Park et al. 2010; Johnstone et al. 2011; Johnstone et al. 2011; Fang et al. 1998; Choi et al. 2003; Arehart et al. 2008, 2010; Gelhausen et al. 2004; Deng et al. 2011; Nakarmi et al. 2009; Nguyen et al. 2016), and most of the entries belong to the mini-band located around $E_C - 0.24$ eV, which thus identifies the energy of the level commonly associated to V_N in gallium nitride.

Table 2.1 Database of traps detected in GaN- and AlGaN-based devices

Type of defect or impurity	Physical origin	Reported E_a (eV)	Detected as	References
Native Defects				
Nitrogen interstitial	N_I-related	1–0.76	EC −	Asghar et al. 2004; Fang et al. 2000; Look et al. 2005; Belahsene et al. 2015; Lee et al. 2019; Boturchuk et al. 2018)*,&, (Polyakov et al. 2016)
		2.42	EV +	Zhang et al. 2015)
Nitrogen antisite	N_{Ga}-related	0.66–0.5	EC −	Fang et al. 2000; Boturchuk et al. 2018; Cho et al. 2003; Cho et al. 2001; Kindl et al. 2009; Umana-Membreno et al. 2008; Stuchlíková et al. 2010; Hacke et al. 1994; Hacke et al. 1994; Chung et al. 2000; Chen et al. 2012; Honda et al. 2012; Tokuda 2016)
		2.48	EV +	Nakano 2018)
Nitrogen vacancy	V_N-related	0.27–0.089	EC −	Look et al. 2005; Cho et al. 2001; Chung et al. 2000; Umana-Membreno et al. 2002; Soh et al. 2004; Park et al. 2010)&, (Cho et al. 2003; Johnstone et al. 2011; Fang et al. 1998)&, (Honda et al. 2012; Choi et al. 2003; Arehart et al. 2008)*, (Cho et al. 2008)&, (Boturchuk et al. 2018; Polyakov et al. 2016; Chen et al. 2012; Park et al. 2010; Johnstone et al. 2011; Arehart et al. 2010; Gelhausen et al. 2004; Deng et al. 2011; Nakarmi et al. 2009; Nguyen et al. 2016)*
		0.4	EC −	Deng et al. 2011)
		3.28	EV +	Zhang et al. 2015)
	V_N or Mg_{Ga}-V_N	3.18	EV +	Nakano 2018)
	V_N complexes	0.35	EC −	Look et al. 2005)
	$(V_N)^{3+}$	0.53	EC −	Nakarmi et al. 2009)
	V_N clusters	0.613	EC −	Johnstone et al. 2011)&
Gallium interstitial	Ga_I complexes	0.91	EC −	Hierro et al. 2002)&
	Ga_I-related	0.8	EC −	Polyakov et al. 2016)*

(continued)

Table 2.1 (continued)

Type of defect or impurity	Physical origin	Reported E_a (eV)	Detected as	References
Gallium vacancy	V_{Ga} or unspecified complexes (main band)	2.85–2.47	EC –	Arehart et al. 2008; Arehart et al. 2010; Polyakov et al. 2004; Calleja et al. 1997; Zhang et al. 2012; Hierro et al. 2000; Henry et al. 2012; Armstrong et al. 2012)*
		1.02, 0.83, and 0.89, 0.9, 0.86	EV +	Belahsene et al. 2015; Polyakov et al. 2016; Zhang et al. 2015)*, (Tokuda 2016)
	V_{Ga}-O	1.12, 0.64, 0.6	EC –	Stuchlíková et al. 2010)
	V_{Ga}-related	0.62	EC –	Cho et al. 2008)
		1.8	EV +	Nakano 2018)
		0.25	EV +	Deng et al. 2011)
	V_N-V_{Ga} complexes	0.24	EC –	Honda et al. 2012)*
Extended defects	Many	1.118–0.24	EC –	Soh et al. 2004)[&], (Stuchlíková et al. 2010)[*,&], (Arehart et al. 2010)[&], (Kindl et al. 2009)[&], (Look et al. 2005)[&], (Fang et al. 2005)[&], (Boturchuk et al. 2018)[*,&], (Galiano et al. 2018)[&], (Tokuda 2016)[&], (Nguyen et al. 2016)[&]
Impurities-Related Defects				
Silicon	Si-related	0.59	EC –	Chen et al. 2003)
		0.4	EC –	Soh et al. 2004)
		0.37	EC –	Soh et al. 2004)
Magnesium	Shallow acceptor level	0.16–0.24	EV +	Nakano 2018)**, (Arehart et al. 2008; Hierro et al. 2000)*, (Armstrong et al. 2012; Sasikumar et al. 2012)
	Mg-H complexes	0.08	EV +	Kogiso et al. 2019)*
		0.62	EC –	Hierro et al. 2000)
	Mg-related	0.355, 0.597	EC –	Umana-Membreno et al. 2002; Polyakov et al. 2010)
	Mg-V_N complexes	0.44	EC –	Gelhausen et al. 2004)
Hydrogen	H-V_{Ga} complexes	2.62–2.47	EC –	Arehart et al. 2008; Zhang et al. 2012)
	C- or H-related	0.578, 0.49	EC –	Wang et al. 1998)*, (Lee et al. 1995)*

(continued)

Table 2.1 (continued)

Type of defect or impurity	Physical origin	Reported E_a (eV)	Detected as	References
Oxygen	O-V_{Ga} complexes or dislocations	1.118, 0.642, 0.599	EC −	Stuchlíková et al. 2010)[&]
	O_N	0.44, 0.01	EC −	Deng et al. 2011; Caesar et al. 2012)*
Iron	Fe-related	0.397, 0.5, 0.57, 0.66, 0.68, 0.72	EC −	Umana-Membreno et al. 2008; Polyakov et al. 2004; Cardwell et al. 2013a; Silvestri et al. 2013)
		2.5	EV +	Aggerstam et al. 2007)
	Fe^{2+} or Fe^{3+} related	0.34	EC −	Heitz et al. 1997)
		3	EV +	Aggerstam et al. 2007)
Carbon	$(C_N)^0$	3.31–3.22	EC −	Hierro et al. 2000)*, (Arehart et al. 2008, 2010; Hierro et al. 2002; Zhang et al. 2012; Henry et al. 2012; Sasikumar et al. 2012; Zhang et al. 2012)*, (Tokuda 2016; Nguyen et al. 2016; Lee et al. 2017)*
		0.29	EV +	Kogiso et al. 2019; Kachi, "The origin of carbon-related carrier compensation in p-type GaN layers grown by MOVPE" 2018)
	$(C_N)^{-1}$	0.8–0.95	EV +	Honda et al. 2012; Kogiso et al. 2019; Lee et al. 2017)*, (Tokuda 2016; Kachi et al. 2018; Huber et al. 2015)
	C_I	1.35–1.2	EC −	Arehart et al. 2008; Arehart et al. 2010; Nguyen et al. 2016; Hierro et al. 2000)*
	C- or H-related	0.578, 0.49, 0.14	EC −	Wang et al. 1998)*, (Lee et al. 1995)*
	C_{Ga}	0.4	EC −	Honda et al. 2012)*

References marked with *,& and ** indicate, respectively, tentative associations, traps related to extended defects, and tentative grouping performed by the authors of this work. Activation energies and related references are reported in the same order. Data re-used from Meneghini et al. (2021) M. Meneghini et al., "GaN-based power devices: Physics, reliability, and perspectives". Journal of Applied Physics 14 November 2021; 130 (18): 181,101. https://doi.org/10.1063/5.0061354, with the permission of AIP Publishing

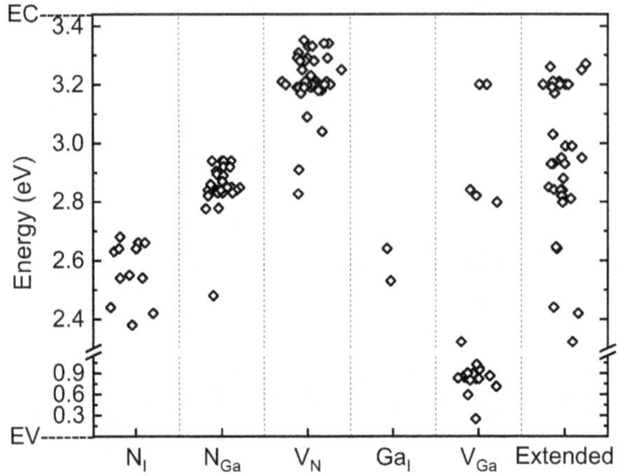

Fig. 2.1 Energy states detected in GaN and associated with native or extended defects. The reference energy level, i.e., E = 0 eV, corresponds to the top of the valence band. Reprinted from Meneghini et al. (2021) Meneghini et al., "GaN-based power devices: Physics, reliability, and perspectives". Journal of Applied Physics 14 November 2021; 130 (18): 181,101. https://doi.org/10.1063/5.0061354, with the permission of AIP Publishing

With respect to the previously described distributions, several outliers are reported in literature: deeper levels located at E_C −0.35 eV (Look et al. 2005), E_C −0.4 eV (Deng et al. 2011), E_C −0.53 eV (Nakarmi et al. 2009), and E_C −0.613 eV (Johnstone et al. 2011) were associated with V_N complexes, simple nitrogen vacancies, triply ionized nitrogen vacancies and clusters of vacancies, respectively.

Additionally, V_N-related levels such as the ones located at E_C −0.19 eV (Cho et al. 2001), at E_C −0.23 eV (Cho et al. 2003, 2001), and at E_C −0.25 eV (Cho et al. 2008) were found to be compatible with extended defects.

Only few papers refer to *gallium interstitials* as the possible physical origin of trap levels detected in GaN. In particular, the authors in Hierro et al. (2002) associate an E_C −0.91 eV level with Ga_I complexes located along dislocations, whereas in Polyakov et al. (2016) a trap located at E_C −0.8 eV detected in a GaN HEMT was tentatively ascribed to gallium interstitials.

Gallium vacancies are often associated with levels located in the E_C −2.42 eV to E_C −2.85 eV band, in particular around E_C −2.6 eV (Belahsene et al. 2015; Polyakov et al. 2016, 2004; Zhang et al. 2015, 2012; Tokuda 2016; Arehart et al. 2008, 2010; Calleja et al. 1997; Hierro et al. 2000; Henry et al. 2012; Armstrong et al. 2012). Among these levels, some have been tentatively associated with oxygen- (Belahsene et al. 2015; Polyakov et al. 2016) or hydrogen- (Arehart et al. 2008; Zhang et al. 2012), containing complexes formed with gallium vacancies. V_{Ga}-O-related defects were associated to the E_C −1.12 eV, the E_C −0.64 eV, and the E_C −0.6 eV levels, found in Stuchlíková et al. (2010).

Finally, considering the traps detected outside the reference energy band for V_{Ga}, we can find at E_C −0.24 eV V_N-V_{Ga} complexes (Honda et al. 2012), as well as simple gallium vacancies at E_C −0.62 eV in Cho et al. (2008), at E_C −1.64 eV in Nakano (2018) and at E_C −3.19 eV in Deng et al. (2011).

As far as *extended defects* are concerned, related traps are reported to cover a wide range of the upper portion of the GaN bandgap, i.e., from E_C −1.118 eV to E_C −0.24 eV (Look et al. 2005; Boturchuk et al. 2018; Kindl et al. 2009; Stuchlíková et al. 2010; Tokuda 2016; Soh et al. 2004; Arehart et al. 2010; Nguyen et al. 2016; Fang et al. 2005; Galiano et al. 2018). Closer to the conduction band edge, from E_C −0.27 eV to E_C −0.17 eV, several of these traps cover the same range of activation energies typically associated to V_N-related defects. Similarly, the band from E_C − 0.641 eV to E_C −0.41 eV finds a good match with the energetic distribution of N_{Ga}-related defects. Curiously, no correspondence with the energy ranges covered by V_{Ga}-related defects was found: in contrast with the behavior of two former types of defects, this indicates that gallium vacancies either tend to behave as non-interacting point defects, or that they are not prone to segregate in highly defective semiconductor regions.

4 Traps Related to Impurities in GaN

Silicon is intentionally introduced in GaN to act as n-type dopant. When occupying a substitutional position in the GaN lattice, Si features a very shallow donor level, whose activation energy ranges from E_C −0.011 eV to E_C −0.028 eV, for moderately doped GaN layers (Götz et al. 1996; Kumar et al. 2012; Irokawa et al. 2005). Reports of deeper levels associated to Si are scarce and only include traps located at E_C − 0.37 eV (Soh et al. 2004), at E_C −0.4 eV (Soh et al. 2004), and at E_C −0.59 eV (Chen et al. 2003).

Magnesium serves as p-type dopant in GaN, owing to the relatively shallow level, typically positioned between 0.12 eV to 0.2 eV above the valence band edge (Tanaka et al. 1994; Nakayama et al. 1996) (Fig. 2.2), which exhibits acceptor characteristics when Mg occupies a substitutional position with respect to Ga. Additional trap states detected near the valence band edge, such as the levels at E_C −3.2 eV in Sasikumar et al. (2012), at E_C −3.22 eV (Arehart et al. 2008; Hierro et al. 2000), at E_C −3.25 eV (Armstrong et al. 2012), and at E_C −3.28 eV (Nakano 2018), can also be associated to Mg_{Ga}, though alternative physical origins are nominally assigned to them.

Considering the records listed in Table 2.1, generic Mg-related defects were assigned to trap levels at E_C −0.355 eV (Umana-Membreno et al. 2002) and at E_C −0.597 eV (Polyakov et al. 2010). Levels at E_C −3.36 eV (Kogiso et al. 2019) and E_C −0.62 eV level in Hierro et al. (2000) were tentatively assigned to Mg-H complexes, whereas an E_C −0.44 eV level was assigned to Mg-V_N complexes in Gelhausen et al. (2004).

Carbon atoms are intentionally introduced in GaN to compensate the intrinsic n-type conductivity of the material by leveraging the formation of a deep acceptor level, whose energy falls in the E_C − 2.64 eV to E_C − 2.49 eV band (Honda et al. 2012; Tokuda 2016; Kogiso et al. 2019; Lee et al. 2017; Kachi et al. 2018; Huber

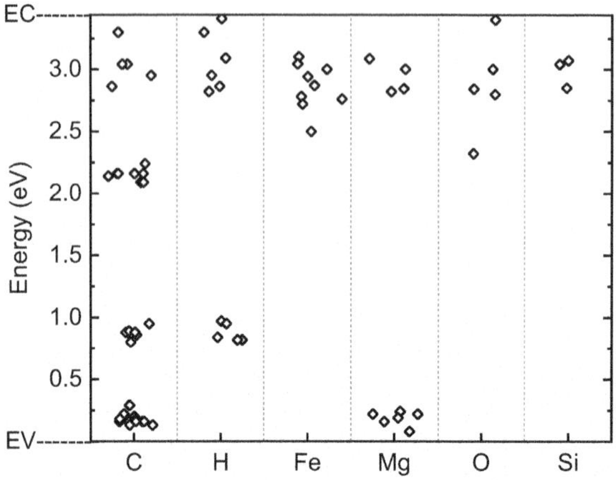

Fig. 2.2 Energy states detected in GaN and associated with impurities-related defects. The reference energy level, i.e.,. E = 0 eV, corresponds to the top of the valence band. The scattering of the detected levels within the lower part of the gap has been magnified with respect to Fig. 2.1. Reprinted from Meneghini et al. (2021) Meneghini et al., "GaN-based power devices: Physics, reliability, and perspectives". Journal of Applied Physics 14 November 2021; 130 (18): 181,101. https://doi.org/10.1063/5.0061354, with the permission of AIP Publishing

et al. 2015), associated with the formation of large concentrations of C_N defects. This process results in the pinning of the Fermi level at about $E_V + 0.9$ eV, and in the subsequent formation of semiconductor layers with very-low conductivity. Also a second band of shallow energy levels, approximately ranging in the $E_C - 3.31$ eV to $E_C - 3.15$ eV band, is often associated to C_N (Tokuda 2016; Arehart et al. 2008, 2010; Nguyen et al. 2016; Hierro et al. 2002, 2000; Zhang et al. 2012; Henry et al. 2012; Kogiso et al. 2019; Sasikumar et al. 2012; Zhang et al. 2012; Lee et al. 2017; Kachi, et al. 2018).

Interstitial C_I is reported to introduce a deep donor level in the upper part of the gap, with activation energies ranging from $E_C - 1.35$ eV to $E_C - 1.2$ eV (Arehart et al. 2008, 2010; Nguyen et al. 2016; Hierro et al. 2000).

Finally, additional shallower levels detected at $E_C - 0.14$ eV (Lee et al. 1995), $E_C - 0.4$ eV (Honda et al. 2012), $E_C - 0.49$ eV (Lee et al. 1995), and $E_C - 0.578$ eV (Wang et al. 1998) have been tentatively ascribed to carbon- or hydrogen-related defects.

Like carbon, also *iron* is used in the gallium nitride technology, especially in the fabrication of HEMTs, to lower the intrinsic conductivity of target semiconductor layers. Several Fe-related deep acceptor levels, positioned in the $E_C - 0.94$ eV to $E_C - 0.34$ eV band, have been detected and reported in literature (Umana-Membreno et al. 2008; Polyakov et al. 2004; Cardwell et al. 2013a; Silvestri et al. 2013; Heitz et al. 1997; Aggerstam et al. 2007). In particular, the latest reports (Horita et al. 2020) indicate that the dominant level associated to Fe in HVPE GaN is represented by a trap positioned at $E_C - 0.58$ eV, and featuring a capture cross-section of $\sigma = 2 \times 10^{-15}$ cm^2, which is likely to be related to Fe_{Ga} sites.

Oxygen is another common impurity introduced in GaN during growth, and/or subsequent treatments. It is found to favor the formation of traps at $E_C - 1.118$ eV, $E_C - 0.642$ eV, and $E_C - 0.599$ eV related to complexes with other defects, such as V_{Ga}-O or dislocations (Stuchlíková et al. 2010) and to O_N in Deng et al. (2011); Caesar et al. 2012).

Like oxygen, also *hydrogen* tends to form complexes with other chemical species. Literature reports associate levels located between $E_C - 2.62$ eV and $E_C - 2.47$ eV to complexes formed with V_{Ga} (Arehart et al. 2008; Zhang et al. 2012), whereas levels at $E_C - 0.578$ eV and at E_C.

5 Compact Modelling of Charge Trapping Effects

An accurate simulation of trap effects starting from first principles can be difficult even for the most powerful software and for the most skilled device physicist. For this reason, it can be worth to simplify the problem by using a compact model. Even though not as accurate and complete as numerical simulators based on Schrödinger-Poisson solvers or non-equilibrium Green's functions (NEGF), compact modelling is still able to provide useful qualitative information while exploiting all the advantages of the compact modelling, i.e., computational speed, ease of use and integration with SPICE circuit simulators. Given that, in a compact model, all the information on the physics of the device has to be defined by the user, the correctness and the accuracy of the results vary significantly depending on the quality of the code.

One of the most widely used compact models for GaN HEMTs is the ASM GaN HEMT model. In the official release, traps are modelled as RC networks or by networks composed of capacitors and ideal current generators, in order to reproduce the exponential behavior typical of carrier emission processes due to thermionic emission (Albahrani et al. 2019). This approach has several issues, arising from the lack of any link between the physical properties of the actual traps in the device and the parameter values set in the model. Moreover, it cannot be used to reproduce the response of defects arranged as energy minibands or with spatial distribution, which have non-ideal exponential behavior (Bisi et al. 2013).

Advancements with respect to the base model were done by Kellogg et al. (2020), who propose to use a 3-pole RC model for matching the stretched exponential response. While this approach leads to an improvement in the quality of the simulation, it is still far from optimal, since a stretched exponential is composed of a theoretically-infinite amount of RC networks (Modolo et al. 2022). Additionally, even in this case, the model parameters are extracted in order to get the best-fit of the experimental data, not based on real and measured physical trap parameters.

This problem was addressed by Li et al. (2018). They still handle the traps as a compensating voltage limiting the external bias, generated by RC elements (see Figs. 2.3), but develop an iterative algorithm to find a more optimized solution with

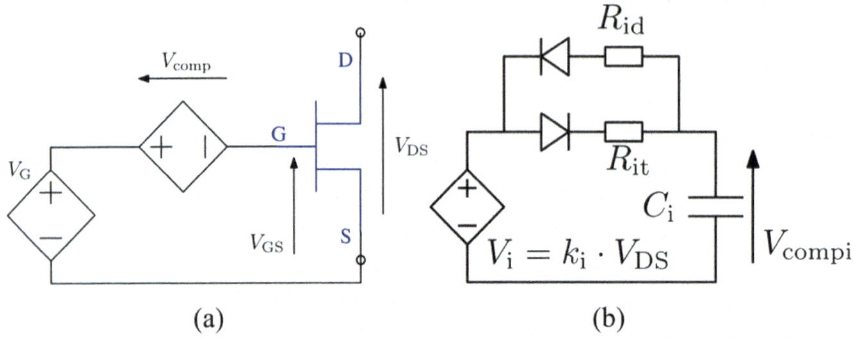

Fig. 2.3 Equivalent circuit used to include trap effects in a compact model, as shown by Li et al. (2018)

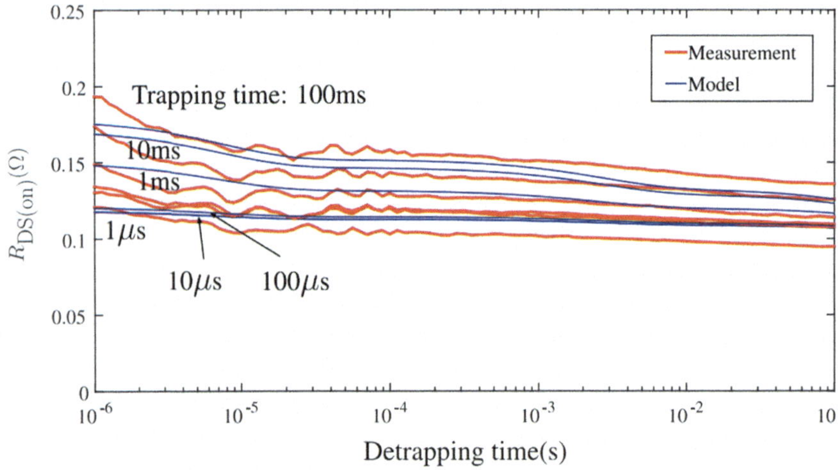

Fig. 2.4 Typical results obtained by compact modelling when non-physical trap parameters are used, as shown by Li et al. (2018)

multiple RC networks. While this innovation improves the quality of the model, results are still far from the experimental data and the matching quality varies significantly depending on the bias and temperature condition under analysis, as shown in Fig. 2.4. The root cause remains the same: the trap model parameters are extrapolated from the experimental data and not related to the defect physical behavior.

Weiser et al. tried to include more physical insight by using a Gaussian distribution of time constants instead of a distribution originated by the fit of the data (Weiser et al. 2021). This type of distribution, while more accurate in some specific cases, is not guaranteed to match any possible defect distribution present in a device under test.

A step forward was done by Modolo et al. (2022). They extract the real distribution of the traps according to the inverse Laplace transform method presented in Modolo et al. (2022), and use the values of the activation energies and cross-sections to compute the values of the RC elements instead of obtaining them through fits of the experimental data. This procedure is based on the Arrhenius law, as shown in Eq. 2.1

$$\tau_i(T) = \frac{1}{T^2}\exp\left(\frac{E_{a,i}}{kT} + ln\left(\frac{h^3}{2\sqrt{3}(2\pi)^{\frac{3}{2}} m_r m_0 k^2 g \sigma_{c,i}}\right)\right) \tag{2.1}$$

Each time constant is then used to extract the RC values at each temperature according to Eq. 2.2, where A_i is the amplitude of the time constant computed by the inverse Laplace analysis.

$$R_i C_i(T) = \tau_i(T)$$

$$R_i(T) = A_i(T) \tag{2.2}$$

This procedure brings the results of the compact modelling to unprecedented levels of accuracy. As shown in Fig. 2.5, the compact model can reproduce the experimental data in a wide temperature range, matching the measurements in a timeframe of more than seven orders of magnitude, both in the filling and in the recovery phase. Since the trap parameters used in the simulation are based on the real activation energy and capture cross-section of the defect and supported by Arrhenius law, the model is fully physically-supported and consistent. Therefore, it can be used to extrapolate the behavior of the device even outside the bias and temperature range used for its calibration, a feat impossible to achieve by usual compact models that rely only on a fitting approach.

6 TCAD Simulations of Charge Trapping Effects

Numerical device simulations based on the drift–diffusion or hydrodynamic transport models are a rapid yet powerful tool to investigate charge trapping effects in GaN transistors and, specifically, to validate hypothesis on their role as performance-, stability-, and reliability-limiting phenomena. In order to practically fulfill this function, a minimum set of physical models needs to be available within the adopted simulation platform and solved self-consistently with the transport model equations that are specific either to the GaN material system, or the transistor type, or the trap physics.

These "mandatory" models, used by various groups (Joshi et al. 2016; Uren et al. 2012; Verzellesi et al. 2014; Bakeroot et al. 2018; Tallarico et al. 2022), include (i) the spontaneous and piezoelectric polarization charges at the interfaces between different III-nitride materials (Ambacher et al. 1999, 2000), (ii) a Schottky contact

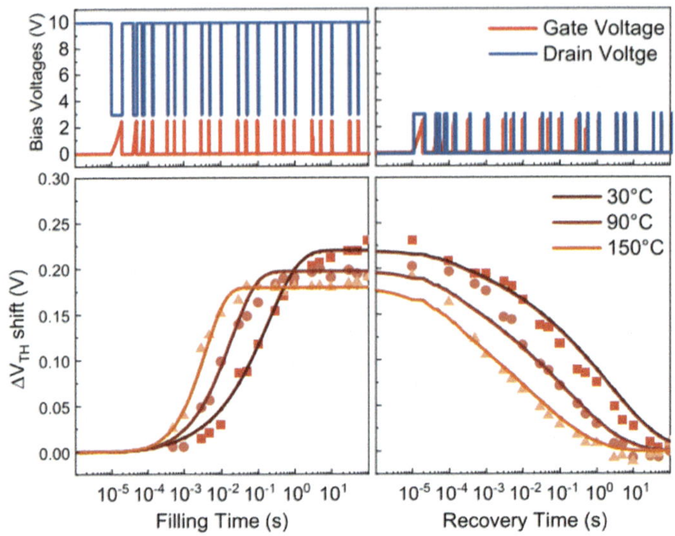

Fig. 2.5 Results of the improved compact model presented in Modolo et al. (2022)

model including both thermionic and field emission mechanisms (Sze and Ng 2006; Ieong et al. 1998), (iii) a trap model based on the Shockley–Read–Hall detailed balance between the different carrier emission and capture processes linking the trap to the conduction and valance bands and allowing the trap-occupation dynamics to be accounted for without any quasi-static approximation (Shockley and Read 1952).

Another aspect that requires particular attention is the way impurities intentionally introduced in the epitaxy are modelled. This is specifically the case of chemical elements adopted to compensate for the unintentional n-type conductivity of GaN in the buffer layer of GaN transistors, rendering the buffer semi-insulating and hence increasing the OFF-state breakdown voltage (V_{BD}). This however comes at the cost of the creation of deep-level traps within the GaN bandgap, in addition to those associated to intrinsic crystallographic defects and therefore potentially at the cost of an increase in detrimental charge trapping effects.

In RF HEMTs, iron (Fe) is generally adopted as a compensating impurity (Silvestri et al. 2013; Heikman et al. 2002; Meneghini et al. 2014). Fe-related traps are acceptor-like levels located about 0.5 eV below the GaN conduction band edge. As far as the spatial distribution of Fe traps is concerned, a realistic model is that of considering Fe concentration constant in the GaN buffer up to the distance from the interface with the barrier corresponding to the depth at which Fe doping is switched off during buffer growth and an exponentially decaying concentration with a slope of one decade every 0.4 μm, to account for the "memory effect" of the incorporation of Fe dopants (Uren et al. 2012).

Carbon (C) doping is instead a common technological solution to reduce buffer conductivity and increase breakdown voltage (V_{BD}) in GaN power transistors (Uren et al. 2012; Bahat Treidel et al. 2010). Collateral, trap-related effects are even stronger than those associated to Fe doping (Bahat Treidel et al. 2010; Alamo and Lee 2019; Chini et al. 2016). Depending on the growth conditions, C atoms can either substitute N or Ga sites, occupy interstitial locations in the crystal, or form complexes with intrinsic defects (Lyons et al. 2021, 2014; Remesh et al. 2020; Matsubara and Bellotti 2017). In typical undoped GaN layers used as buffer in HEMTs, the position of the Fermi level is such that both acceptor and donor traps can form. Several works, discussing either simulation or experimental results, have indicated the occurrence of partial "auto-compensation" between the dominant, deep acceptor traps generally attributed to C_N levels and the concomitantly introduced (i.e., non pre-existing) shallow donors, reducing the *effective* concentration of acceptor traps well below the level of the introduced C concentration (especially in case of extrinsic C doping) (Verzellesi et al. 2014; Chini et al. 2016; Zagni et al. 2019; Rackauskas et al. 2018; Uren et al. 2017; Koller et al. 2018; Joshi et al. 2019; Fariza et al. 2016). The way we accounted for C doping in all of our works is actually based on this partial donor–acceptor compensation model and is consistent with the current understanding of the energy levels associated with C introduction into GaN (Zimmermann et al. 2021). While early DFT calculations based on LDA and GGA approximations predicted the substitutional C_N state to be a shallow acceptor (Wright 2002), more accurate DFT calculations using the HSE functional consistently predicted the C_N acceptor level in the 0.9–1.1 eV range above the valence band maximum (Lyons et al. 2010). This is in good agreement with extensive experimental findings, including *(i)* measurement of thermally activated reduction in the resistivity of carbon-doped GaN (Richter et al. 2020); *(ii)* observation of p-type conductivity and its thermal activation energy (Iwinska et al. 2016); *(iii)* a hole trap detection by DLTS and minority carrier transient spectroscopy (Honda et al. 2012); *(iv)* results from photocapacitance, pulsed photoionization, and DLOS (Hierro et al. 2000); *(v)* yellow luminescence studies (Honda et al. 2012; Yamada et al. 2018).

7 Charge Trapping Effects in GaN RF Transistors

7.1 Barrier Versus Buffer Trapping Effects

In RF HEMTs both barrier and buffer traps can lead to non-negligible current-collapse effects, but the associated physical mechanisms are different and their peculiarities can effectively be analyzed by means of numerical device simulations.

In 0.5 mm Schottky-gate AlGaN/GaN HEMTs on SiC substrates, drain current transient (DCT) spectroscopy experiments yielded four different trap signatures. One trap with an energy of 0.52 eV (T1) could be attributed to the Fe impurities introduced in the GaN buffer layer to compensate for the unintentional n-type conductivity. The other three traps with energies 0.80 eV (T2), 0.87 eV (T3), and 0.76 eV (T4) were assigned to the AlGaN barrier, based on numerical device simulations allowing to dependably reproduce specific features of the associated drain-current waveforms.

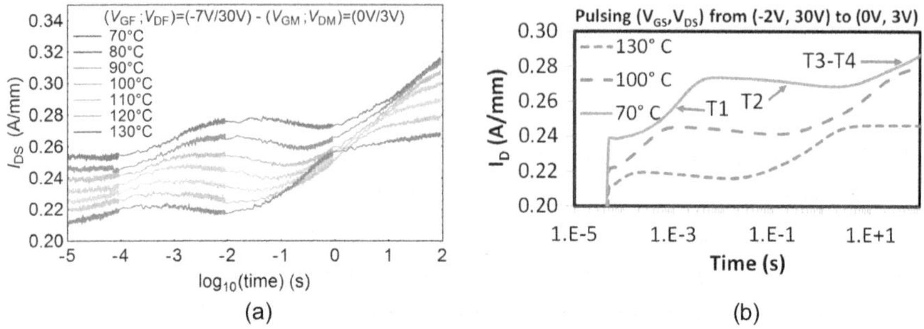

Fig. 2.6 Experimental **a** and simulated **b** temperature-dependent drain current transients from a Schottky-gate AlGaN/GaN HEMTs. Device is pulsed from off-state to on-state conditions

The simulated drain-current transients in response to the application of a double, gate-source, and drain-source pulse from OFF-to-ON-state conditions are compared to the corresponding experimental curves in Fig. 2.6a and b. As can be noted, simulations are able to capture the key aspects of the actual device behavior, i.e., (1) the overall shape of the transient characterized by an increasing drain-current waveform after application of the turn-on double pulse ("low-pass-filter"-like behavior), (2) the presence of separated transients, associated with the different identified traps T1-T2-T3-T4 with amplitudes, types (increasing or decreasing) and time constants at the different temperatures comparable with those obtained experimentally.

Figure 2.7 shows contour plots of the two-dimensional distributions of the negative charge density associated with the Fe-related buffer trap T1 for a temperature of 70 °C. As can be noted, prior to the application of the OFF-to-ON double-pulse ($t = 10^{-7}$ s to 10^{-5} s), the negative trapped charge density peaks in the GaN buffer far from the AlGaN/GaN interface, indicating that T1 traps are "fed" by drain-source buffer leakage electrons under the OFF-state phase. At $t = 10^{-6}$ s, the device has been turned on and the electron current density now peaks at the AlGaN/GaN interface. The T1 trapped charge density, however, cannot change immediately in the buffer owing to the finite trap emission time. As a matter of fact, it decreases in Fig. 2.2 over the 10^{-4}–10^{-2} s time span, consistently with Fig. 2.6 (see curve for T = 70 °C). After $t = 10^{-2}$ s, no significant changes take place for the T1 charge density in the buffer anymore, as the T1 related transient is finished, see Fig. 2.6 (T = 70 °C).

Figure 2.8 shows the simulated negative trapped charge density at T = 70 °C associated with barrier traps (T2-T3-T4) as a function of position along a horizontal cut in the barrier layer at different times during the turn-on transient shown in Fig. 2.6. In this case, barrier traps are "fed" by gate injected electrons during the OFF-state phase and consistently the negative trapped charge density is maximum at $t = 10^{-7}$ s. After application of the turn-on pulse at $t = 5 \times 10^{-5}$ s, the trapped charge remains "frozen" until 10^{-4} s due to the long emission time constants of barrier traps T2-T3-T4. In the 10^{-4} to 10^{-1} s interval, the dominant effect is the negative trapped charge increase associated with the T2 trap and localized at the source end of the barrier region under the gate. This peculiar aspect is due to the reduction of the depletion

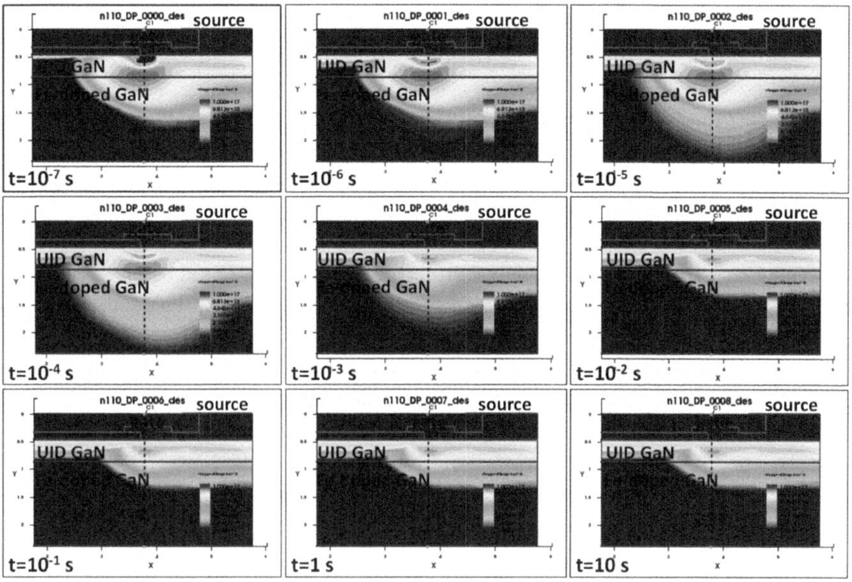

Fig. 2.7 Simulated negative trapped charge density distributions in the buffer for T = 70 °C at different times during the turn-on transient shown in Fig. 2.6

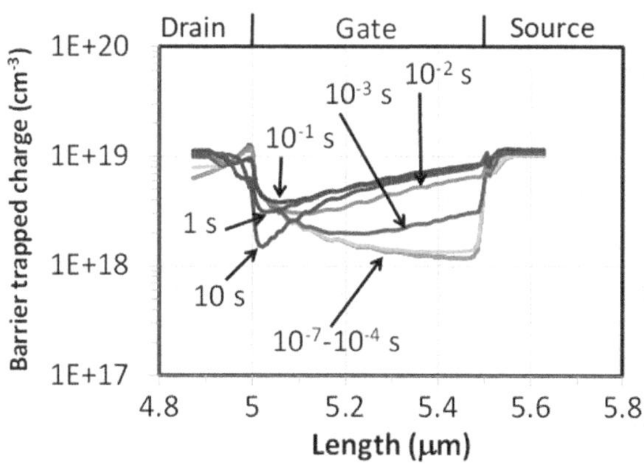

Fig. 2.8 Simulated negative trapped charge density associated with barrier traps vs length along a horizontal cut in the barrier layer for T = 70 °C at different times during the turn-on transient shown in Fig. 2.6

region in the gate-source region, stimulating electron capture into barrier traps. Afterwards, consistently with the drain-current transient in Fig. 2.6, the "standard" barrier trap behavior resumes, leading to a negative trapped charge decrease as a result of electron emission by barrier traps T3 and T4 especially in the drain end region of the barrier layer under the gate.

To summarize, buffer traps are "fed" by buffer punch-through current under OFF-state and emit electrons as the device is turned ON. Barrier traps are "fed" by gate-injected electrons in OFF-state phase. Turn-on transient can be dominated either (i) by electron emission from barrier traps due to reduction of gate electron injection, or (ii) by electron capture mainly localized on the source side owing to contraction of the depletion region.

7.2 Temperature-Independent Current Collapse Effects

There are cases where current-collapse effects are not characterized by an Arrhenius-like temperature dependence that points to trapping as the responsible mechanism. An example of this behavior was analyzed in Zagni et al. (2244), with both measurements and simulations of 0.15 mm gate-length AlGaN/GaN RF HEMTson SiC substrates. The GaN buffer is doped with Feimpurities to reduce the drain-source leakage current and to increase breakdown voltage. The threshold voltage (V_{TH}) is 2.2 V.

DCTs were acquired at temperatures ranging from 30 to 70 °C with 10 °C steps, to characterize the activation energy of the device turn-on process after the application of an OFF-state, 100 s trap filling pulse with ($V_{GS,fill}$, $V_{DS,fill}$) = (V_{TH}−1, 56) V. The measurement conditions were (V_{GS},V_{DS}) = (−1, 2) V. Figure 2.9a shows the results of DCT monitoring. As can be seen, single-exponential behavior is detected with time constants in the 10^{-3}–5×10^{-2} s range depending on temperature, which is typical of pure electron emission from point defects associated with Fe-doping (Cioni et al. 2021; Zagni et al. 2021; Cardwell et al. 2013b). At each temperature, DCTs time constants were extracted by fitting the transients with stretched exponential functions (Bisi et al. 2013), and determined as the point where the $dI_D/d[\log_{10}(t_{meas})]$ signals peaked, see Fig. 2.9b. The time constants are then used to extract the activation energy (E_A) of the thermally activated emission by linear fitting of the Arrhenius plot as shown in the inset of Fig. 2.9b. Extracted E_A = 0.55 eV is typical of Fe-associated traps (Meneghini et al. 2021; Bisi et al. 2013; Cioni et al. 2021). As shown in Fig. 2.9, however, besides this process there is also another one characterized by time constants of about 1–2 s and negligibly accelerated by temperature.

We thus performed numerical device simulations to confirm the role of Fe-related traps and to gain more insights into the nature of the second, T-independent process shown in Fig. 2.9. Fe traps in the buffer were modelled as acceptor-like states 0.55 eV below GaN conduction band, with a capture cross-section for electrons of 1.7×10^{-15} cm^2. Analogous to the actual samples, Fe trap concentration of 1×10^{18} cm^{-3} was uniform up to 0.6 μm below the AlGaN/GaN interface, from which concentration exponentially decayed by one decade per 0.4 μm (Uren et al. 2012; Zagni et al. 2021). Interestingly, the slow temperature-independent process characterized in Fig. 2.9 could be successfully reproduced by the simulations by assuming that this phenomenon was induced by electron injection from the gate into a thin (2 nm) interface layer between the passivation and the AlGaN barrier. By including a large density of shallow levels with a very small electron capture cross-section, the electron transport through the thin interface layer, possibly governed by a hopping mechanism, is mimicked.

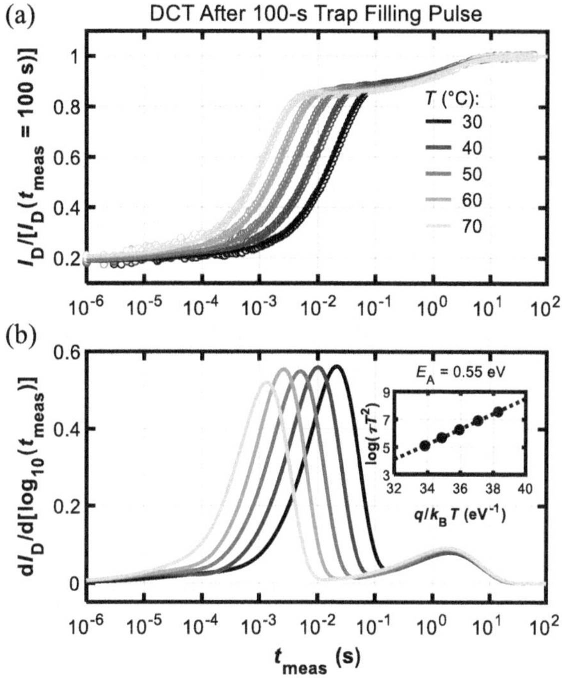

Fig. 2.9 **a** DCTs measured at different temperatures (see legend) after the long 100 s trap filling pulse (dots are data and lines are fitting functions). Current is normalized to final value, i.e., $I_D(t_{meas} = 100\ s)$. **b** $dI_D/d[\log10(t_{meas})]$ used for the extraction of the time constants (τ) to build the Arrhenius plot (see inset) that yields $E_A = 0.55$ eV. In addition to this, another T-independent process is observed (Adapted from Zagni et al. (2244), copyright 2022 by the authors of the cited paper, licensed under CC BY 4.0)

Simulated DCTs, carried out under with the same conditions used to characterize the DUTs, are shown in Fig. 2.10. The good agreement between simulations and experimental data shows the soundness of the modelling framework, validates the role of Fe traps in the observed current-collapse effects and suggests a route for a full understanding of the slow recovery process.

To summarize, we investigated the current dispersion in short gate-length AlGaN/GaN HEMTs for 5G applications. Pulsed I–V characterization revealed the absence of pure gate-lag effects but significant current-collapse effects under double gate and drain voltage pulsing. Comparison of measured and simulated DCTs carried out after a long 100 s trap-filling pulse allowed determining that current collapse was mainly due to Fe-related traps in the GaN buffer and to surface charging effects induced by gate electron injection and transport through the passivation/barrier interface.

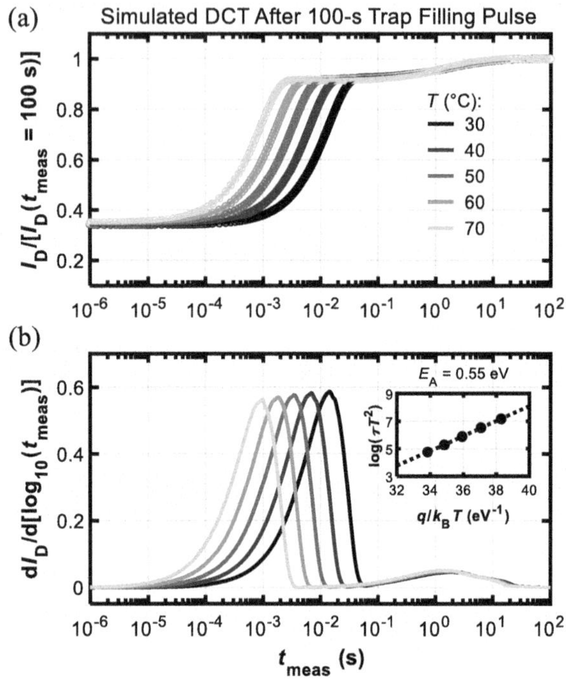

Fig. 2.10 **a** Simulated DCTs at different temperatures (see legend) after the long 100 s trap filling pulse (dots are data and lines are fitting functions). Current is normalized to final value, i.e., $I_D(t_{meas} = 100\ s)$. **b** $dI_D/d[\log10(t_{meas})]$ used for the extraction of the time constants (τ) used to build the Arrhenius plot (see inset) that yields $E_A = 0.55$ eV. Both processes found in Fig. 2.4 are reproduced by the simulations (Adapted from Zagni et al. (2244), copyright 2022 by the authors of the cited paper, licensed under CC BY 4.0)

7.3 Charge Trapping Mechanisms in Fe- and C-Doped Buffers

To master short-channel effects of GaN RF HEMT technologies due to leakage in the buffer, Fe is the most commonly used compensating impurity. However, as a consequence of its wide adoption in power GaN HEMTs for switching converters, C doping is also being investigated as a possible alternative to Fe.

In (Santi et al. 2022) we characterized and simulated Schottky-gate AlGaN/GaN HEMTs with 0.15 μm gate length and 4 μm drain-source gap. Wafers having either Fe- or C-doped buffers were considered. To account for hot-electron effects, the hydrodynamic transport model was employed. In agreement with (Uren et al. 2012), Fe doping was modelled by 0.56–eV electron traps. C doping instead was modelled by including deep acceptor traps at $E_V + 0.9$ eV (i.e., 2.5 eV under E_C) corresponding to C_N states, partially compensated by shallow donors (Verzellesi et al. 2014; Chini et al. 2016; Zagni et al. 2021).

Fe and C doping were found to have a different impact on the negative charge build-up in the buffer under off-state stress. Figure 2.11 shows the contour plots of the negative trapped charge within the buffer layer for the Fe-doped (a) and the C-doped device (b). In the former case, the negative charge corresponds to trapped electrons provided by the source-drain leakage current, consistently with what was

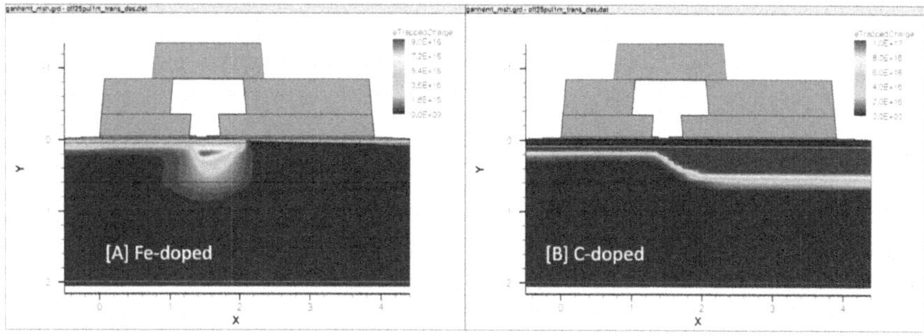

(a) (b)

Fig. 2.11 Contour plots of the negative charge build-up within the buffer layer at $V_{DS} = 25V$ under off-state conditions. For Fe-doped buffer **a** the negative charge corresponds to trapped electrons while for the C-doped one **b** the negative charge corresponds to negatively ionized acceptors due to emission of holes

previously discussed. For the latter case instead, the negative charge corresponds to negatively ionized acceptors stemming from hole emission by the C_N states.

The negative charge build-up in the case of Fe-doped device is localized beneath the gate at the edge towards the drain contact, see Fig. 2.11a. A relatively small portion of the source-drain spacing is involved, hence negligible R_{ON} variations occur. Moreover, the negative charge build-up is rapidly reabsorbed after the stress bias is removed. This happens thanks to the relatively short time constant characterizing electron emission from the Fe states to the conduction band, in the order of 10^{-2} s.

Conversely, the negative charge build-up in the case of C-doped device is spread along the whole gate-drain access region, as shown in Fig. 2.11b. This leads to a significant impact on the device R_{ON}. Additionally, this effect is recovered slowly after stress bias removal because of the large $0.9 - eV$ activation energy (i.e., energy offset from valence band) characterizing $C_N$states, resulting in 10^2–10^3 s hole emission and capture time constants. This observation agrees with the "hole-redistribution" model (Zagni et al. 2021). According to the model, during off-state stress holes are predominantly emitted from C_N states within the gate-drain depleted region of the buffer to move to the gate-source and the bottom regions of the buffer. There, they get captured by the available C_N states. To recover the pre-stress conditions, holes need to be re-emitted from the traps and then to move back to be eventually re-captured in the gate-drain region. The model is described more in detail in Sec. 2.2.

7.4 Fe-Traps Influence on Time-Dependent Breakdown

In GaN RF HEMTs, the intentional incorporation of Fe traps into the buffer has been observed to cause not only current-collapse effects but also time-dependent breakdown voltage (V_{BD}), which can have serious implications for device reliability

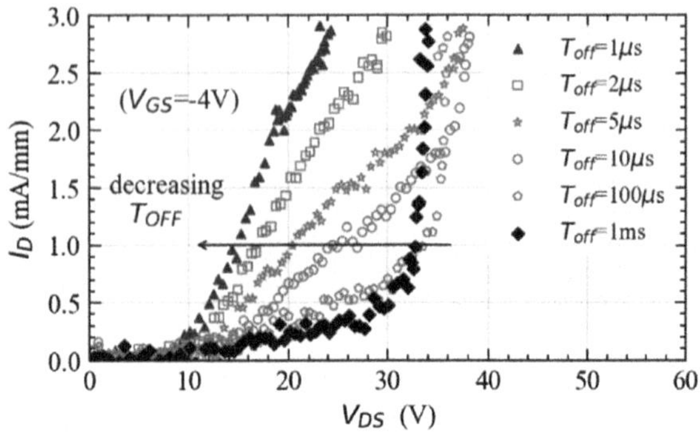

Fig. 2.12 Experimental breakdown characteristics measured for different T_{OFF} in the 1 μs to 1 ms time range at $V_{GS} = -4$ V. The breakdown voltage reduces while decreasing T_{OFF} (Adapted from Cioni et al. (2022) © 2022 IEEE)

(Zhang et al. 2020). This phenomenon was studied in Cioni et al. (2022) using scaled AlGaN/GaN RF HEMTs (with $L_G = 0.1$ μm) on SiC substrates. During the experiment, pulsed I–V characterization was performed on the devices with varying pulse duration (T_{OFF}) in the off-state condition, and the drain current (I_D) was measured to reconstruct the I_D-V_{DS} breakdown curve up to a 3 mA/mm current density. The characterization was then repeated for different T_{OFF} values ranging from 1 μs to 1 ms, and the corresponding V_{BD} was extracted for each T_{OFF} value. The results indicated a significant decrease in V_{BD} for shorter pulse durations (Cioni et al. 2022), as depicted in Fig. 2.12.

To aid in understanding the experimental findings, numerical device simulations were conducted, representing Fe-traps as deep acceptors located approximately 0.5 eV below the GaN conduction band edge. These traps had a concentration of 10^{18} cm^{-3} in the GaN buffer from the SiC substrate interface up to 0.6 μm away from the AlGaN/GaN interface, after which the iron concentration decayed with a slope of one decade every 0.4 μm.

The numerical simulations showed that the peak of ionized Fe traps was situated below the gate edge towards the drain terminal of the device (Uren et al. 2012). This region also corresponded to where the electric field (E) reached its highest value (Rajasingam et al. 2004). Consequently, the time-dependent occupancy of Fe traps affected the concentration of free electrons in the high field region, potentially influencing V_{BD}. Specifically, if the electron concentration was sufficiently high, the high electric field would lead to a significant generation rate due to impact ionization. For short pulses, incomplete ionization of Fe traps resulted in a relatively high concentration of free electrons. On the other hand, longer T_{OFF} led to a higher concentration of trapped electrons in Fe traps, causing a reduction in the density of free electrons in the high field region. As a result, the impact ionization generation rate decreased for a given V_{DS}. This interpretation was supported by the breakdown curves simulated at different T_{OFF} values, as shown in Fig. 2.13. The behavior of the simulated I–V curves aligned with the experimental observations, demonstrating that V_{BD} decreased as T_{OFF} decreased.

In summary, the breakdown voltage (V_{BD}) of scaled 0.1 μm GaN HEMTs obtained through pulsed I–V characterization was found to be influenced by the off-state pulse

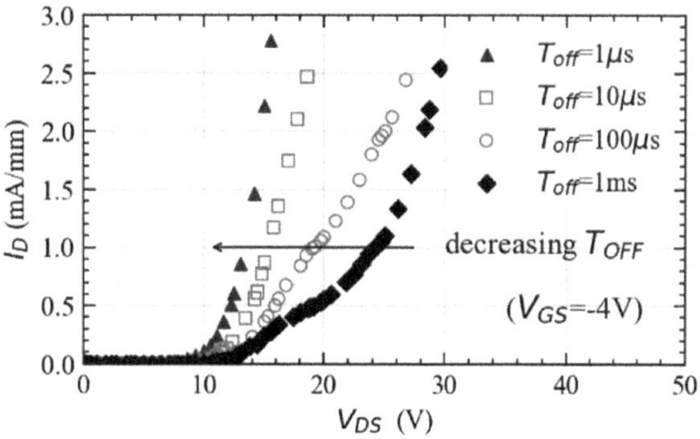

Fig. 2.13 Simulated breakdown characteristics obtained for $V_{GS} = -4$ V and T_{OFF} in the time range between 1 μs and 1 ms. The breakdown voltage reduces while decreasing T_{OFF} (Adapted from Cioni et al. (2022) © 2022 IEEE)

duration. Shorter pulse durations resulted in a decrease in V_{BD}. This decrease in V_{BD} was attributed to a reduced occupation of Fe traps purposely introduced into the buffer layer. The decrease in trap occupation led to an increase in the source-drain leakage current, subsequently raising the impact-ionization rate. Therefore, the time-dependent behavior of V_{BD} in these GaN HEMTs is closely linked to the occupancy of Fe traps and their impact on the device's leakage current and breakdown characteristics.

7.5 Effect of Trap-Filling Bias on the Extraction of the Time Constant of Drain Current Transients

DCT spectroscopy is a widely used method for identifying deep levels within GaN-based devices and measuring the charge emission rate (e_n) from these levels (Bisi et al. 2013). However, the emission time constants ($\tau_e = 1/e_n$) obtained through this method can be influenced by the trap filling conditions used during the measurement process (Joh and Alamo 2011). This, in turn, affects the extraction of the capture cross-section from the Arrhenius plot associated with the trap state (Angelotti et al. 2020).

In (Zagni et al. 2021), researchers investigated these aspects in AlGaN/GaN HEMTs grown on SiC substrates with specific gate dimensions. Temperature-dependent DCT measurements revealed a trap signature with an activation energy (E_A) of 0.52 eV, consistent with known Fe-related buffer traps in the literature (Cardwell et al. 2013b; Axelsson et al. 2016). These traps are involved in the capture and emission of electrons from and to the conduction band. To simulate this behavior, Fe-related traps in the GaN buffer were modelled as deep acceptor traps located at 0.52 eV below the conduction band edge. The density of Fe-traps was assumed to be constant in the bulk of the GaN (10^{18} cm^{-3}) and exponentially reduced towards the AlGaN/GaN interface with a rate of 0.4 μm/decade, accounting for the "memory effect" of the Fe dopant incorporation process.

To investigate the trap dynamics, different trap filling conditions were applied both experimentally and in the simulations. A fixed $V_{GS,fill}$ of − 3.8 V was used, while $V_{DS,fill}$ varied in the range between 15 and 60 V. The time constant (τ) was extracted as the peak of the $dI_D/d\log_{10}t$ signal. It is worth noting that, according to the SRH theory (Schroder 2005), the measured τ can be expected to coincide with the emission time constant (τ_e) only if, under the adopted measurement conditions, electron capture is negligible compared to electron emission. When capture is not negligible, the time evolution of the trapped charge (and thus of the emitted charge) deviates from the dynamics expected for a pure emission process. As a result, the time constant extracted from the resulting DCT will differ from the inverse of the trap emission rate (e_n).

This discrepancy was found to be the cause of the observed dependence of τon $V_{DS,fill}$, shown in Fig. 2.14. At low $V_{DS,fill}$, electron capture by Fe traps during the filling pulse was relatively small, allowing traps to continue capturing electrons during the DCT measurement phase. This concurrent contribution of emission and capture processes reduced the extracted τ. In contrast, at high $V_{DS,fill}$, traps were fully filled at the end of the filling pulse, and the capture contribution during the measurement transient vanished. Consequently, the extracted τapproached its theoretical value ($1/e_n$), resulting in a flattening of the τcurve. This observation indicated that only in the flat region of τ the trap emission rate (e_n) could be accurately estimated from the DCT time constant. The importance of this finding lies in the fact that the dependence of the extracted τon $V_{DS,fill}$ directly affects the estimation of e_n, potentially leading to errors in the extraction of the trap cross-section from the Arrhenius plot if this dependence is not properly considered.

In summary, the apparent emission time (τ) from deep levels in AlGaN/GaN HEMTs is affected by the trap filling conditions during DCT measurements. Particularly, the drain bias applied during the filling pulse showed a significant effect on τ. The explanation for this dependence lies in the non-negligible contribution of

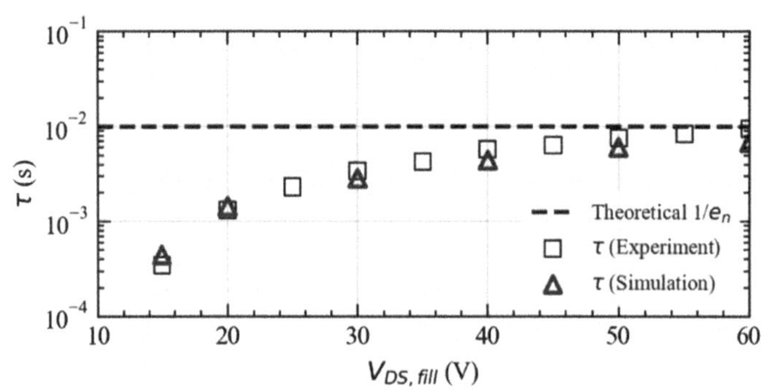

Fig. 2.14 Emission time constants (τ) extracted from measured and simulated DCTs at several $V_{DS,fill}$ levels and $V_{GS,fill}$ = -3.8 V. A variation on $V_{DS,fill}$ produces a readable variation on τ for $V_{DS,fill}$ below 40 V, while data points collected at higher $V_{DS,fill}$ approach the inverse of the trap emission rate ($1/e_n$) (Adapted from Zagni et al. (2021) © IEEE)

trap capture, causing to deviate from the inverse of the trap emission rate. Numerical simulations supported this interpretation, showing that the reduction in τ was a consequence of the small variation in trapped charge during the filling pulse if the applied $V_{DS,fill}$ was not large enough to induce almost complete trap filling.

8 Charge Trapping Effects in GaN Power Devices

8.1 Donor–Acceptor Compensation in C Doped HEMTs

Proper modelling of C-related trap states in GaN transistors is essential for device simulations aimed at investigating charge trapping in the doped buffer and breakdown effects related to source-drain leakage current. Therefore, to calibrate device simulations for a given technology and reproduce the characteristics of real devices, the concentration of acceptor states ($N_{C,A}$) and donor states ($N_{C,D}$), as well as their compensation ratio, i.e., $CR = N_{C,D}/N_{C,A}$, must all be accurately determined (Koller et al. 2021).

By means of calibrated device simulations, ref. (Zagni et al. 2021) analyzed CR impact on both off-state breakdown voltage (V_{BD}) and current-collapse magnitude (ΔI_{CC}), unveiling the dependence of V_{BD} and ΔI_{CC} on the total ($N_{C,TOT} = N_{C,A} + N_{C,D}$) and effective ($N_{C,EFF} = N_{C,A} - N_{C,D}$) C-related trap concentrations in the buffer, respectively. The simulations were calibrated on experimental data obtained from the characterization of AlGaN/GaN power HEMTs on SiC substrates (Bahat Treidel et al. 2010).

Figure 2.15 shows V_{BD}vs$N_{C,TOT}$ for three different CRs. As $N_{C,TOT}$ increase, V_{BD} first increases and then saturates, independent of CR. This behavior is consistent with the fact that V_{BD} increases as the electric field peak at the gate edge decreases due to the increased $N_{C,A}$ (negative charge). As a matter of fact, doping with acceptor impurities (like Fe and C) is employed in power GaN HEMTs for this specific end. When the electric-field peak moves from the gate edge to the drain contact, further increasing $N_{C,A}$ no longer influences V_{BD}, which then saturates as shown in Fig. 2.15. Two other considerations can be drawn from Fig. 2.15:

(1) the maximum saturation V_{BD} value ($V_{BD,max}$) is a non-monotonic function of CR;
(2) to reproduce the experimental V_{BD} value in Bahat Treidel et al. (2010), CR of about 50% must be assumed.

The non-monotonic dependence of $V_{BD,max}$ on CR is explained as follows. Once the electric field peak has moved to the drain contact, increasing $N_{C,D}$ (at fixed $N_{C,A}$)makes the electric field profile more uniform, hence $V_{BD,max}$ increases with CR. On the other hand, increasing CR beyond a certain value increases the source-drain punch-through current across the buffer, because the higher $N_{C,D}$ increases the conductivity of the buffer. This leads to $V_{BD,max}$ reduction at high CR.

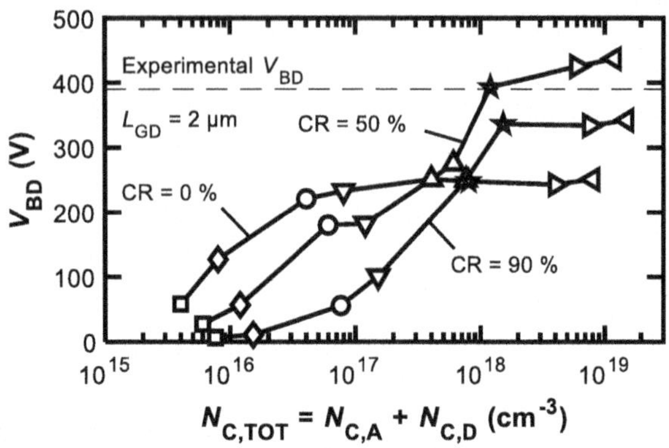

Fig. 2.15 V_{BD} vs. $N_{C,TOT}$ for CR = {0, 50, 90}%. The same symbols correspond to the same $N_{C,A}$ at different $N_{C,D}$ depending on CR.(Adapted from Zagni et al. (2021) copyright 2021 by the authors of the cited paper, licensed under CC BY 4.0)

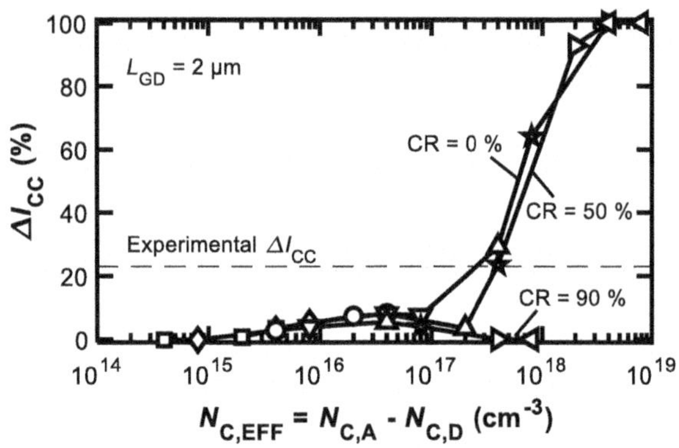

Fig. 2.16 ΔI_{CC} vs $N_{C,EFF}$ for CR = {0, 50, 90}%. The same symbols correspond to the same $N_{C,A}$ at different $N_{C,D}$ depending on CR (Adapted from Zagni et al. (2021) copyright 2021 by the authors of the cited paper, licensed under CC BY 4.0)

We now consider ΔI_{CC} plotted against $N_{C,EFF}$ in Fig. 2.16 for the same CR values used in Fig. 2.15. As shown in Fig. 2.16, regardless of CR for $N_{C,EFF}$ smaller than 10^{17} cm^{-3}ΔI_{CC} remains small. On the other hand, unless CR is very large (90%), for higher $N_{C,EFF}$ ΔI_{CC} increases steeply and exceeds 60%. This value is considerably larger than those reported for state-of-the-art C-doped GaN power HEMTs for $N_{C,EFF} > 10^{18}$ cm^{-3}, which is instead typical in extrinsically doped devices (i.e., using C precursors).

These results indicate that it is unrealistic to assume that high concentrations of C dopants could translate to C_N acceptors only. In fact, if this were the case, real devices would be completely overwhelmed by effects such as current collapse and dynamic R_{ON} increase. This observation agrees with previous simulations results obtained by considering CR = 0% (i.e., acceptors only) with $N_{C,A}$ of about the order of the nominal C density (i.e., ~ 10^{18}–10^{19} cm^{-3}) that led to a large overestimation of current-collapse effects measured in actual devices of different technologies (Verzellesi et al. 2014; Chini et al. 2016; Meneghesso et al. 2014; Iucolano et al. 2016).

To summarize, by combining results shown in Figs. 2.15 and 2.16 we observe that the V_{BD} and ΔI_{CC} values measured in the device under study can be reproduced with a single set of parameters (and specifically with the same $N_{C,TOT}$ and $N_{C,EFF}$) only when considering a CR of about 50%. More generally, our results indicate that a non-negligible part of incorporated C atoms results in donor-like levels or contributes to donor-like defect-impurity centers, thus compensating up to a significant degree the dominant acceptor traps introduced by C doping.

8.2 «Hole Redistribution» Model for Dynamic R_{ON} Effects

Reference (Zagni et al. 2021) reports on a model to explain the physics of the dynamic R_{ON} effects in C-doped GaN HEMTs. The negative charge accumulation in acceptor traps-related to C doping in the buffer-is generally assumed to be the main factor determining current collapse and dynamic R_{ON} degradation during OFF-state stress (Meneghesso et al. 2016; Bisi et al. 2014; Fabris et al. 2019; Moens et al. 2015a; Tanaka et al. 2013). Electron capture into buffer traps is generally blamed on for the negative charge build-up affecting the 2DEG (Bisi et al. 2014; Fabris et al. 2019; Moens et al. 2015a; Tanaka et al. 2013). Experimental results of the R_{ON} stress and recovery transients in power MIS-HEMTs during front- and back-gate test conditions reported in Meneghesso et al. (2016) show that each transient (R_{ON} increase and recovery) under both conditions is thermally activated with a single $E_A \approx 0.9$ eV. Inthe same report (Meneghesso et al. 2016), the fact that both stress and recovery transients were thermally activated with the 0.9-eV E_A was attributed to the observed correlation of the thermally activated stress with the temperature-accelerated buffer leakage current (Bisi et al. 2014). On the other hand, in the literature the acceleration with temperature of the recovery transients was attributed to the vertical/lateral leakage paths (also thermally activated) that lead to the discharge of C-related buffer traps (Uren et al. 2017; Moens et al. 2015b; Meneghini et al. 2018). In (Zagni et al. 2021),an alternative, comprehensive explanation of the results in Meneghesso et al. (2016) is proposed: the 0.9–eV activation energy of both stress and recovery transients stems from the dominant acceptor trap energy level associated with C in the buffer, as a result of hole emission, redistribution, and re-trapping in the buffer itself. This is the so-called "hole-redistribution" model. The interpretation was based on the results of calibrated numerical device simulations.

The Arrhenius plot extracted from both experimental and simulated stress/ recovery R_{ON} transients is shown in Fig. 2.17. Lines are the linear fitting of the data showing that both experiments and simulations are characterized by the same $E_A \approx 0.9$ eV.

Stress is performed with two different biasing conditions:

(1) by applying a negative bias to the gate contact and a large positive bias to the drain contact (with source and substrate contacts grounded). This is referred to as front-gate OFF-state stress (FGOS).

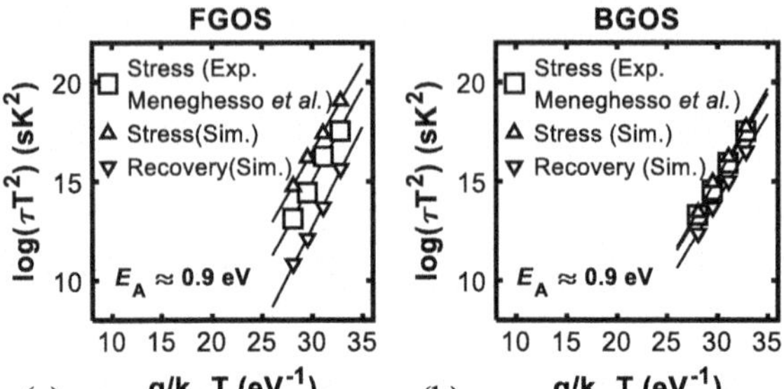

(a) (b)

Fig. 2.17 Arrhenius plot for the simulated **a** FGOS and **b** BGOS transients as well as the relative recovery processes. Experimental data from Meneghesso et al. (2016) of stress processes are reported for comparison. Lines are the linear fitting of the data showing that both experiments and simulations are characterized by the same $E_A \approx 0.9$ eV. (Adapted from Zagni et al. (2021) © IEEE)

(2) by applying a negative bias to the substrate contact (with all other contacts grounded). This is referred to as back-gate OFF-state stress (BGOS)

The explanation of the proposed mechanism underlying R_{ON} stress and recovery transients is as follows. During FGOS (or BGOS) stress, the net negative trapped charge-obtained as the difference between ionized acceptors and donors –increases in the gate-drain region of the buffer near the interface with the channel layer because holes are being emitted from the 0.9–eV C-related acceptor traps to the valence band. This is consistent with the observed R_{ON} increase during stress transient. The hole emission process is thermally activated with $E_A = 0.9$ eV. The emitted holes drift towards the gate–source region at the bottom of the buffer. Holes get partially trapped there and thus discharge the negatively charged acceptor traps. The situation is inverted during recovery: holes are emitted from C-related acceptors in the gate-source (bottom) region of the buffer and drift back towards the gate-drain (top) region. Here, the holes get re-trapped by the acceptor states from where they were emitted during stress in the first place. This is consistent with the R_{ON} decrease during recovery. The hole re-trapping process during recovery is also thermally activated with $E_A = 0.9$ eV, since the re-trapped holes in the gate-source (top) part of the buffer need to be first emitted from the C-related acceptor traps in the gate-drain (bottom) region of the buffer. If the recovery phase is long enough, it restores completely the pre-stress value of R_{ON}. This holds true only if emitted holes do not leak out from the device contacts or recombine with electrons. These phenomena are actually likely to occur either in pre-breakdown conditions, when applying very large negative gate bias, due to large electron leakage currents in the buffer, or in ON-state stress conditions, where channel hot electrons are injected into the buffer (Santi et al. 2022).

To summarize, the "hole redistribution" model solves the puzzle of the activation energy being the same in both stress/recovery phases and is seamlessly related to the commonly accepted model for C doping in GaN, i.e., a dominant acceptor trap level at about 0.9 eV from the valence band edge (E_V) (Lyons et al. 2014), that renders the GaN buffer a weakly p-type region. Note that the model describes accurately the dynamics of stress/recovery transients only up to drain bias for which electron leakage current through the GaN buffer is negligible thanks to the effective suppression by C-related impurities.

8.3 Non-monotonic, Dynamic R_{ON} Degradation

A widely reported-yet not definitely assessed-result in the literature is the non-monotonic dependence of R_{ON} on the off-state drain voltage ($V_{DS,OFF}$), that has indeed been observed in several AlGaN/GaN power HEMT technologies (Moens et al. 2015b; Canato et al. 2020; Karboyan et al. 2018; Li et al. 2020; Meneghesso et al. 2016; Uren et al. 2015). Some of the data from the literature are collected and shown in Fig. 2.18. While dynamic R_{ON} degradation is generally attributed to dispersion effects due to buffer traps, different interpretations for the observed partial R_{ON} recovery for large $V_{DS,OFF}$ have been proposed:

(1) Balancing of negative and positive buffer charge storage (Moens et al. 2015b) induced by leakage paths between the 2DEG and the carbon-doped buffer. These leakage paths are modelled either with heavily p-doped shorts (Uren et al. 2017, 2015; Karboyan et al. 2018) or with band-to-band hopping transport (Uren and Kuball 2021),
(2) Ionization of donor traps leading to positive charge generation in the UID GaN channel layer (Yang et al. 2019),
(3) Charge redistribution within the device structure due to increasing vertical drain-to-substrate leakage (Meneghini et al. 2018),
(4) Neutralization of ionized acceptors (or ionization of donors) due to positive charge accumulation at the interface between the carbon-doped buffer and the strain-relief layers/nucleation layer (Uren et al. 2017, 2015; Meneghini et al. 2018; Uren and Kuball 2021; Moens et al. 2017).

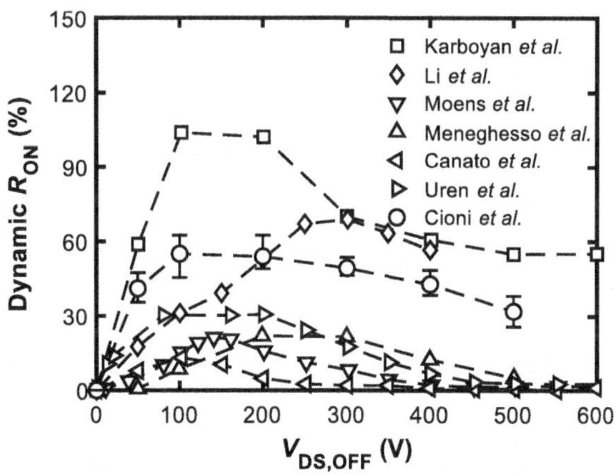

Fig. 2.18 Dynamic R_{ON} vs $V_{DS,OFF}$ data found in the literature (Canato et al. 2020; Karboyan et al. 2018; Li et al. 2020; Meneghesso et al. 2016; Uren et al. 2015; Moens et al. 2015c) and obtained from measurements in Cioni et al. (2021) (see legend). The partial R_{ON}recovery behavior for large $V_{DS,OFF}$ is observed on different GaN power HEMT technologies (Adapted from Cioni et al. (2021) © IEEE)

In (Cioni et al. 2021), a quantitative interpretation for the observed partial recovery of dynamic R_{ON} vs $V_{DS,OFF}$ in packaged p-GaN gate AlGaN/GaN HEMTs grown on a p-type Si substrate has been proposed. The interpretation was based on the results of calibrated numerical device simulations. The model for the vertical leakage through the substrate and nucleation layers was implemented after (Cornigli et al. 2015; Borga et al. 2020). Recent Monte-Carlo calculations reported in Bellotti and Bertazzi (2012) provided the impact-ionization coefficients for the Cynoweth's law for both electrons and holes in the GaN layers. The parameters used can be found in Zagni et al. (2019).In both experiments and simulations, OFF-state stress was held for 1000 s at different $V_{DS,OFF}$ biases. R_{ON} was evaluated at the end of the stress by rapidly sweeping the bias to drive the device in its triode region ($V_{GS} = 6$ V; $V_{DS} = 0.5$ V). Simulation results are shown in Fig. 2.19 along with measurement data points, finding an overall good agreement for the $V_{DS,OFF}$ range under investigation. Particularly, simulations capture the observed dynamic R_{ON} decrease occurring for $V_{DS,OFF} \geq 300$ V. Importantly, the simulations can accurately reproduce the experimental results only when including the hole generation mechanism (i.e., impact ionization). Simulation results suggest that it is the partial neutralization of negatively charged acceptors in the buffer by the generated holes (that get trapped by these same states) that causes R_{ON} to decrease. On the other hand, the partial R_{ON} recovery is not captured by the simulations if the impact ionization model is not activated, because the negatively ionized acceptors monotonically increase with $V_{DS,\,OFF}$.

The results discussed here unveil the need for a hole generation mechanism to reduce R_{ON} increase at large V_{DS} stress. The interpretation for partial R_{ON} recovery finds validation by an independent experimental analysis carried out in Tanaka et al. (2015), in which the reduced R_{ON} dispersion is obtained as a consequence of the reduction of the negative charge in the gate-drain access region thanks to hole injection through an ad-hoc p-type injector integrated into the device structure.

To summarize, a quantitative interpretation of the non-monotonic dynamic R_{ON} behavior with OFF-state stress bias in p-GaN gate AlGaN/GaN power HEMTs has

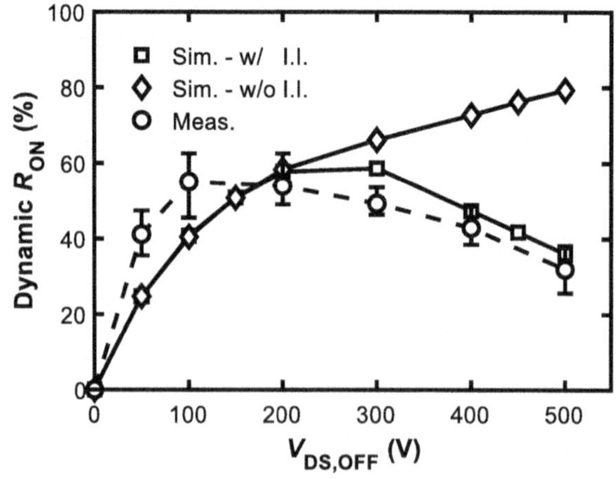

Fig. 2.19 Comparison between simulated and measured dynamic R_{ON} vs $V_{DS,OFF}$ (see legend). Simulations were carried out with and without Impact Ionization (I.I.). Stress time is $t_{OFF} = 1000$ s for both experiments and simulations. (Adapted from Cioni et al. (2021) © IEEE)

been provided. We attribute the dynamic R_{ON} increase to hole de-trapping from buffer acceptor traps within the gate-drain access region during OFF-state stress (see also Sect. 3). The observed partial R_{ON} recovery at large OFF-state stress is instead associated with the partial neutralization of negatively ionized acceptors by holes generated by a high electric field mechanism.

8.4 Bidirectional NBTI Effects in GaN MIS-HEMTs

Although typical GaN High Electron Mobility Transistors (HEMTs) are normally-on devices, for commercial applications it is more favorable to have normally-off operation. One technology option enabling this kind of operation is the fully recessed metal–insulator-semiconductor (MIS) -HEMT (Alamo and Lee 2019). As the acronym suggests, this technology integrates an insulator layer in the gate stack of the transistor which however leads to threshold-voltage (V_T) instability issues related to the defective oxide/semiconductor interface (Alamo and Lee 2019). Even in normally-off HEMTs investigating the V_T stability after negative bias temperature instabilities stress (NBTI) is important, since $V_{GS} \ll V_T$ is conventionally employed to prevent from false turn-on when pulsing at large drain voltages (Hua et al. 2018).In addition, high negative gate bias during NBTI test can be considered as a proxy condition for the OFF-state under pulse-mode operation. This applies because under both conditions comparable drain-gate voltages are achieved.

NBTI has been widely investigated in the literature. Depending on the sign of the observed V_T shifts (ΔV_T), different NBTI trends have been observed:

(1) $\Delta V_T < 0$ V (Meneghini et al. 2016; Guo and Alamo 2017; Viey et al. 2019);
(2) $\Delta V_T > 0$ V (Hua et al. 2018; He et al. 2018; Hua et al. 2018);
(3) ΔV_T of both signs (Guo and Alamo 2016; Sang et al. 2015).

$\Delta V_T < 0$ V due to NBTI is generally associated with the emission of trapped electrons from traps located either in the gate insulator or at the oxide/semiconductor interface. On the other hand, a comprehensive model for $\Delta V_T > 0$ V after NBTI stress has not been developed yet. Also, a consistent explanation for $\Delta V_T > 0$ due to NBTI would require to take into account similar results obtained after off-state stress, that have been ascribed to a mechanism involving hole generation (Hua et al. 2018, 2020).

Reference (Zagni et al. 2021) reports on an extensive characterization of V_T stability in AlGaN/GaN MIS-HEMTs, incorporating a 20 nm fully recessed (i.e., with complete barrier removal) Al_2O_3 gate dielectric that partially penetrates into the undoped GaN channel (typically of about 10 nm). 2DEG suppression when no gate bias is applied (leading to $V_T > 0$ V) is thus obtained. Several stress conditions were considered by varying gate bias ($V_{GS,STR}$), stress time (t_{STR}), and temperature (T). ΔV_T of both signs were observed depending on $V_{GS,STR}$ magnitude, thermally activated with different activation energies (E_A). Here we discuss two different $V_{GS,STR}$ magnitude, i.e., moderate ($V_{GS,STR} = -7$ V) and high ($V_{GS,STR} = -17$ V), which lead

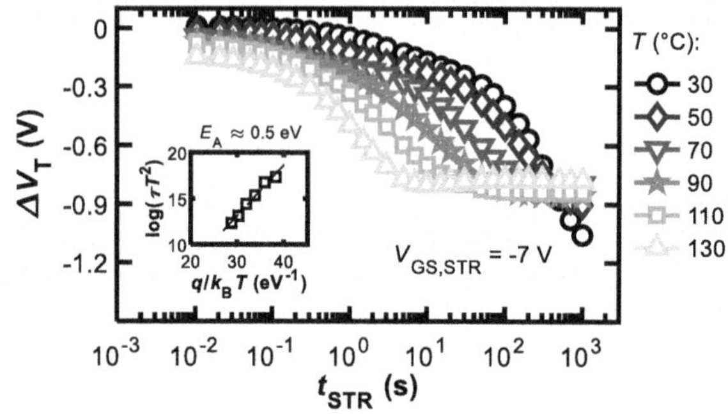

Fig. 2.20 Measured ΔV_T verses t_{STR} for different temperatures (T = 30–130°C, ΔT = 20 °C) at $V_{GS,STR} = -7$ V. Inset: Arrhenius plot of the stress data showing that $E_A \approx 0.5$ eV.(Adapted from Zagni et al. (2021) © IEEE)

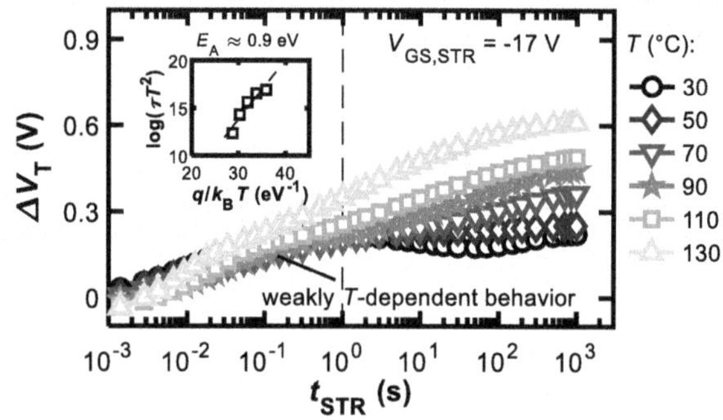

Fig. 2.21 Measured ΔV_T verses t_{STR} for different temperatures (T = 30–130°C, ΔT = 20 °C) at $V_{GS,STR} = -17$ V. Inset: Arrhenius plot of the stress data showing that $E_A \approx 0.9$ eV.(Adapted from Zagni et al. (2021) © IEEE)

to negative and positive$\boldsymbol{\Delta} V_T$, respectively. The measured $\boldsymbol{\Delta} V_T$vs $t_{STR}$is shown in Figs. 2.20 and 2.21 for the moderate and high stress conditions, respectively.

As shown in Fig. 2.20, under moderate stress conditions V_T shifts negatively (i.e., becomes less than the initial value, V_{T0}) and then at $t_{STR} = 1000$ s it saturates to ≈–0.8 V or ≈–1 V. This well-known behavior is generally attributed to the emission of trapped electrons in the gate oxide or at the interface with the semiconductor (Guo and Alamo 2017, 2016). The process is thermally activated (Meneghini et al. 2016), and for the devices tested in this work, we extracted $E_A \approx 0.5$ eV, see Fig. 2.15. The time constants (τs) were extracted on the full stress measurement range, i.e., 1 ms-1000 s. Another possible explanation for the observed negative $\boldsymbol{\Delta} V_T$ is the de-ionization of acceptor traps in the buffer (related to C-doping) as a consequence of the applied negative gate stress (Viey et al. 2019).

Figure 2.21 shows the results of the high stress conditions, that leads V_T to shift positively (i.e., V_T becomes larger than V_{T0}). Interestingly, this drift is weakly dependent on temperature for t_{STR} up to 1 s, whereas it exhibits a strong temperature dependence for higher stress times and we extracted $E_A \approx 0.9$ eV, see Fig. 2.21. The fast $\boldsymbol{\Delta} V_T > 0$ V (i.e., $t_{STR} < 1$ s) is due to electrons injected by the gate terminal that trap in the gate oxide (He et al. 2018). Conversely, the thermally activated $\boldsymbol{\Delta} V_T$ increase is induced by a slower process with $E_A \approx 0.9$ eV, that can be correlated with the increased negatively ionized0.9-eV C_N acceptor trap concentration in the buffer after hole emission from the same traps.

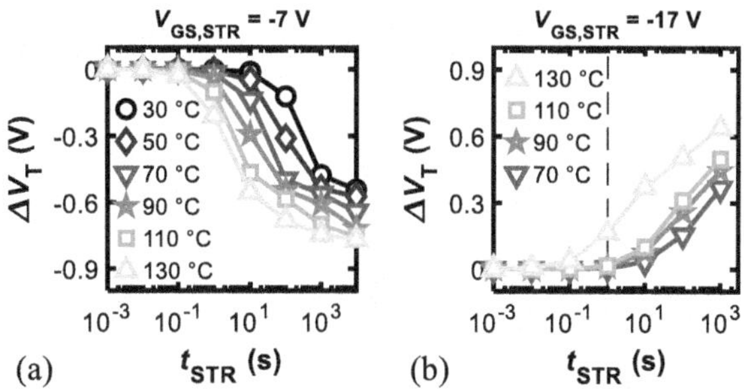

Fig. 2.22 Simulated ΔV_T after **a** moderate and **b** high $V_{GS,STR}$ verses t_{STR} at different temperatures (see legend). The negative and positive ΔV_T with different gate bias stress-as well as its temperature-dependent behavior-observed in the tested devices can be correctly captured

This interpretation is validated by simulations shown in Fig. 2.22, capturing the bidirectional $\boldsymbol{\Delta} V_T$ for the two stress conditions considered.

To summarize, the negative V_T shiftunder moderate gate bias stress (i.e., $V_{GS,STR} = -7$ V) was associated with the thermally activated emission of electron (with $E_A \approx 0.5$ eV) from interface and/or oxide traps. On the other hand, the positive V_T shift under high stress (i.e., $V_{GS,STR} = -17$ V)and for long stress time was explained by the increased negatively ionized acceptor trap density in the buffer. The interpretation, validated by simulations, is reinforced by the same 0.9-eV activation energy of both the V_T shift and the widely accepted C_N energy level.

9 Conclusion

Charge trapping effects still represent a limit for the performance and reliability of GaN transistor technologies and need to be carefully characterized, understood, and modelled. After reviewing the different deep levels that have been detected in GaN and describing the possible compact and TCAD approaches that be adopted to model charge trapping effects, a selection of cases where modelling has substantially contributed to the understanding of charge trapping effects has been presented in this chapter for both GaN RF and GaN power transistors. For RF transistors, these case studies include (i) the different mechanisms by which barrier and buffer traps charge up and cause the current collapse, (ii) the possible contribution to current collapse stemming from non-Arrhenius, surface charging effects induced by gate electron injection and transport through the passivation/barrier interface; (iii) the different features of buffer charge trapping effects depending on whether the buffer is doped with Fe or C; (iv) time-dependent breakdown effects linked to the occupancy of Fe traps in the GaN buffer, (v) the dependence of the trap emission time on the trap filling conditions during DCT measurements caused by the applied bias is not large enough to induce almost complete trap filling. For power transistors, analyzed cases are: (i) the necessity to assume that a non-negligible part of incorporated C atoms, used to doped the GaN buffer, result in donor-like levels thus compensating up to a significant degree the dominant acceptor traps introduced by C doping; (ii) the "hole-redistribution model" explaining the dynamic R_{ON} effects in C-doped

GaN HEMTs, as a result of hole emission, redistribution and capture within the buffer layer; (iii) a quantitative interpretation of the observed partial R_{ON} recovery at large OFF-state stress as attributed to neutralization of negatively ionized C-related acceptors by holes generated by a high electric field mechanism; (iv) the explanation of the bidirectional threshold-voltage shifts observed in GaN MIS-HEMTs as a result of different mechanisms involving surface and buffer traps.

References

Aggerstam T, Pinos A, Marcinkevičius S, Linnarsson M, Lourdudoss S (2007) Electron and hole capture cross-sections of Fe acceptors in GaN: Fe epitaxially grown on sapphire. J Electron Mater 36(12):1621–1624. https://doi.org/10.1007/s11664-007-0202-9

Albahrani SA, Mahajan D, Hodges J, Chauhan YS, Khandelwal S (2019) ASM GaN: Industry standard model for GaN RF and power Devices-Part-II: Modeling of charge trapping. IEEE Trans Electron Devices 66(1):87–94. https://doi.org/10.1109/TED.2018.2868261

Alkauskas A, McCluskey MD, Van De Walle CG (2016) Tutorial: Defects in semiconductors - Combining experiment and theory. J Appl Phys 119(18):181101. https://doi.org/10.1063/1.4948245

Ambacher O, Smart J, Shealy JR, Weimann NG, Chu K, Murphy M, Schaff WJ, Eastman LF, Dimitrov R, Wittmer L, Stutzmann M, Rieger W, Hilsenbeck J (1999) Two-dimensional electron gases induced by spontaneous and piezoelectric polarization charges in N- and Ga-face AlGaN/GaN heterostructures. J Appl Phys 85(6):3222–3233. https://doi.org/10.1063/1.369664

Ambacher O, Foutz B, Smart J, Shealy JR, Weimann NG, Chu K, Murphy M, Sierakowski AJ, Schaff WJ, Eastman LF, Dimitrov R, Mitchell A, Stutzmann M (2000) Two dimensional electron gases induced by spontaneous and piezoelectric polarization in undoped and doped AlGaN/GaN heterostructures. J Appl Phys 87(1):334–344. https://doi.org/10.1063/1.371866

Angelotti AM, Gibiino GP, Santarelli A, Florian C (2020) Experimental characterization of charge trapping dynamics in 100-nm AlN/GaN/AlGaN—on-Si HEMTs by wideband transient measurements. IEEE Trans Electron Devices 67(8):3069–3074. https://doi.org/10.1109/TED.2020.3000983

Arehart AR, Homan T, Wong MH, Poblenz C, Speck JS, Ringel SA (2010) Impact of N- and Ga-face polarity on the incorporation of deep levels in n-type GaN grown by molecular beam epitaxy. Appl Phys Lett 96(24):242112. https://doi.org/10.1063/1.3453660

Arehart AR, Corrion A, Poblenz C, Speck JS, Mishra UK, DenBaars SP, Ringel SA (2008) Comparison of deep level incorporation in ammonia and rf-plasma assisted molecular beam epitaxy n-GaN films. Phys Status Solidi C 5(6):1750–1752. https://doi.org/10.1002/pssc.200778622

Armstrong A, Henry TA, Koleske DD, Crawford MH, Lee SR (2012) Quantitative and depth-resolved deep level defect distributions in InGaN/GaN light emitting diodes. Opt. Express, OE 20(106):A812–A821. https://doi.org/10.1364/OE.20.00A812

Asghar M, Muret P, Beaumont B, Gibart P (2004) Field dependent transformation of electron traps in GaN p–n diodes grown by metal–organic chemical vapour deposition. Mater Sci Eng, B 113(3):248–252. https://doi.org/10.1016/j.mseb.2004.09.001

Axelsson O, Gustafsson S, Hjelmgren H, Rorsman N, Blanck H, Splettstoesser J, Thorpe J, Roedle T, Thorsell M (2016) Application relevant evaluation of trapping effects in AlGaN/GaN HEMTs with Fe-doped buffer. IEEE Trans Electron Devices 63(1):326–332. https://doi.org/10.1109/TED.2015.2499313

Bahat Treidel E, Brunner F, Hilt O, Cho E, Würfl J, Trankle G (2010) AlGaN/GaN/GaN:C back-barrier HFETs with breakdown voltage of over 1 kV and low RON× A. IEEE Trans Electron Devices 57(11):3050–3058. https://doi.org/10.1109/TED.2010.2069566

Bakeroot B, Stockman A, Posthuma N, Stoffels S, Decoutere S (2018) Analytical model for the threshold voltage of p- (Al)GaN high-electron-mobility transistors. IEEE Trans Electron Devices 65(1):79–86. https://doi.org/10.1109/TED.2017.2773269

Belahsene S, Al Saqri NA, Jameel D, Mesli A, Martinez A, De Sanoit J, Ougazzaden A, Salvestrini JP, Ramdane A, Henini M (2015) Analysis of deep level defects in GaN p-i-n diodes after beta particle irradiation. Electronics 4(4):4. https://doi.org/10.3390/electronics4041090.

Bellotti E, Bertazzi F (2012) A numerical study of carrier impact ionization in Al xGa 1-xN. J Appl Phys 111(10):103711. https://doi.org/10.1063/1.4719967

Bisi D, Meneghini M, De Santi C, Chini A, Dammann M, Bruckner P, Mikulla M, Meneghesso G, Zanoni E (2013) Deep-level characterization in GaN HEMTs-Part I: Advantages and limitations of drain current transient measurements. IEEE Trans Electron Devices 60(10):3166–3175. https://doi.org/10.1109/TED.2013.2279021

Bisi D, Meneghini M, Marino FA, Marcon D, Stoffels S, Van Hove M, Decoutere S, Meneghesso G, Zanoni E (2014) Kinetics of buffer-related RON-increase in GaN-on-silicon MIS-HEMTs. IEEE Electron Device Lett 35(10):1004–1006. https://doi.org/10.1109/LED.2014.2344439

Borga M, De Santi C, Stoffels S, Bakeroot B, Li X, Zhao M, Van Hove M, Decoutere S, Meneghesso G, Meneghini M, Zanoni E (2020) Modeling of the vertical leakage current in AlN/Si heterojunctions for GaN power applications. IEEE Trans Electron Devices 67(2):595–599. https://doi.org/10.1109/TED.2020.2964060

Boturchuk I, Scheffler L, Larsen AN, Julsgaard B (2018) Evolution of electrically active defects in n-GaN during heat treatment typical for ohmic contact formation. Phys Status Solidi (A) 215(9):1700516. https://doi.org/10.1002/pssa.201700516

Buffolo M, Caria A, Piva F, Roccato N, Casu C, De Santi C, Trivellin N, Meneghesso G, Zanoni E, Meneghini M (2022) Defects and Reliability of GaN-Based LEDs: Review and Perspectives. Phys Status Solidi (A) 219(8):2100727. https://doi.org/10.1002/pssa.202100727.

Caesar M, Dammann M, Polyakov V, Waltereit P, Bronner W, Baeumler M, Quay R, Mikulla M, Ambacher O (2012) Generation of traps in AlGaN/GaN HEMTs during RF- and DC-stress test. In: 2012 IEEE International Reliability Physics Symposium (IRPS), p CD.6.1-CD.6.5. https://doi.org/10.1109/IRPS.2012.6241883

Calleja E, Sánchez FJ, Basak D, Sánchez-García MA, Muñoz E, Izpura I, Calle F, Tijero JMG, Sánchez-Rojas JL, Beaumont B, Lorenzini P, Gibart P (1997) Yellow luminescence and related deep states in undoped GaN. Phys Rev B 55(7):4689–4694. https://doi.org/10.1103/PhysRevB.55.4689

Canato E, Meneghini M, De Santi C, Masin F, Stockman A, Moens P, Zanoni E, Meneghesso G (2020) OFF-state trapping phenomena in GaN HEMTs: Interplay between gate trapping, acceptor ionization and positive charge redistribution. Microelectron Reliab 114(May):113841. https://doi.org/10.1016/j.microrel.2020.113841

Cardwell DW, Sasikumar A, Arehart AR, Kaun SW, Lu J, Keller S, Speck JS, Mishra UK, Ringel SA, Pelz JP (2013) Spatially-resolved spectroscopic measurements of Ec—0.57 eV traps in AlGaN/GaN high electron mobility transistors. Appl Phys Lett 102(19):193509. https://doi.org/10.1063/1.4806980

Cardwell DW, Sasikumar A, Arehart AR, Kaun SW, Lu J, Keller S, Speck JS, Mishra UK, Ringel SA, Pelz JP (2013) Spatially-resolved spectroscopic measurements of Ec -0.57 eV traps in AlGaN/GaN high electron mobility transistors. Appl Phys Lett 102(19):0–4. https://doi.org/10.1063/1.4806980

Chen XD, Huang Y, Fung S, Beling CD, Ling CC, Sheu JK, Lee ML, Chi GC, Chang SJ (2003) Deep level defect in Si-implanted GaN n+-p junction. Appl Phys Lett 82(21):3671–3673. https://doi.org/10.1063/1.1578167

Chen S, Honda U, Shibata T, Matsumura T, Tokuda Y,. Ishikawa K, Hori M, Ueda H, Uesugi T, Kachi T (2012) As-grown deep-level defects in n-GaN grown by metal–organic chemical vapor deposition on freestanding GaN. J Appl Phys 112(5). https://doi.org/10.1063/1.4748170

Chini A, Meneghesso G, Meneghini M, Fantini F, Verzellesi G, Patti A, Iucolano F (2016) Experimental and numerical analysis of hole emission process from carbon-related traps in GaN buffer layers. IEEE Trans Electron Devices 63(9):3473–3478. https://doi.org/10.1109/TED.2016.2593791

Cho HK, Kim KS, Hong C-H, Lee HJ (2001) Electron traps and growth rate of buffer layers in unintentionally doped GaN. J Cryst Growth 223(1):38–42. https://doi.org/10.1016/S0022-0248(00)00982-9

Cho HK, Kim CS, Hong C-H (2003) Electron capture behaviors of deep level traps in unintentionally doped and intentionally doped n-type GaN. J Appl Phys 94(3):1485–1489. https://doi.org/10.1063/1.1586981

Cho HK, Khan FA, Adesida I, Fang Z-Q, Look DC (2008) Deep level characteristics in n-GaN with inductively coupled plasma damage. J Phys d: Appl Phys 41(15):155314. https://doi.org/10.1088/0022-3727/41/15/155314

Choi KJ, Jang HW, Lee J-L (2003) Observation of inductively coupled-plasma-induced damage on n-type GaN using deep-level transient spectroscopy. Appl Phys Lett 82(8):1233–1235. https://doi.org/10.1063/1.1557316

Chung HM, Chuang WC, Pan YC, Tsai CC, Lee MC, Chen WH, Chen WK, Chiang CI, Lin CH, Chang H (2000) Electrical characterization of isoelectronic In-doping effects in GaN films grown by metalorganic vapor phase epitaxy. Appl Phys Lett 76(7):897–899. https://doi.org/10.1063/1.125622

Cioni M, Zagni N, Selmi L, Meneghesso G, Meneghini M, Zanoni E, Chini A (2021) Electric field and self-heating effects on the emission time of iron traps in GaN HEMTs. IEEE Trans Electron Devices 68(7):3325–3332. https://doi.org/10.1109/TED.2021.3081613

Cioni M, Zagni N, Iucolano F, Moschetti M, Verzellesi G, Chini A (2021) Partial recovery of dynamic R_{ON} versus OFF-state stress voltage in p-GaN Gate AlGaN/GaN power HEMTs. IEEE Trans Electron Devices 68(10):4862–4868. https://doi.org/10.1109/TED.2021.3105075

Cioni M, Zagni N, Chini A (2022) Fe-traps influence on time-dependent breakdown voltage in 0.1-μm GaN HEMTs for 5G applications. In: 2022 IEEE International Reliability Physics Symposium (IRPS), Dallas, TX, USA, p 11B.3–1–11B.3–5. https://doi.org/10.1109/IRPS48227.2022.9764502

Cornigli D, Reggiani S, Gnani E, Gnudi A, Baccarani G, Moens P, Vanmeerbeek P, Banerjee A, Meneghesso G (2015) Numerical Investigation of the Lateral and Vertical Leakage Currents and Breakdown Regimes in GaN-on-Silicon Vertical Structures. In: IEEE International Electron Devices Meeting (IEDM), pp 109–112

del Alamo JA, Lee ES (2019) Stability and reliability of lateral GaN power field-effect transistors. IEEE Trans Electron Devices 66(11):4578–4590. https://doi.org/10.1109/TED.2019.2931718

Deng Q, Wang X, Xiao H, Wang C, Yin H, Chen H, Lin D, Jiang L, Feng C, Li J, Wang Z, Hou X (2011) Comparison of as-grown and annealed GaN/InGaN : Mg samples. J Phys d: Appl Phys 44(34):345101. https://doi.org/10.1088/0022-3727/44/34/345101

Fabris E, Meneghini M, De Santi C, Borga M, Meneghesso G, Zanoni E, Kinoshita Y, Tanaka K, Ishida H, Ueda T (2019) Hot-Electron Effects in GaN GITs and HD-GITs: A Comprehensive Analysis. In: IEEE International Reliability Physics Symposium Proceedings, Monterey, CA, pp 1–6. https://doi.org/10.1109/IRPS.2019.8720472

Fang Z-Q, Look DC, Kim W, Fan Z, Botchkarev A, Morkoç H (1998) Deep centers in n-GaN grown by reactive molecular beam epitaxy. Appl Phys Lett 72(18):2277–2279. https://doi.org/10.1063/1.121274

Fang Z-Q, Look DC, Kim DH, Adesida I (2005) Traps in AlGaN/GaN/SiC heterostructures studied by deep level transient spectroscopy. Appl Phys Lett 87(18):182115. https://doi.org/10.1063/1.2126145

Fang Z-Q, Polenta L, Hemsky JW, Look DC (2000) Deep centers in as-grown and electron-irradiated n-GaN. In: Presented at the IEEE Semiconducting and Semi-Insulating Materials Conference, SIMC, vol 2000, pp 35–42. https://doi.org/10.1109/SIM.2000.939193

Fariza A, Lesnik A, Bläsing J, Hoffmann MP, Hörich F, Veit P, Witte H, Dadgar A, Strittmatter A (2016) On reduction of current leakage in GaN by carbon-doping. Appl Phys Lett 109(21):212102. https://doi.org/10.1063/1.4968823

Galiano K, Deitz JI, Carnevale SD, Gleason DA, Paul PK, Zhang Z, McSkimming BM, Speck JS, Ringel SA, Grassman TJ, Arehart AR, Pelz JP (2018) Spatial correlation of the EC-0.57 eV trap state with edge dislocations in epitaxial n-type gallium nitride. J Appl Phys 123(22):224504. https://doi.org/10.1063/1.5022806

Gelhausen O, Phillips MR, Goldys EM, Paskova T, Monemar B, Strassburg M, Hoffmann A (2004) Dissociation of H-related defect complexes in Mg-doped GaN. Phys Rev B 69(12):125210. https://doi.org/10.1103/PhysRevB.69.125210

Götz W, Johnson NM, Chen C, Liu H, Kuo C, Imler W (1996) Activation energies of Si donors in GaN. Appl Phys Lett 68(22):3144–3146. https://doi.org/10.1063/1.115805

Guo A, Del Alamo JA (2016) Negative-bias temperature instability of GaN MOSFETs. In: IEEE International Reliability Physics Symposium Proceedings, pp 4A11–4A16. https://doi.org/10.1109/IRPS.2016.7574526

Guo A, Del Alamo JA (2017) Unified mechanism for positive- and negative-bias temperature instability in GaN MOSFETs. IEEE Trans Electron Devices 64(5):2142–2147. https://doi.org/10.1109/TED.2017.2686840

Hacke P, Detchprohm T, Hiramatsu K, Sawaki N, Tadatomo K, Miyake K (1994) Analysis of deep levels in n-type GaN by transient capacitance methods. J Appl Phys 76(1):304–309. https://doi.org/10.1063/1.357144

He J, Hua M, Zhang Z, Chen KJ (2018) Performance and V TH stability in E-Mode GaN fully recessed MIS-FETs and partially recessed MIS-HEMTs with LPCVD-SiN x /PECVD-SiN x gate dielectric stack. IEEE Trans Electron Devices 65(8):3185–3191. https://doi.org/10.1109/TED.2018.2850042

Heikman S, Keller S, DenBaars SP, Mishra UK (2002) Growth of Fe doped semi-insulating GaN by metalorganic chemical vapor deposition. Appl Phys Lett 81(3):439–441. https://doi.org/10.1063/1.1490396

Heitz R, Maxim P, Eckey L, Thurian P, Hoffmann A, Broser I, Pressel K, Meyer BK (1997) Excited states of Fe^{3+} in GaN. Phys Rev B 55(7):4382–4387. https://doi.org/10.1103/PhysRevB.55.4382

Henry TA, Armstrong A, Kelchner KM, Nakamura S, DenBaars SP, Speck JS (2012) Assessment of deep level defects in m-plane GaN grown by metalorganic chemical vapor deposition. Appl Phys Lett 100(8):082103. https://doi.org/10.1063/1.3687700

Hierro A, Kwon D, Ringel SA, Hansen M, Speck JS, Mishra UK, DenBaars SP (2000) Optically and thermally detected deep levels in n-type Schottky and p+-n GaN diodes. Appl Phys Lett 76(21):3064–3066. https://doi.org/10.1063/1.126580

Hierro A, Ringel SA, Hansen M, Speck JS, Mishra UK, DenBaars SP (2000) Hydrogen passivation of deep levels in n–GaN. Appl Phys Lett 77(10):1499–1501. https://doi.org/10.1063/1.1290042

Hierro A, Arehart AR, Heying B, Hansen M, Mishra UK, DenBaars SP, Speck JS, Ringel SA (2002) Impact of Ga/N flux ratio on trap states in n-GaN grown by plasma-assisted molecular-beam epitaxy. Appl Phys Lett 80(5):805–807. https://doi.org/10.1063/1.1445274

Honda U, Yamada Y, Tokuda Y, Shiojima K (2012) Deep levels in n-GaN doped with carbon studied by deep level and minority carrier transient spectroscopies. Jpn J Appl Phys 51(4S):04DF04. https://doi.org/10.1143/JJAP.51.04DF04

Horita M, Narita T, Kachi T, Suda J (2020) Identification of origin of EC –0.6 eV electron trap level by correlation with iron concentration in n-type GaN grown on GaN freestanding substrate by metalorganic vapor phase epitaxy. Appl Phys Express 13(7):071007. https://doi.org/10.35848/1882-0786/ab9e7c

Hua M, Wei J, Bao Q, Zheng Z, Zhang Z, He J, Chen KJ (2018) Hole-induced threshold voltage shift under reverse-bias stress in E-mode GaN MIS-FET. IEEE Trans Electron Devices 65(9):3831–3838. https://doi.org/10.1109/TED.2018.2856998

Hua M, Wei J, Bao Q, Zhang Z, Zheng Z, Chen KJ (2018) Dependence of VTH stability on gate-bias under reverse-bias stress in E-mode GaN MIS-FET. IEEE Electron Device Lett 39(3):413–416. https://doi.org/10.1109/LED.2018.2791664

Hua M, Yang S, Wei J, Zheng Z, He J, Chen KJ (2020) Hole-induced degradation in E-mode GaN MIS-FETs: Impact of substrate terminations. IEEE Trans Electron Devices 67(1):217–223. https://doi.org/10.1109/TED.2019.2954282

Huber M, Silvestri M, Knuuttila L, Pozzovivo G, Andreev A, Kadashchuk A, Bonanni A, Lundskog A (2015) Impact of residual carbon impurities and gallium vacancies on trapping effects in AlGaN/GaN metal insulator semiconductor high electron mobility transistors. Appl Phys Lett 107(3):032106. https://doi.org/10.1063/1.4927405

Ieong M, Solomon PM, Laux SE, Wong H-SP, Chidambarrao D (1998) Comparison of raised and schottky source/drain MOSFETs using a novel tunneling contact model. In: International Electron Devices Meeting 1998. Technical Digest (Cat. No.98CH36217), pp 733–736. https://doi.org/10.1109/IEDM.1998.746461

Irokawa Y, Fujishima O, Kachi T, Nakano Y (2005) Electrical activation characteristics of silicon-implanted GaN. J Appl Phys 97(8):083505. https://doi.org/10.1063/1.1863458

Iucolano F, Parisi A, Reina S, Patti A, Coffa S, Meneghesso G, Verzellesi G, Fantini F, Chini A (2016) Correlation between dynamic Rdsou transients and Carbon related buffer traps in AlGaN/GaN HEMTs. In: IEEE International Reliability Physics Symposium (IRPS), pp CD21–CD24. https://doi.org/10.1109/IRPS.2016.7574586

Iwinska M, Piotrzkowski R, Litwin-Staszewska E, Sochacki T, Amilusik M, Fijalkowski M, Lucznik B, Bockowski M (2016) Highly resistive C-doped hydride vapor phase epitaxy-GaN grown on ammonothermally crystallized GaN seeds. Appl Phys Express 10(1):011003. https://doi.org/10.7567/APEX.10.011003

Joh J, Del Alamo JA (2011) A current-transient methodology for trap analysis for GaN high electron mobility transistors. IEEE Trans Electron Devices 58(1):132–140. https://doi.org/10.1109/TED.2010.2087339

Johnstone DK, Ahoujja M, Yeo YK, Hengehold RL, Guido L (2011) Deep centers and their capture barriers in MOCVD-grown GaN. MRS Online Proc Libr 692(1):271. https://doi.org/10.1557/PROC-692-H2.7.1

Joshi V, Soni A, Tiwari SP, Shrivastava M (2016) A comprehensive computational modeling approach for AlGaN/GaN HEMTs. IEEE Trans Nanotechnol 15(6):947–955. https://doi.org/10.1109/TNANO.2016.2615645

Joshi V, Tiwari SP, Shrivastava M (2019) Part I: Physical insight into carbon-doping-induced delayed avalanche action in GaN buffer in AlGaN/GaN HEMTs. IEEE Trans Electron Devices 66(1):561–569. https://doi.org/10.1109/TED.2018.2878770

Narita (成田哲生) T, Tomita (冨田一義) K, Tokuda (徳田豊) Y, Kogiso (小木曽達也) T, Horita (堀田昌宏) M, Kachi (加地徹) T (2018) The origin of carbon-related carrier compensation in p-type GaN layers grown by MOVPE. J Appl Phys 124(21):215701. https://doi.org/10.1063/1.5057373

Karboyan S, Uren MJ, Manikant JW, Pomeroy, Kuball M (2018) On the origin of dynamic Ron in commercial GaN-on-Si HEMTs. Microelectron Reliab 81:306–311. https://doi.org/10.1016/j.microrel.2017.10.006

Kellogg K, Khandelwal S, Dunleavy L, Wang J (2020) Characterization of thermal and trapping time constants in a GaN HEMT. In: 2020 94th ARFTG Microwave Measurement Symposium (ARFTG), Jan. 2020, pp. 1–4. https://doi.org/10.1109/ARFTG47584.2020.9071731

Kindl D, Hubík P, Krištofik J, Mareš JJ, Výborný Z, Leys MR, Boeykens S (2009) Deep defects in GaN/AlGaN/SiC heterostructures. J Appl Phys 105(9):093706. https://doi.org/10.1063/1.3122290

T. Kogiso, T. Narita, H. Yoshida, Y. Tokuda, K. Tomita, and T. Kachi, "Characterization of hole traps in MOVPE-grown p-type GaN layers using low-frequency capacitance deep-level transient spectroscopy," *Jpn. J. Appl. Phys.*, vol. 58, no. SC, p. SCCB36, May 2019, https://doi.org/10.7567/1347-4065/ab0408.

Koller C, Pobegen G, Ostermaier C, Pogany D (2018) Effect of carbon doping on charging/discharging dynamics and leakage behavior of carbon-doped GaN. IEEE Trans Electron Devices 65(12):5314–5321. https://doi.org/10.1109/TED.2018.2872552

Koller C, Lymperakis L, Pogany D, Pobegen G, Ostermaier C (2021) Mechanism leading to semi-insulating property of carbon-doped GaN: Analysis of donor acceptor ratio and method for its determination. J Appl Phys 130(18):185702. https://doi.org/10.1063/5.0060912

Kumar M, Bhat TN, Roul B, Rajpalke MK, Kalghatgi AT, Krupanidhi SB (2012) Carrier concentration dependence of donor activation energy in n-type GaN epilayers grown on Si (111) by plasma—assisted MBE. Mater Res Bull 47(6):1306–1309. https://doi.org/10.1016/j.materresbull.2012.03.016

Lee WI, Huang TC, Guo JD, Feng MS (1995) Effects of column III alkyl sources on deep levels in GaN grown by organometallic vapor phase epitaxy. Appl Phys Lett 67(12):1721–1723. https://doi.org/10.1063/1.115028

Lee I-H, Polyakov AY, Smirnov NB, Zinovyev RA, Bae K-B, Chung T-H, Hwang S-M, Baek JH, Pearton SJ (2017) Changes in electron and hole traps in GaN-based light emitting diodes from near-UV to green spectral ranges. Appl Phys Lett 110(19):192107. https://doi.org/10.1063/1.4983556

M. Lee, C. W. Ahn, T. K. O. Vu, H. U. Lee, E. K. Kim, and S. Park, "First observation of electronic trap levels in freestanding GaN crystals extracted from Si substrates by hydride vapour phase epitaxy," *Sci Rep*, vol. 9, no. 1, Art. no. 1, May 2019, https://doi.org/10.1038/s41598-019-43583-y.

Li K, Evans PL, Johnson CM (2018) Characterisation and modeling of gallium nitride power semiconductor devices dynamic on-state resistance. IEEE Trans Power Electron 33(6):5262–5273. https://doi.org/10.1109/TPEL.2017.2730260

Li S, Yang S, Han S, Sheng K (2020) Investigation of temperature-dependent dynamic RONof GaN HEMT with hybrid-drain under hard and soft switching. In: Proceedings of the International Symposium on Power Semiconductor Devices and ICs, vol 2020, pp 306–309. https://doi.org/10.1109/ISPSD46842.2020.9170048

Look DC, Fang Z-Q, Claflin B (2005) Identification of donors, acceptors, and traps in bulk-like HVPE GaN. J Cryst Growth 281(1):143–150. https://doi.org/10.1016/j.jcrysgro.2005.03.035

Lyons JL, Janotti A, Van De Walle CG (2014) Effects of carbon on the electrical and optical properties of InN, GaN, and AlN. Phys Rev B-Condens Matter Mater Physics 89(3):1–8. https://doi.org/10.1103/PhysRevB.89.035204

Lyons JL, Wickramaratne D, Van de Walle CG (2021) A first-principles understanding of point defects and impurities in GaN. J Appl Phys 129(11):111101. https://doi.org/10.1063/5.0041506

Lyons JL, Janotti A, Van De Walle CG (2010) Carbon impurities and the yellow luminescence in GaN. Appl Phys Lett 97(15). https://doi.org/10.1063/1.3492841

Matsubara M, Bellotti E (2017) A first-principles study of carbon-related energy levels in GaN. I. Complexes formed by substitutional/interstitial carbons and gallium/nitrogen vacancies. J Appl Phys 121(19):195701. https://doi.org/10.1063/1.4983452

Meneghesso G, Meneghini M, Rossetto I, Bisi D, Stoffels S, Van Hove M, Decoutere S, Zanoni E (2016) Reliability and parasitic issues in GaN-based power HEMTs: a review. Semicond Sci Technol 31(9):093004. https://doi.org/10.1088/0268-1242/31/9/093004

Meneghesso G, Silvestri R, Meneghini M, Cester A, Zanoni E, Verzellesi G, Pozzovivo G, Lavanga S, Detzel T, Haberlen O, Curatola G (2014) Threshold voltage instabilities in D-mode GaN HEMTs for power switching applications. In: IEEE International Reliability Physics Symposium (IRPS), pp 6–10. https://doi.org/10.1109/IRPS.2014.6861109

Meneghesso G, Meneghini M, Silvestri R, Vanmeerbeek P, Moens P, Zanoni E (2016) High voltage trapping effects in GaN-based metal-insulator-semiconductor transistors. Jpn J Appl Phys 55(1). https://doi.org/10.7567/JJAP.55.01AD04.

Meneghini M, Rossetto I, Bisi D, Stocco A, Chini A, Pantellini A, Lanzieri C, Nanni A, Meneghesso G, Zanoni E (2014) Buffer traps in Fe-doped AlGaN/GaN HEMTs: Investigation of the physical properties based on pulsed and transient measurements. IEEE Trans Electron Devices 61(12):4070–4077. https://doi.org/10.1109/TED.2014.2364855

Meneghini M, Rossetto I, Bisi D, Ruzzarin M, Van Hove M, Stoffels S, Wu T-LL, Marcon D, Decoutere S, Meneghesso G, Zanoni E (2016) Negative bias-induced threshold voltage instability in GaN-on-Si power HEMTs. IEEE Electron Device Lett 37(4):474–477. https://doi.org/10.1109/LED.2016.2530693

Meneghini M, Tajalli A, Moens P, Banerjee A, Zanoni E, Meneghesso G (2018) Trapping phenomena and degradation mechanisms in GaN-based power HEMTs. Mater Sci Semicond Process 78:118–126. https://doi.org/10.1016/j.mssp.2017.10.009

Meneghini M, De Santi C, Abid I, Buffolo M, Cioni M, Khadar RA, Nela L, Zagni N, Chini A, Medjdoub F, Meneghesso G, Verzellesi G, Zanoni E, Matioli E (2021) GaN-based power devices: Physics, reliability, and perspectives. J Appl Phys 130(18):181101. https://doi.org/10.1063/5.0061354

Meneghini M, Chowdhury S, Derluyn J, Medjdoub F, Ji D, Chun J, Kabouche R, De Santi C, Zanoni E, Meneghesso G (2023) GaN-Based Lateral and Vertical Devices. In: Rudan M, Brunetti R, Reggiani S (eds) Springer Handbook of Semiconductor Devices. Cham: Springer International Publishing, pp 525–578 (2023). https://doi.org/10.1007/978-3-030-79827-7_15

Modolo N, De Santi C, Baratella G, Bettini A, Borga M, Posthuma N, Bakeroot B, You S, Decoutere S, Bevilacqua A, Neviani A, Meneghesso G, Zanoni E, Meneghini M (2022) Compact modeling of nonideal trapping/detrapping processes in GaN power devices. IEEE Trans Electron Devices 69(8):4432–4437. https://doi.org/10.1109/TED.2022.3184622

Modolo N, De Santi C, Minetto A, Sayadi L, Prechtl G, Meneghesso G, Zanoni E, Meneghini M (2022) Trap-state mapping to model GaN transistors dynamic performance. Sci Rep 12(1):1–10. https://doi.org/10.1038/s41598-022-05830-7

Moens P, Vanmeerbeek P, Banerjee A, Guo J, Liu C, Coppens P, Salih A, Tack M, Caesar M, Uren MJ, Kuball M, Meneghini M, Meneghesso G, Zanoni E (2015) On the impact of carbon-doping on the dynamic Ron and off-state leakage current of 650V GaN power devices. In: Proceedings of the International Symposium on Power Semiconductor Devices and ICs, vol 2015, pp 37–40. https://doi.org/10.1109/ISPSD.2015.7123383

Moens P, Banerjee A, Uren MJ, Meneghini M, Karboyan S, Chatterjee I, Vanmeerbeek P, Casar M, Liu C, Salih A, Zanoni E, Meneghesso G, Kuball M, Tack M (2015) Impact of buffer leakage on intrinsic reliability of 650V AlGaN/GaN HEMTs. In: 2015 IEEE International Electron Devices Meeting (IEDM), p 35.2.1–35.2.4. https://doi.org/10.1109/IEDM.2015.7409831

Moens P, Banerjee A, Uren MJ, Meneghini M, Karboyan S, Chatterjee I, Vanmeerbeek P, Cäsar M, Liu C, Salih A, Zanoni E, Meneghesso G, Kuball M, Tack M, Casar M, Liu C, Salih A, Zanoni E, Meneghesso G, Kuball M, Tack M (2015) Impact of buffer leakage on intrinsic reliability of 650V AlGaN/GaN HEMTs. In: IEEE International Electron Devices Meeting (IEDM), p 35.2.1–35.2.4. https://doi.org/10.1109/IEDM.2015.749831

Moens P, Uren MJ, Banerjee A, Meneghini M, Padmanabhan B, Jeon W, Karboyan S, Kuball M, Meneghesso G, Zanoni E, Tack M (2017) Negative dynamic ron in AlGaN/GaN power devices. In: IEEE International Symposium on Power Semiconductor Devices and IC's (ISPSD), Sapporo, Japan, pp 97–100. https://doi.org/10.23919/ISPSD.2017.7988902

Nakano Y. (2018) Deep-level defects in homoepitaxial p-type GaN. J Vac Sci & Technol A 36(2). https://doi.org/10.1116/1.5017867

Nakarmi ML, Nepal N, Lin JY, Jiang HX (2009) Photoluminescence studies of impurity transitions in Mg-doped AlGaN alloys. Appl Phys Lett 94(9):091903. https://doi.org/10.1063/1.3094754

Nakayama H, Hacke P, Khan MRH, Detchprohm T, Hiramatsu KHK, Sawaki NSN (1996) Electrical Transport Properties of p-GaN. Jpn J Appl Phys 35(3A):L282. https://doi.org/10.1143/JJAP.35.L282

Nguyen XS, Hou HW, De Mierry P, Vennéguès P, Tendille F, Arehart AR, Ringel SA, Fitzgerald EA, Chua SJ (2016) Deep level traps in semi-polar n-GaN grown on patterned sapphire substrate by metalorganic vapor phase epitaxy. Phys Status Solidi (B) 253(11):2225–2229. https://doi.org/10.1002/pssb.201600364

Park YS, Lee M, Jeon K, Yoon IT, Shon Y, Im H, Park CJ, Cho HY, Han M-S (2010) Deep level transient spectroscopy in plasma-assisted molecular beam epitaxy grown Al0.2Ga0.8N/GaN interface and the rapid thermal annealing effect. Appl Phys Lett 97(11):112110. https://doi.org/10.1063/1.3491798.

Polyakov AY, Smirnov NB, Govorkov AV, Shlensky AA, Pearton SJ (2004) Influence of high-temperature annealing on the properties of Fe doped semi-insulating GaN structures. J Appl Phys 95(10):5591–5596. https://doi.org/10.1063/1.1697616

Polyakov AY, Smirnov NB, Govorkov AV, Markov AV, Sun Q, Zhang Y, Yerino CD, Ko T-S, Lee I-H, Han J (2010) Electrical properties and deep traps spectra of a-plane GaN films grown on r-plane sapphire. Mater Sci Eng, B 166(3):220–224. https://doi.org/10.1016/j.mseb.2009.11.030

Polyakov AY, Smirnov NB, Turutin AV, Shemerov IS, Ren F, Pearton SJ, Johnson JW (2016) Deep traps and instabilities in AlGaN/GaN high electron mobility transistors on Si substrates. J Vac Sci Technol, B 34(4):041216. https://doi.org/10.1116/1.4953347

Rackauskas B, Uren MJ, Stoffels S, Zhao M, Decoutere S, Kuball M (2018) Determination of the self-compensation ratio of carbon in AlGaN for HEMTs. IEEE Trans Electron Devices 65(5):1838–1842. https://doi.org/10.1109/TED.2018.2813542

Rajasingam S, Pomeroy JW, Kuball M, Uren MJ, Martin T, Herbert DC, Hilton KP, Balmer RS (2004) Micro-Raman temperature measurements for electric field assessment in active AlGaN-GaN HFETs. IEEE Electron Device Lett 25(7):456–458. https://doi.org/10.1109/LED.2004.830267

Remesh N, Mohan N, Raghavan S, Muralidharan R, Nath DN (2020) Optimum carbon concentration in GaN-on-silicon for breakdown enhancement in AlGaN/GaN HEMTs. IEEE Trans Electron Devices 67(6):2311–2317. https://doi.org/10.1109/TED.2020.2989421

Richter E, Beyer FC, Zimmermann F, Gärtner G, Irmscher K, Gamov I, Heitmann J, Weyers M, Tränkle G (2020) Growth and properties of intentionally carbon-doped GaN layers. Cryst Res Technol 55(2):1900129. https://doi.org/10.1002/crat.201900129

Sang F, Wang M, Zhang C, Tao M, Xie B, Wen CP, Wang J, Hao Y, Wu W, Shen B (2015) Investigation of the threshold voltage drift in enhancement mode GaN MOSFET under negative gate bias stress. Jpn J Appl Phys 54(4):044101. https://doi.org/10.7567/JJAP.54.044101

De Santi C, Buffolo M, Meneghesso G, Zanoni E, Meneghini M (2021) Dynamic performance characterization techniques in gallium nitride-based electronic devices. Crystals 11(9). https://doi.org/10.3390/cryst11091037

De Santi C, Zanoni E, Meneghini M, Meneghesso G, Rampazzo F, Gao VZ, Sharma C, Chiocchetta F, Verzellesi G, Chini A, Cioni M, Zagni N, Lanzieri C, Pantellini A, Peroni M, Latessa L (2022) Role of carbon in dynamic effects and reliability of 0.15-um AlGaN/GaN HEMTs for RF power amplifiers. In: Gallium Nitride Materials and Devices XVII, San Francisco, United States, p 56. https://doi.org/10.1117/12.2609666

Sasikumar A, Arehart A, Kolluri S, Wong MH, Keller S, DenBaars SP, Speck JS, Mishra UK, Ringel SA (2012) Access-region defect spectroscopy of DC-stressed N-polar GaN MIS-HEMTs. IEEE Electron Device Lett 33(5):658–660. https://doi.org/10.1109/LED.2012.2188710

Sasikumar A, Arehart A, Ringel SA, Kaun S, Wong MH, Mishra UK, Speck JS (2012) Direct correlation between specific trap formation and electric stress-induced degradation in MBE-grown AlGaN/GaN HEMTs. In: 2012 IEEE International Reliability Physics Symposium (IRPS), p 2C.3.1–2C.3.6. https://doi.org/10.1109/IRPS.2012.6241780

Schroder DK (2005) "Defects," in Semiconductor Material and Device Characterization, Hoboken, NJ, USA: John Wiley & Sons, Inc., pp 251–317. https://doi.org/10.1002/0471749095.ch5

Shockley W, Read WT (1952) Statistics of the recombinations of holes and electrons. Phys Rev 87(5):835–842. https://doi.org/10.1103/PhysRev.87.835

Silvestri M, Uren MJ, Kuball M (2013) Iron-induced deep-level acceptor center in GaN/AlGaN high electron mobility transistors: Energy level and cross section. Appl Phys Lett 102(7):073501. https://doi.org/10.1063/1.4793196

Soh CB, Chua SJ, Lim HF, Chi DZ, Liu W, Tripathy S (2004) Identification of deep levels in GaN associated with dislocations. J Phys: Condens Matter 16(34):6305. https://doi.org/10.1088/0953-8984/16/34/027

Stuchlíková Ľ, Šebok J, Rybár J, Petrus M, Nemec M, Harmatha L, Benkovská J, Kováč J, Škriniarová J, Lalinský T, Paskiewicz R, Tlaczala M (2010) Investigation of deep energy levels in heterostructures based on GaN by DLTS. In: The Eighth International Conference on Advanced Semiconductor Devices and Microsystems, pp 135–138. https://doi.org/10.1109/ASDAM.2010.5666319

Sze SM, Ng KK (2006) Physics of Semiconductor Devices. Hoboken, NJ, USA: John Wiley & Sons, Inc. https://doi.org/10.1002/0470068329

Tallarico AN, Millesimo M, Bakeroot B, Borga M, Posthuma N, Decoutere S, Sangiorgi E, Fiegna C (2022) TCAD modeling of the dynamic VTH Hysteresis under fast sweeping characterization in p-GaN gate HEMTs. IEEE Trans Electron Devices 69(2):507–513. https://doi.org/10.1109/TED.2021.3134928

Tanaka T, Watanabe A, Amano H, Kobayashi Y, Akasaki I, Yamazaki S, Koike M (1994) P-type conduction in Mg-doped GaN and Al0.08Ga0.92N grown by metalorganic vapor phase epitaxy. Appl Phys Lett 65(5):593–594. https://doi.org/10.1063/1.112309

Tanaka K, Ishida M, Ueda T, Tanaka T (2013) Effects of Deep Trapping States at High Temperatures on Transient Performance of AlGaN/GaN Heterostructure Field-Effect Transistors. Jpn J Appl Phys 52(4S):04CF07. https://doi.org/10.7567/JJAP.52.04CF07

Tanaka K, Morita T, Umeda H, Kaneko S, Kuroda M, Ikoshi A, Yamagiwa H, Okita H, Hikita M, Yanagihara M, Uemoto Y, Takahashi S, Ueno H, Ishida H, Ishida M, Ueda T (2015) Suppression of current collapse by hole injection from drain in a normally-off GaN-based hybrid-drain-embedded gate injection transistor. Appl Phys Lett 107(16). https://doi.org/10.1063/1.4934184.

Tokuda Y (2016) (Invited) DLTS Studies of defects in n-GaN. ECS Trans 75(4):39. https://doi.org/10.1149/07504.0039ecst

Umana-Membreno GA, Dell JM, Hessler TP, Nener BD, Parish G, Faraone L, Mishra UK (2002) 60Co gamma-irradiation-induced defects in n-GaN. Appl Phys Lett 80(23):4354–4356. https://doi.org/10.1063/1.1483390

Umana-Membreno GA, Parish G, Fichtenbaum N, Keller S, Mishra UK, Nener BD (2008) Electrically Active Defects in GaN layers grown with and without Fe-doped buffers by metal-organic chemical vapor deposition. J Electron Mater 37(5):569–572. https://doi.org/10.1007/s11664-007-0313-3

Uren MJ, Moreke J, Kuball M (2012) Buffer design to minimize current collapse in GaN/AlGaN HFETs. IEEE Trans Electron Devices 59(12):3327–3333. https://doi.org/10.1109/TED.2012.2216535

Uren MJ, Caesar M, Karboyan S, Moens P, Vanmeerbeek P, Kuball M (2015) Electric field reduction in C-doped AlGaN/GaN on Si high electron mobility transistors. IEEE Electron Device Lett 36(8):826–828. https://doi.org/10.1109/LED.2015.2442293

Uren MJ, Karboyan S, Chatterjee I, Pooth A, Moens P, Banerjee A, Kuball M (2017) 'Leaky Dielectric' model for the suppression of dynamic R_{ON} in carbon-doped AlGaN/GaN HEMTs. IEEE Trans Electron Devices 64(7):2826–2834. https://doi.org/10.1109/TED.2017.2706090

Uren MJ, Kuball M (2021) Impact of carbon in the buffer on power switching GaN-on-Si and RF GaN-on-SiC HEMTs. Jpn J Appl Phys 60(SB):SB0802. https://doi.org/10.35848/1347-4065/abdb82

Verzellesi G, Morassi L, Meneghesso G, Meneghini M, Zanoni E, Pozzovivo G, Lavanga S, Detzel T, Häberlen O, Curatola G (2014) Influence of buffer carbon doping on pulse and AC behavior of insulated-gate field-plated power AlGaN/GaN HEMTs. IEEE Electron Device Lett 35(4):443–445. https://doi.org/10.1109/LED.2014.2304680

Viey AG, Vandendaele W, Jaud M-A, Cluzel J, Barnes J-P, Martin S, Krakovinsky A, Gwoziecki R, Plissonnier M, Gaillard F, Modica R, Iucolano F, Meneghini M, Zanoni E, Meneghesso G, Ghibaudo G (2019) Investigation of nBTI degradation on GaN-on-Si E-mode MOSc-HEMT. In: IEEE International Electron Devices Meeting (IEDM), San Francisco, CA, USA, p 4.3.1–4.3.4. https://doi.org/10.1109/IEDM19573.2019.8993588

Wang CD, Yu LS, Lau SS, Yu ET, Kim W, Botchkarev AE, Morkoç H (1998) Deep level defects in n-type GaN grown by molecular beam epitaxy. Appl Phys Lett 72(10):1211–1213. https://doi.org/10.1063/1.121016

Weiser MCJ, Hückelheim J, Kallfass I (2021) A novel approach for the modeling of the dynamic ON-state resistance of GaN-HEMTs. IEEE Trans Electron Devices 68(9):4302–4309. https://doi.org/10.1109/TED.2021.3098498

Wright AF (2002) Substitutional and interstitial carbon in wurtzite GaN. J Appl Phys 92(5):2575–2585. https://doi.org/10.1063/1.1498879

Yam FK, Low LL, Oh SA, Hassan Z, Yam FK, Low LL, Oh SA, Hassan Z (2011) Gallium nitride: An overview of structural defects. In: Optoelectronics -Materials and Techniques, IntechOpen. https://doi.org/10.5772/19878.

Yamada H, Chonan H, Takahashi T, Yamada T, Shimizu M (2018) Deep-level traps in lightly Si-doped n-GaN on free-standing m-oriented GaN substrates. AIP Adv 8(4):045311. https://doi.org/10.1063/1.5011362

Yang S, Han S, Sheng K, Chen KJ (2019) Dynamic On-resistance in gan power devices: mechanisms, characterizations, and modeling. IEEE J Emerg Sel Top Power Electron 7(3):1425–1439. https://doi.org/10.1109/JESTPE.2019.2925117

Zagni N, Puglisi FM, Pavan P, Chini A, Verzellesi G (2019) Insights into the off-state breakdown mechanisms in power GaN HEMTs. Microelectron Reliab 100–101:113374. https://doi.org/10.1016/j.microrel.2019.06.066

Zagni N, Cioni M, Chini A, Iucolano F, Puglisi FM, Pavan P, Verzellesi G (2021) Mechanisms underlying the bidirectional V_T shift after negative-bias temperature instability stress in carbon-doped fully recessed AlGaN/GaN MIS-HEMTs. IEEE Trans Electron Devices 68(5):2564–2567. https://doi.org/10.1109/TED.2021.3063664

Zagni N, Chini A, Puglisi FM, Pavan P, Verzellesi G (2021) On the modeling of the donor/acceptor compensation ratio in carbon-doped GaN to univocally reproduce breakdown voltage and current collapse in lateral GaN power HEMTs. Micromachines 12(6):709. https://doi.org/10.3390/mi12060709

Zagni N, Chini A, Puglisi FM, Meneghini M, Meneghesso G, Zanoni E, Pavan P, Verzellesi G (2021) 'Hole redistribution' model explaining the thermally activated RON stress/recovery transients in carbon-doped AlGaN/GaN power MIS-HEMTs. IEEE Trans Electron Devices 68(2):697–703. https://doi.org/10.1109/TED.2020.3045683

Zagni N, Cioni M, Chini A (2021) Effect of trap-filling bias on the extraction of the time constant of drain current transients in AlGaN/GaN HEMTs. In: 2021 IEEE 8th Workshop on Wide Bandgap Power Devices and Applications (WiPDA), Redondo Beach, CA, USA, pp 231–235. https://doi.org/10.1109/WiPDA49284.2021.9645115

Zagni N, Verzellesi G, Chini A (2022) Temperature-independent current dispersion in 0.15 μm AlGaN/GaN HEMTs for 5G applications. Micromachines 13(12):2244. https://doi.org/10.3390/mi13122244.

Zhang Z, Hurni CA, Arehart AR, Yang J, Myers RC, Speck JS, Ringel SA (2012) Deep traps in nonpolar m-plane GaN grown by ammonia-based molecular beam epitaxy. Appl Phys Lett 100(5):052114. https://doi.org/10.1063/1.3682528

Zhang Z, Hurni CA, Arehart AR, Speck JS, Ringel SA (2012) Influence of V/III growth flux ratio on trap states in m-plane GaN grown by ammonia-based molecular beam epitaxy. Appl Phys Lett 101(15):152104. https://doi.org/10.1063/1.4759037

Zhang Z, Arehart AR, Kyle ECH, Chen J, Zhang EX, Fleetwood DM, Schrimpf RD, Speck JS, Ringel SA (2015) Proton irradiation effects on deep level states in Mg-doped p-type GaN grown by ammonia-based molecular beam epitaxy. Appl Phys Lett 106(2):022104. https://doi.org/10.1063/1.4905783

Zhang R, Kozak JP, Song Q, Xiao M, Liu J, Zhang Y (2020) Dynamic breakdown voltage of GaN power HEMTs. In: 2020 IEEE International Electron Devices Meeting (IEDM), p 23.3.1–23.3.4. https://doi.org/10.1109/IEDM13553.2020.9371904.

Zimmermann F, Beyer J, Röder C, Beyer FC, Richter E, Irmscher K, Heitmann J (2021) Current status of carbon-related defect luminescence in GaN. Phys Status Solidi (A) 218(20):2100235. https://doi.org/10.1002/pssa.202100235

Recent Developments and Applications of High Electron Mobility Transistors

Kalyan Biswas, Rachita Ghoshhajra, and Angsuman Sarkar

Abstract Several material systems, including AlGaAs/GaAs, AlGaN/GaN, InP/AlInAs/InGaAs, GaAs/AlInAs/InGaAs, etc., may be employed to develop HEMTs. HEMTs with low bandgap channel materials are most suitable for high-speed, low-power applications, whereas those with wide bandgap channel materials are excellent for high-power applications. Several industries have demonstrated many circuits with varying levels of complexity using HEMTs. According to survey report published by Spherical Insights & Consulting, the global size of HEMT market is expected to reach US$7.51 billion by 2022, and the global high electron mobility transistor market trade is expected to reach US$16.82 billion by 2032. A lot of research has been done on the development of high power electronic devices (HEMTs) because they are a promising alternative to conventional transistors, more promising and used in wireless applications. HEMT devices are receiving consideration for their greater electron transport and are competing with traditional field-effect transistors (FETs) and replacing them with outstanding performance at very larger frequency, enhanced power density and improved proficiency. This chapter highlights the applications of GaN HEMTs for RF electronics & digital electronics, integrated circuits and quantum computing applications.

Keywords MOSHEMT · GaN HEMT · InP HEMT · Integrated circuits · RF electronics · Quantum computing

K. Biswas (✉)
ECE Department, MCKV Institute of Engineering, Howrah, West Bengal, India
e-mail: kalyan.b.2006@ieee.org

R. Ghoshhajra
CSE Department, MCKV Institute of Engineering, Howrah, West Bengal, India

A. Sarkar
ECE Department, Kalyani Government Engineering College, Kalyani, West Bengal, India

D. Nirmal and J. Ajayan (eds.), *Modeling of AlGaN/GaN High Electron Mobility Transistors*, Springer Tracts in Electrical and Electronics Engineering,
https://doi.org/10.1007/978-981-97-7506-4_3

1 Introduction

The need for massive data generation, real-time video streaming, and faster downloads on mobile devices has fuelled the requirements for faster wireless communication. In power conversion at RF frequencies, GaAs and InP-based III-V semiconductors have been tried to outperform silicon-based devices since the innovation of the Gunn diode in 1962 (Barratt et al. 2009). A lot of research has been done on the development of high-power electronic devices (HEMTs) because they are a promising alternative to conventional transistors, more promising and used in wireless applications. HEMT devices are receiving consideration for their greater electron transport and are competing with traditional field-effect transistors (FETs) and replacing them with outstanding performance at very larger frequency, enhanced power density and improved proficiency. Starting with GaAs MESFETs, Heterojunction Bipolar Transistors (HBTs) and High Electron Mobility Transistors (HEMTs) devices are extending the limits of RF in several frequency bands (from high frequency to mmWave) and at all levels of power. This chapter is planned to give an overview of some popular and mostly used HEMT device structures along with working principles and simulation methodologies with graphical illustrations. From the beginning, the high mobility and confined nature of 2D electron gas demonstrated that modulation doping may be used to make very fast and efficient electronic devices with high scaling potential and very good short channel effects. Fujitsu introduces the first HEMT IC-based Ring Oscillator with enhancement-mode switching and depletion-mode load HEMTs having a gate length of 1.7 μm and assess logic performance capability. Soon the advantages of HEMTs in terms of speed-power performance become evident. Few companies have demonstrated many circuits with varying levels of complexity. According to survey report published by Spherical Insights & Consulting, the global size of HEMT market is expected to reach US$7.51 billion by 2022, and the global high electron mobility transistor market trade is expected to reach US$16.82 billion by 2032 (https://www.globenewswire.com/news-release/2023/06/01/2680054/0/en/Global-High-Electron-Mobility-Transistor-HEMT-Market-Size-To-Exceed-USD-16-82-Billion-By-2032-CAGR-of-8-4.html).

2 Working Principle of HEMTs

In HEMTs, heterojunctions are made of different semiconductors with different band gaps that require the valence band (VB) and conduction band (CB) to be bent across the materials to have a continuous level. The combination of band gap energies is very important when creating heterostructures using wide and narrow band gap materials. Depending on the orientations of the bandgap of materials, there exist different forms of structure dependent on material combinations. Various probable alignments of band gaps are presented in Fig. 1 below such as (a) "Straddling", (b) "Staggered",

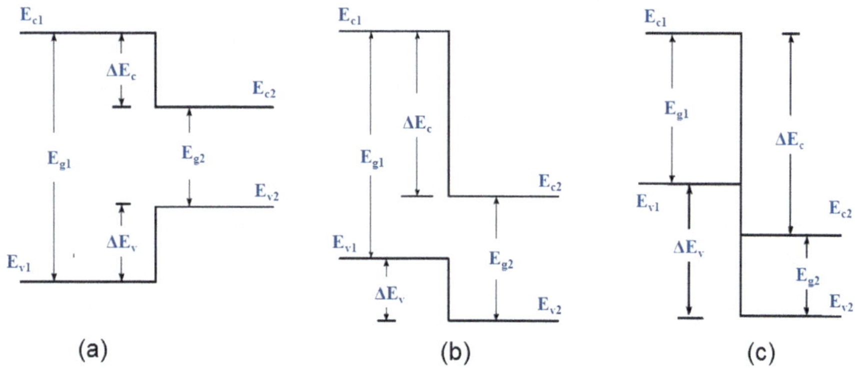

Fig. 1 Different possible band gap alignments

and (c) "Broken". Heterojunctions can also be categorized according to the electrical conductivity of the material creating the heterojunction. According to conductivity, there exist two types of heterojunction, one is isotype and the other is anisotype. Isotype heterojunction means heterojunction formed between semiconductors of the same conductivity (N–n or p-P); where uppercase letters represent materials with larger band gaps and lowercase letters represent materials with a lower line. However, an anisotropic heterojunction is made from semiconductor materials with different conductivity (N-p or n-P). Straddling type heterojunction is highly important for producing high mobility in HEMT channel. The offset in conduction band is shaped depending on the difference in electron affinity of the semiconductor materials used.

When creating a heterojunction, the bending of the energy band should have a continuous level. Diffusion of electrons from doped material having larger bandgap happens to neighbouring small bandgap materials with lower states of energy. The movement of electrons thus creates a potential difference which generates an electric field amongst the concerned materials. Electron drift is caused due to this electric field to restore the conduction band (CB) of the material with larger bandgap. This drift/diffusion process will continue until they become stable with each other and form an equilibrium intersection. It's very important to remember that the undoped material with smaller bandgap has a surplus of majority carriers. Lower bandgap material has no donor atoms to generate catering since it is undoped thus results in high mobility. The most unique feature of HEMTs is the mismatch between the valence and conduction bands, which can be adjusted to control the charge flow in and out of the device. Due to the diffusion of carriers, accumulation of electrons occurs in narrow band gap materials at the junction of the two regions. Hence, high currents are produced in these devices due to the electron accumulation. The high electron density region is created by electron deposition, also known as two-dimensional electron gas (2DEG). Figure 2 provides a schematic representation of the overall energy band diagram of a HEMT heterojunction (Muhammad Navid Anjum et al. 2017).

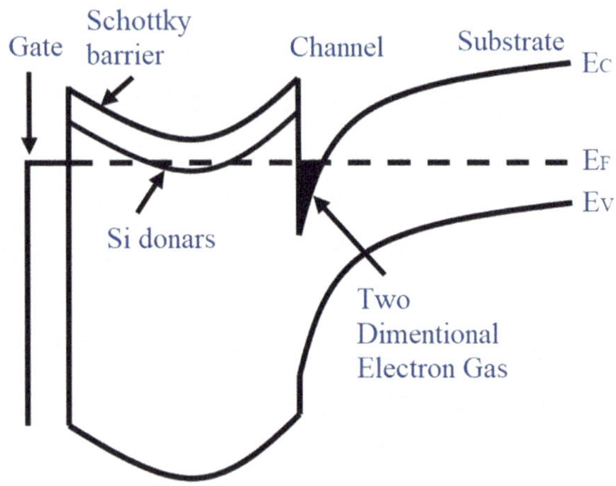

Fig. 2 A schematics of HEMTs' band diagram

3 Common HEMT Structures

The principle of high electron mobility is in the device can be realized by different combinations and orientations. A few common material combinations are described in this section.

3.1 GaAs HEMTs

A typical GaAs-based HEMT model is shown in Fig. 3. The material pair used for this device is AlGaAs and GaAs. Basically, the arrangement has a semi-insulating substrate GaAs on which a thin layer of GaAs is grown first; then an undoped AlGaAs layer, then a layer of doped AlGaAs, and finally another thin layer of GaAs. Usual thickness of these layers is marked in the figure. The transistor is completed by depositing aluminium and placing two ohmic contacts, called source and drain and finally create a Schottky barrier to act as a gate. To isolate most of the carriers from ionized impurities, a rapid transition is made from wide bandgap AlGaA material and lower bandgap material GaAs when wide bandgap material is doped. Accordingly, a channel is shaped at the junction of the GaAs/AlGaAs heterojunction.

3.2 GaN HEMT

GaN-based HEMTs are constructed using the same fabrication process as GaAs-based HEMTs, as shown in Fig. 4. However, AlGaN/GaN HEMTs do not require

Fig. 3 Schematic diagram of a GaAs HEMT structure

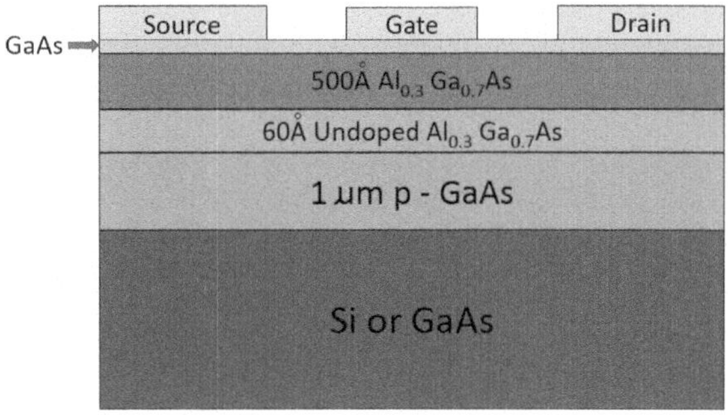

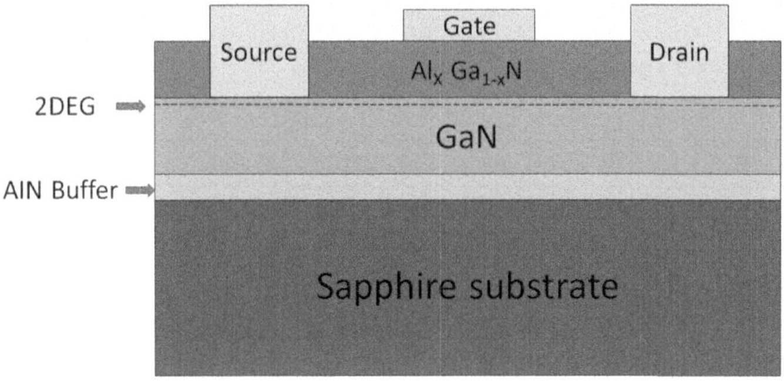

Fig. 4 Schematic view of a conventional GaN-HEMT

deliberate doping. Instead, electrons come from their surface states due to the polarization phenomenon in wurtzite-structured GaN. The accumulation of free carriers creates a high carrier density at the interface, creating a 2DEG channel.

3.3 InP-Based HEMTs

InP HEMTs provide lower electron effective mass at the InGaAs channel layer compared to traditional GaAs-based HEMTs. These HEMTs have a large transition (about 0.5 eV) between the channel layer and the adjacent InAlAs layer. So InP-based HEMTs exhibit a strong, high electron saturation rate and high electron concentration. A schematic diagram of the InP HEMT is shown in Fig. 5. InP-HEMTs exhibit good noise immunity and gain performance at microwave frequencies. At higher temperatures, these properties are further enhanced.

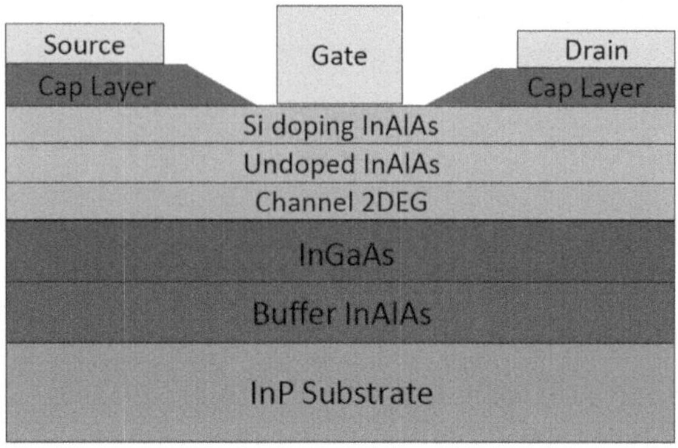

Fig. 5 Typical InP HEMT structure

3.4 MOS-HEMTs

Recently, high electron mobility transistors (HEMTs) have received great attention due to their greater electron transport which confirms extraordinary speed and finds applications in high power devices. The HEMT devices were invented by Mimura of Fujitsu Laboratories Ltd, Kobe, Japan (Mimura 2002; Liu et al. 2013).

To get rid of the need for high-frequency, low-noise, high-power-density applications, traditional MOSFETs, and MESFETs have had to be manufactured with very short channel lengths so that most carriers face minimal impurity scattering and reduce performance degradation. There are design and performance limitations that require a large saturation current and a high value of transconductance, which can be obtained by high doping level. In order to overcome these limitations, HEMT devices integrate heterojunctions made in between two dissimilar bandgap materials where to avoid impurity scattering electrons are restricted in a quantum well.

A commonly used material combination is GaAs with AlGaAs, although there is wide-ranging variation, which depends on the application of the device. The main features of this device are high 2DEG concentration, high transport speed, and high breakdown field at the heterojunction, so it has better performance compared to AlGaN/GaN materials.

However, HEMT gates have a Schottky barrier that is not particularly effective at preventing gate leakage, especially for scaled devices. Hence, in recent times, people have started to search for more insulating methods between the metal junction and the semiconductor layer. In most cases, oxides are utilized as the insulators, producing a metal–oxide–semiconductor HEMT (MOS-HEMT) structure. A schematic view of the MOS-HEMT device cross-section is shown in Fig. 6. Al_2O_3 (Liu et al. 2013; Hori et al. 2013; Chou et al. 2014) is chosen as oxide layer very frequently and used in MOSHEMT devices because of its large bandgap, large breakdown field, better thermal and chemical stability, and simplicity in deposition. Very good performance of MOSHEMT devices is reported by researchers where they have used TiO_2 (Dong et al. 2010), HfO_2 (Lee et al. 2014), Pr_2O_3 (Chiu et al. 2008), and SiO_2 (Hsu et al. 2016) as the gate dielectric to dodge the aforesaid limitation.

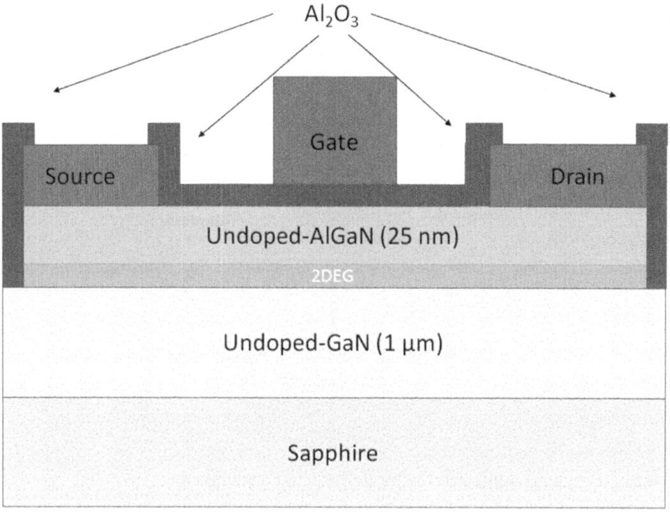

Fig. 6 Cross-section of Al2O3/AlGaN/GaN MOS-HEMT

4 Performance Analysis: TCAD Approach

In practical use, most electronic devices must eventually be integrated into a system to be effective. For example, AlGaN/GaN HEMTs are included in MMIC PAs to act as a major power source. While device-level research is important for the development of functional devices, it is equally important to understand the level of use. A simulation model should be developed to ensure that the device-circuit interaction is fully studied, especially for new devices under investigation (Jian et al. 2021). With the increasing acceptance of HEMT technology for high frequency and high power applications, researchers have shown increasing interest in this technology in recent years. Due to the lack of test results available for these materials, designers need a reliable tool that can predict the material's performance in a short period of time before making it available in commercial use. Therefore, physics-based device simulators are very important for the design and performance analysis of these semiconductor devices. Although researchers have published many analytical models in the literature to study and analyse the features of HEMTS and improve device performance, TCAD simulator tools are easier to use (Sarkar 2016). Physics-based models are developed taking into account material-specific properties such as polarization effects, energy band diagrams, piezoelectricity, and more. The developed specific models have a set of primary equations that connect the carrier densities and the electrostatic potential using some simulation domain. These important equations derived from Maxwell's equations include Poisson's equations, transport equations, continuity equations, etc. Distribution of carrier densities as a result of charge transport along with generation and recombination process are obtained by solving those equations. In TCAD simulation software, the model is divided into many small elements

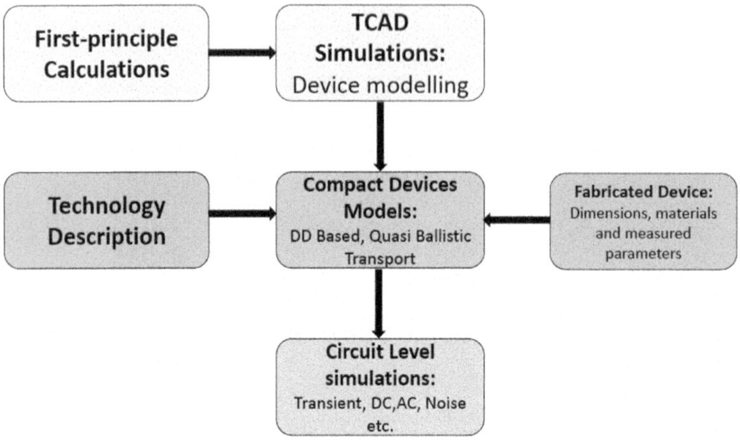

Fig. 7 Overview of general approach for modelling and simulation

and grid points are called "nodes". Different electrical parameters are calculated in these nodes of the device such as current, voltage, transconductance, resistance, etc. depending on requirements and model capability. These parameters are calculated by the TCAD software using the numerical simulations. This element size of the mesh is very crucial and plays a significant role in the accuracy of simulation results. If the grid density is increased for simulation model, the accuracy of the calculation will increase, but the simulation time will increase in turn. Therefore, the compromise between the accuracy and simulation time is crucial to realize the benefits of physical simulation. Different methodology used by TCAD software for simulation in different domains is shown in Fig. 7.

TCAD software provides various transport models to determine the behaviour of the system (Sarkar 2016) like Drift Diffusion, Mobility, SRH, Thermodynamic, Impact ionization, polarization, and many more. To effectively calculate/estimate the properties of the device, researchers need to include such parameters during simulation model set up. Popular TCAD software can often provide the different physical models required for modelling (Biswas et al. 2023). Mixed model simulation approach helps to study the circuit performance of the device under study. Impact of different device parameters on circuit may be studied easily using mixed-mode simulation approach. The compact design of the device paves way large-scale circuit simulations with lots of interconnected transistors, which enables easy implementation of circuit-device co-design for the HEMT devices. A general overview of simulation approaches used in HEMT device simulation is illustrated in Fig. 7.

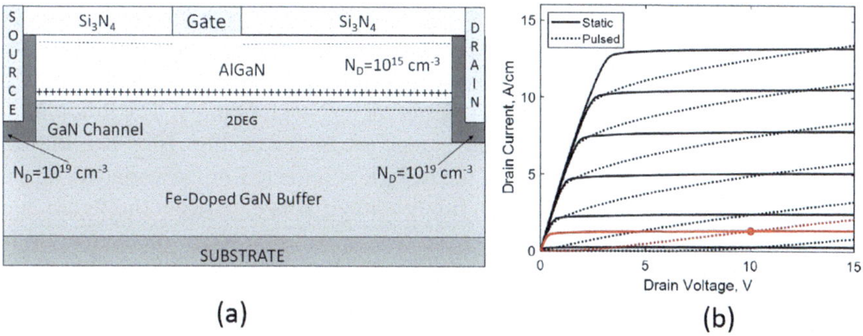

Fig. 8 **a** Cross-sectional view of A GAN HEMT device **b** I_d-V_d curve of the device

4.1 Device Performance Analysis

TCAD simulation may be carried out in device level for the HEMT device considering suitable material combinations and structures. The cross-sectional view of a AlGaN/GaN HEMT structure and the simulation result is shown in Fig. 8 (Catoggio et al. 2023). The length of the gate is 150 nm, the source to gate distance is 0.8 μm, and the drain to gate distance is 2 μm. The AlGaN layer that acts as barrier layer is 15 nm thick with a 25% Al mole fraction, while the undoped GaN channel region is 5 nm thick. 2 μm deep GaN buffer layer is also used. Simulated output characteristics are shown in Fig. 8b. It has been shown that the threshold voltage of the device is V_{th} = 2.5 V and the saturation current at V_D = 10 V is I_{Dss} = 1.3 A/mm (Catoggio et al. 2023).

4.2 Circuit Performance Analysis

Mixed mode simulation capability of TCAD software can be used for circuit performance estimation of the device using device circuit co-design. The impact of the device performance parameters for its usefulness in the circuit was estimated using the device in a cascode generator (Biswas et al. 2018). A schematic diagram of the InGaAs double gate device structure and simulated Vg-Id curve of the device is shown in Fig. 9. The mole fraction of Indium in the compound semiconductor is varied to obtain different curves. A cascode amplifier is basically a two-stage circuit designed using two transistors. The circuit diagram of the two-stage cascode amplifier where one of the MOSFETs (M_2) is connected with the reference voltage V_r and the input is used at the gate terminal of the MOSFET M_1 is shown in Figure X.10. The schematic circuit diagram of a two-stage cascode amplifier is shown in Fig. 10a where one of the MOSFETs indicated by M_2 is biased by the reference voltage V_r and in the gate terminal of MOSFET M_1, input is applied. In the circuit, for one transistor, the mode of operation is a common source and for the other is a common gate configuration. This combination has advantages like higher gain or higher bandwidth, higher input–output isolation, better slew rate, stability and more flexibility.

The device has a width of 1 μm and a reference voltage of 0.6 V. The differential gain of the amplifier is simulated and plotted as shown in Fig. 10b. The impact of different Indium concentrations in the device is reflected in the variation of the differential gain curve as a function of input voltage. It is clear from the figure that the maximum gain of the circuit using the device with the lower mole fraction is greater than the circuit using device with the larger indium percentage.

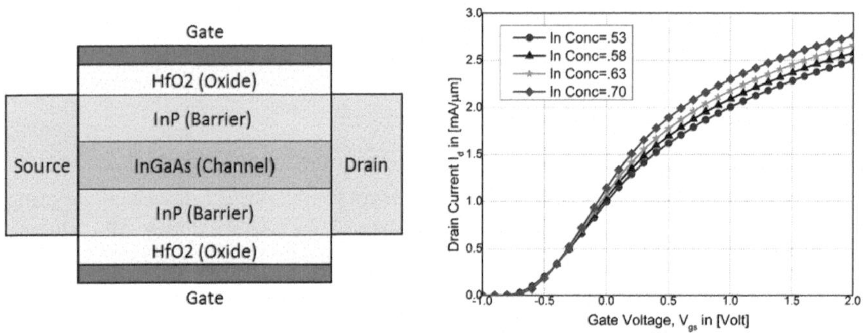

(a) (b)

Fig. 9 **a** Schematic diagram of a double gate InGaAs MOSHEMT

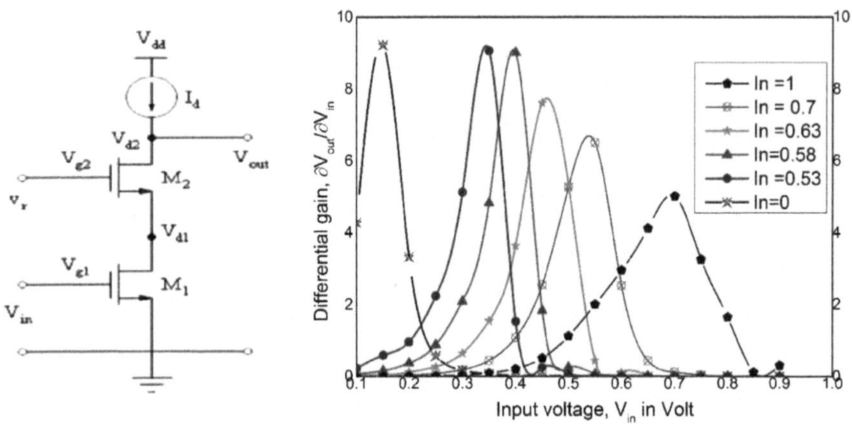

(a) (b)

Fig. 10 **a** Cascode amplifier **b** differential gain variation with input voltage due to different Indium mole fraction

5 Applications

5.1 GaN for RF Electronics

Compared to GaAs, Si, and SiGe, GaN technology has the best features for RF power applications such as satellite communications and 5G mobile phones. Fifth generation (5G) and beyond wireless communication, base station transmitters need signal modulation bandwidths (>100 t0 400). A similar pattern has been observed for Ku and Ka band satellite communications also. Power amplifier is the most potent source of energy starvation in radio transmitters, so increasing its effectiveness is a major research topic (Ma et al. 2017). Various PA architectures have been proposed in recent times for upcoming generations of cellular communication.

5.2 GaN for Digital Electronics

III-N alloys have excellent properties that enable advanced technology with channel lengths below 10 nm. Compared to Si and other semiconductors, III-N semiconductors have a wide bandgap, which reduces band-to-band tunnelling and gate-induced current leakage, leading to a reduction of current and therefore low lower. Using atomic quantum transport simulations, the performance of 5 nm gate-length GaN nMOS nanowire field-effect transistors (GaN-NWnFETs) in various geometries around the cross-sectional scalability limit was investigated and reported (Lu et al. 2019). Circles, triangles and hexagonal shapes were found to have higher current than square shapes, even though their lesser surface area. The better performance of circular, triangular and hexagonal cross-sections can be directly attributed to the greater effective mass of the carriers, which prevent direct source to drain tunnelling.

5.3 GaN Integrated Circuits

GaN's high critical electric field, combined with its high transport characteristics, allows GaN power transistors to have shorter drift region and smaller gate width than silicon or SiC transistors. This results in reduced gate capacitance and higher switching frequency from conventional power switches at the same operating voltage. To take advantage of the gate reduction and high switching efficiency of GaN power transistors, it is necessary to reduce the parasitic inductance of the power switch/transistor and gate driver circuit. Lot of research has been done on the new all-GaN doping technology, which allows the integration of high-performance n-channel and p-channel GaN-advanced type transistors on the same chip without epitaxial regrowth or special processing (Chowdhury et al. 2019; Bader et al. 2020; Zheng et al. 2021).

5.4 *GaN for Quantum Computing Applications*

Quantum computing is considered the next great revolution in technology. From recently published reports on research and development made by many universities and major industry players, the so-called "quantum superiority" has been realized and success has been achieved. Although each of Si-Ge, III-V, and III-N platforms has its own strengths and limitations, the III-N material platform stands out as a unique material that combines superconducting, semiconducting, and piezoelectric/ferroelectric materials in the same epitaxial group. Because of the wurtzite crystal structure, many III-N materials such as III-N and its alloys exhibit piezoelectricity and even ferroelectricity (such as ScAlN). Alloys such as AlN and ScAlN give rise to high-performance MEMS resonators, which are increasingly used for phonon state matching and even quantum memories (Gokhale et al. 2020; Satzinger et al. 2018; Cheng et al. 2020; Yan et al. 2018).

6 Conclusion

The prospect of GaN-based devices and technologies in electronics, radio frequency, and digital and quantum electronics, with their potential for integration was presented in this chapter. For power electronic devices, views from high to low power applications, vertical and lateral, 2D and 3D and multi-channel devices, and discussion on GaN material strength and toughness were discussed. The potential of GaN for digital PAs and the integration of software, hardware, and artificial intelligence could lead to improved systems and systems for future mmWave operations. A vision to go beyond the traditional power and RF applications of GaN and usher in a new era in GaN-based digital and quantum electronics was also proposed.

References

Muhammad Navid Anjum A, Sharadindu K, Farhana A, Md Khosru AQDM (2017) High electron mobility transistors: performance analysis, research trend and applications. In: book: Different Types of Field-Effect Transistors-Theory and Applications Chapter: 3, Publisher: InTech. https://doi.org/10.5772/67796

Mimura T (2002) The early history of the high electron mobility transistor (HEMT). IEEE Trans Microw Theory Tech 50(3):780–782

Liu H, Chou B, Hsu W, Lee C, Sheu J, Ho C (2013) Enhanced AlGaN/GaN MOS-HEMT Performance by Using Hydrogen Peroxide Oxidation Technique. IEEE Trans Electron Devices 60(1):213–220

Hori Y, Yatabe Z, Hashizume T (2013) Characterization of interface states in Al2O3/AlGaN/GaN structures for improved performance of high-electron-mobility transistors. J Appl Phys 114:244503

Chou B-Y, Hsu W-C, Liu H-Y, Lee C-S, Wu Y-S, Sun W-C, Wei S-Y, Yu S-M, Chiang M-H (2014) Investigations of AlGaN/GaN MOS-HEMT with Al2O3deposition by ultrasonic spray pyrolysis method. Semicond Sci Technol 30(1)

Dong Z, Wang J, Wen CP, Gong D, Li Y, Yu M et al (2010) High Breakdown AlGaN/GaN MOSHEMT with Thermal Oxidized Ni/ti as Gate Insulator Solid State Electron 54:1339–1342

Lee C-S, Liao Y-H, Chou B-Y, Liu H-Y, Hsu W-C (2014) Composite HfO2/Al2O3-dielectric AlGaAs/InGaAs MOS-HEMTs by using RF sputtering/ozone water oxidation Superlattice. Microst 72:194–203

Chiu H-C, Yang C-W, Lin Y-H, Lin R-M, Chang L-B, Horng K-Y (2008) Device characteristics of AlGaN/GaN MOS-HEMTs using high-praseodymium oxide layer IEEE Trans Electron Dev. 55:3305–3309

Hsu C-H, Shih W-C, Lin Y-C, Hsu H-T, Hsu H-H, Huang Y-X, et al. (2016) Improved linearity and reliability in GaN metal–oxide–semiconductor high-electron-mobility transistors using nanolaminate La2O3/SiO2 gate dielectric Jpn. J Appl Phys 55:04EG04

Jian ZA, Mohanty S, Ahmadi E (2021) Switching performance analysis of 3.5 kV Ga2O3 power FinFETs. IEEE Trans Electron Devices 68(2):672–678. https://doi.org/10.1109/TED.2020.3043988

Sarkar CK (2016) Technology computer aided design: Simulation for VLSI MOSFET. CRC Press

Biswas K, Ghoshhajra R, Sarkar A (2023) High electron mobility transistor: Physics-based TCAD simulation and performance analysis. In: Lenka TR, Nguyen HPT (eds) HEMT Technology and Applications. Springer Tracts in Electrical and Electronics Engineering. Springer, Singapore. https://doi.org/10.1007/978-981-19-2165-0_12

Catoggio E, DonatiGuerrieri S, Bonani F (2023) TCAD Modeling of GaN HEMT output admittance dispersion through trap rate equation green's functions. Electronics 12:2457. https://doi.org/10.3390/electronics12112457

Biswas K, Sarkar A, Sarkar CK (2018) Effect of varying Indium concentration of InGaAs channel on device and circuit performance of nanoscale double gate heterostructure MOSFET. Micro NanoLett. 13:690–694. https://doi.org/10.1049/mnl.2017.0884

Ma R, Teo KH, Shinjo S, Yamanaka K, Asbeck PM (2017) A GaN PA for 4G LTE-Advanced and 5G: Meeting the Telecommunication Needs of Various Vertical Sectors Including Automobiles, Robotics, Health Care, Factory Automation, Agriculture, Education, and More. IEEE microwave magazine 18(7):77–85. https://doi.org/10.1109/MMM.2017.2738498.

Lu CS, Chowdhury N, Povolotskyi M, Klimeck G, Mohamed M, Palacios T (2019) Superior performance of 5-nm gate length GaN nanowire nFET for digital logic applications. IEEE Electron Device Lett 40:874–877

Chowdhury N, Lemettinen J, Xie Q, Zhang Y, Rajput NS, Xiang P, Cheng K, Suihkonen S, Then HW, Palacios T (2019) P-channel GaN transistor based on p-GaN/AlGaN/GaN on Si. IEEE Electron Device Lett 40:1036–1039

Bader SJ, Lee H, Chaudhuri R, Huang S, Hickman A, Molnar A, Xing HG, Jena D, Then HW, Chowdhury N et al (2020) Prospects for wide bandgap and ultrawide bandgap CMOS devices. IEEE Trans Electron Devices 67:4010–4020

Zheng Z, Song W, Zhang L, Yang S, Wei J, Chen KJ (2021) Monolithically integrated GaN ring oscillator based on high-performance complementary logic inverters. IEEE Electron Device Lett 42:26–29

Gokhale VJ, Downey BP, Katzer DS, Nepal N, Lang AC, Stroud RM, Meyer DJ (2020) Epitaxial bulk acoustic wave resonators as highly coherent multi-phonon sources for quantum acoustodynamics. Nat Commun 11:2314

Satzinger KJ, Zhong YP, Chang H-S, Peairs GA, Bienfait A, Chou M-H, Cleland AY, Conner CR, Dumur É, Grebel J, Gutierrez I, November BH, Povey RG, Whiteley SJ, Awschalom DD, Schuster DI, Cleland AN (2018) Quantum control of surface acoustic-wave phonons. Nature 563:661–665

Cheng R, Wright J, Xing HG, Jena D, Tang HX (2020) Epitaxial niobium nitride superconducting nanowire single-photon detectors. Appl Phys Lett 117:132601

Yan R, Khalsa G, Vishwanath S, Han Y, Wright J, Rouvimov S, Katzer DS, Nepal N, Downey BP, Muller DA, Xing HG, Meyer DJ, Jena D (2018) GaN/NbN epitaxial semiconductor/superconductor heterostructures. Nature 555:183–189

Barratt CA (2009) iii–v semiconductors, a history in rf applications. ECS Transactions 19(3):79–84

Global High Electron Mobility, 2032. https://www.globenewswire.com/news-release/2023/06/01/2680054/0/en/Global-High-Electron-Mobility-Transistor-HEMT-Market-Size-To-Exceed-USD-16-82-Billion-By-2032-CAGR-of-8-4.html

Neural Network-Based GaN HEMT Modelling Techniques

J. Ajayan, D. Nirmal, and B. Mounika

Abstract The AlGaN/GaN HEMT models can be classified into three main groups: the artificial neural network (ANN)-based model, the physical model and the empirical model. The empirical model employs empirical equations to describe nonlinear device elements like the gate capacitance and the drain current. On the other hand, the physical model specifies the characteristics of a device using physical equations or theory. The neural network (ANN)-based model is one more kind of GaN HEMT model. It employs neural networks (NNs) to precisely characterize nonlinear elements like HEMT gate capacitances and nonlinear drain current. Accuracy and outstanding nonlinear fitting capability are the major benefits of ANN-based models. Recently, AlGaN/GaN HEMTs have demonstrated the highest Johnson figure of merit (JFOM), increased frequency response (higher fmax), high power density, high breakdown voltage, best power device figure of merit and low dielectric constant, which makes it into a highly promising transistor technology for multiple applications, including the existing 5G/communication systems, satellite communications, next generation of 5G networks, military and commercial and avionics.

Keywords Artificial neural networks · 5G Networks · Electrothermal model · Large signal model · Machine learning

J. Ajayan (✉) · B. Mounika
Department of Electronics and Communication Engineering, SR University, Warangal, Telangana, India
e-mail: email2ajayan@gmail.com

D. Nirmal
Department of Electronics and Communication Engineering, Karunya Institute of Technology and Sciences, Coimbatore, Tamilnadu, India

D. Nirmal and J. Ajayan (eds.), *Modeling of AlGaN/GaN High Electron Mobility Transistors*, Springer Tracts in Electrical and Electronics Engineering,
https://doi.org/10.1007/978-981-97-7506-4_4

1 Introduction

Several significant breakthroughs in solid-state devices have been driven by the rising demand for high-power amplifiers in wireless communication systems. AlGaN/GaN HEMT is emerging as promising technology for developing power amplifiers (Mounika and Ajayan 2023a, 2023b; Angen Franklin et al. 2023; Kumar et al. 2022a, 2022b; Mounika et al. 2022; Arivazhagan et al. 2021; Fletcher et al. 2022; Husna Hamza et al. 2022, 2021; Murugapandiyan et al. 2021a, 2021b, 2022, 2017; Arivazhagan et al. 2021, 2019; Mounika et al. 2022; Ajayan et al. 2021, 2022; Augustine Fletcher et al. 2020). Power amplifier design requires precise HEMT modelling over the operational frequency bandwidth in order to forecast the device's power performance. Neural network modelling techniques meet the requirements for a straightforward and precise device modelling method, which is not only desirable but essential in the design of power amplifiers. Neural networks have the ability to generalize patterns from data, learn from data, and model nonlinear properties. These features make neural network techniques an outstanding tool in modelling of GaN HEMTs. The ANN (artificial neural network) techniques can be used to model the small signal behaviour of GaN HEMTs. In 2001, Sang Yun Lee et al. (Nirmal et al. 2018) adopted ANN modelling technique for the first time for the design of X-Band Power Amplifier implemented using GaN HEMT.

2 Neural Network Modelling (Nirmal et al. 2018)

A neural network is a system made up of interconnected neurons that may be taught to predict the responses to new inputs by being exposed to a specific input pattern together with the desired output. The processing element (neuron shown in Fig. 1) serves as the central processing unit of the network and is the fundamental unit for developing the neural network. The n-dimensional vector x_1, x_2, $x_3 \ldots x_n$ represents the inputs of the neuron. The neuron adds a threshold value (θ) and creates a weighted sum of the input vector's components. The output of the summer is then passed through a activation function to obtain the output. Because of their interconnections with each other, neuron function as a sizable parallel network. It is configured as a MLP (multi-layered system known as multilayer perceptron). Between the output and input layers, there exists a hidden layer that is the central part of the neural network. Based on the desired output and complexity of the input response, the number of neurons at each layer and hidden layers may vary (Fig. 2).

In order to model the small signal behaviour of transistors, an MLP with only one hidden layer having the same number of neurons as the output layer is required. Four neurons are present in the input layer of neuron (Fig. 1), and its inputs can be voltages (Vgs and Vds), input power (Pin) and frequency (f). The output is an S-matrix with phase and magnitude values. Let W and V represent the weighting matrix between input and hidden layers and hidden and output layers, respectively. Then,

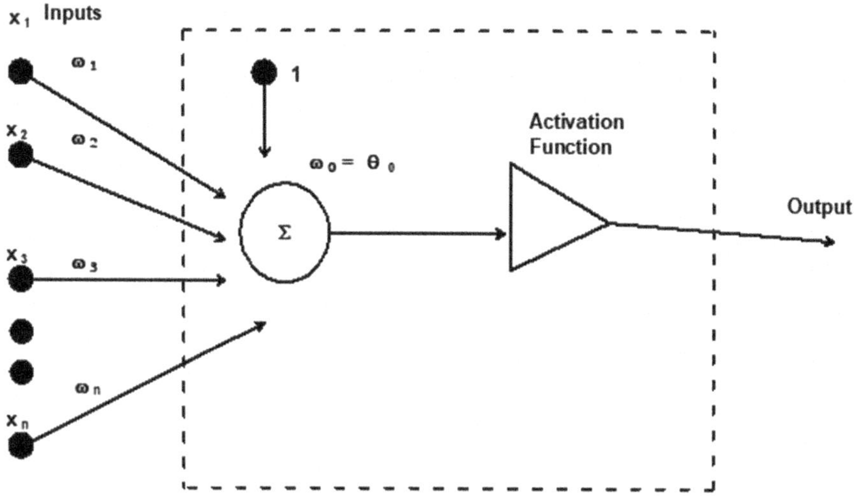

Fig. 1 Schematic of a neuron model (Nirmal et al. 2018)

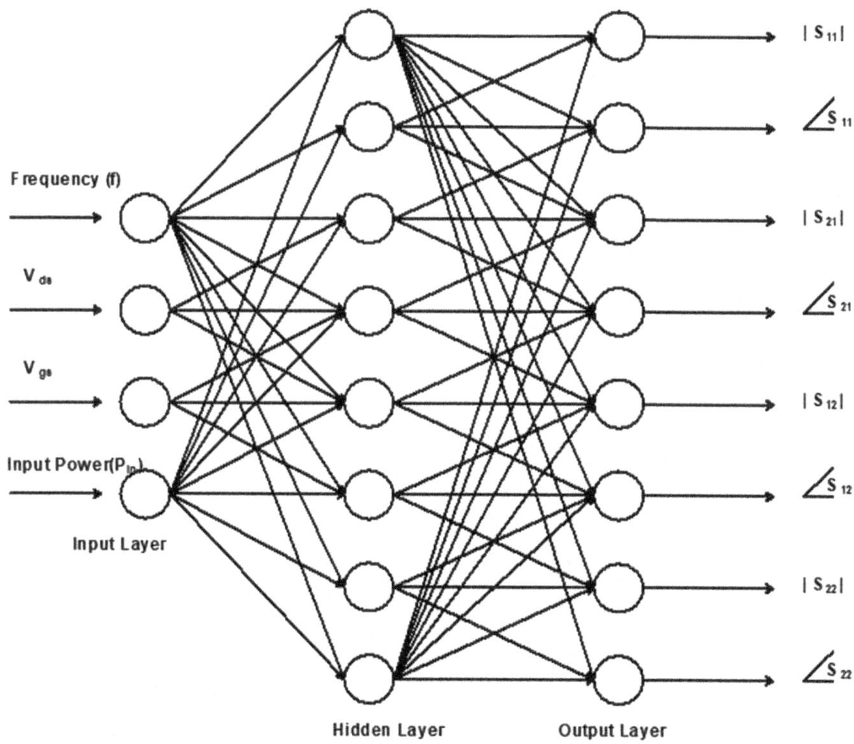

Fig. 2 Configuration of neural network for S-parameter model (Nirmal et al. 2018)

$$W = [\omega_{nq}] N_x Q \quad (1)$$

n ϵ (1, 2, 3, N = number of input neurons = 4)

$$V = [v_{qp}] Q_x P \quad (2)$$

p ϵ (1, 2, 3, ………..P = number of neurons in the hidden layer = 8)

q ϵ (1, 2, 3, ………..Q = number of neurons in the output layer = 8).

Here, ω_{nq} indicates the weighting value between qth hidden layer neuron and nth input layer neuron and v_{qp} indicates the weighting value between qth hidden layer neuron and pth output layer neuron.

3 Training Process (Nirmal et al. 2018)

The calculation of the node threshold values and the weighting matrix in the MLP is referred to as the training process in order to minimize the error function ($E(\omega_{nq}, v_{qp}, \theta_i, \beta_i)$).

Here, the mean square error (E) over all the neurons in the output layer can be computed as

$$E(\omega_{nq}, v_{qp}, \theta_i, \beta_i) = \sum_{k=1}^{n_s} \left[\frac{1}{2} \sum_{i=1}^{8} \left(y_i^k - z_i^k \right)^2 \right] \quad (3)$$

n_s = Number of data samples

z_i^k and y_i^k are measured or desired and computed responses of the ith output node at the f_k (training frequency).

θ_i = Input activation function.

β_i = Output activation function.

Back propagation algorithm can be employed to train the neural network. The mean square error function can be adopted to modify the threshold and weight values in the MLP with the help of the following equations.

$$(\omega_{nq})^k + 1 = (\omega_{nq})^k - \eta \frac{\partial E^k}{\partial \omega_{nq}} + \alpha \left[(\omega_{nq})^k - (\omega_{nq})^{k-1} \right] \quad (4)$$

$$(v_{qp})^k + 1 = (v_{qp})^k - \eta \frac{\partial E^k}{\partial v_{qp}} + \alpha \left[(v_{qp})^k - (v_{qp})^{k-1} \right] \quad (5)$$

$$(\theta_h)^k + 1 = (\theta_h)^k - \eta \frac{\partial E^k}{\partial \theta_h} + \alpha \left[(\theta_h)^k - (\theta_h)^{k-1} \right] \quad (6)$$

$$(\beta_i)^k + 1 = (\beta_i)^k - \eta \frac{\partial E^k}{\partial \beta_i} + \alpha\left[(\beta_i)^k - (\beta_i)^{k-1}\right] \tag{7}$$

η = learning rate

α = momentum coefficient.

4 ANN-Based Electrothermal Model

In 2016, An-Dong Huang et al., (Augustine Fletcher et al. 2019) developed an electrothermal model for GaN HEMTs with the consideration of trapping effects. Consider a 16-element equivalent circuit of GaN HEMT (Fig. 3), which can be used for a small signal model.

The internal elements of the GaN HEMT are determined analytically by averaging over frequency after the exterior portion of the device has been de-embedded. These internal elements can then be used to rebuild the intrinsic Y-parameters. Then, one of the error references is considered to be the differences between the de-embedded and rebuilt intrinsic Y-parameters. Intrinsic Y-parameter error reference can be expressed as

$$\varepsilon_{ij}^{sub1}(V_{gs}, V_{ds}, f_k) = \frac{\left|\mathrm{Re}\left(Y_{ij}^{\mathrm{int\text{-}de}} - Y_{ij}^{\mathrm{int\text{-}cal}}\right)\right|^p + \left|\mathrm{Im}\left(Y_{ij}^{\mathrm{int\text{-}de}} - Y_{ij}^{\mathrm{int\text{-}cal}}\right)\right|^p}{\max\left(\left|Y_{ij}^{\mathrm{nt\text{-}de}}\right|^p\right)} \tag{8}$$

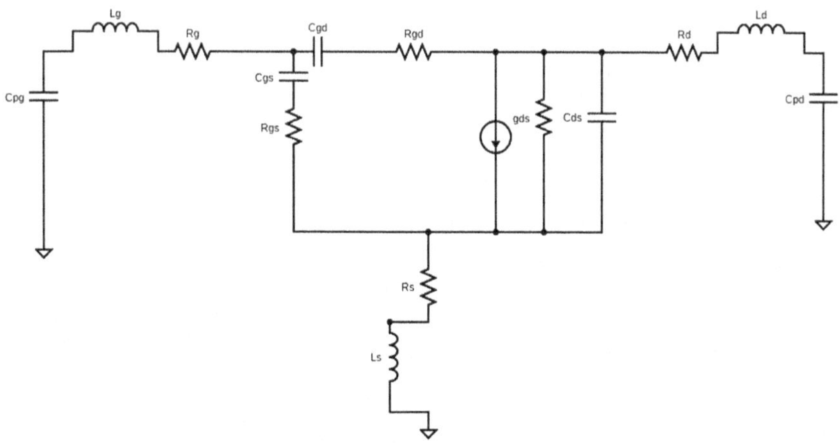

Fig. 3 16-Element equivalent circuit of GaN HEMT, which can be used for small signal model (Augustine Fletcher et al. 2019)

$$\varepsilon_{ij}^{sub1} = \frac{1}{4N_f N_{bias}} \sum_{bias} \left(\sum_{i,j=1}^{2} \sum_{k=1}^{N_f} \varepsilon_{ij}^{sub1}(V_{gs}, V_{ds}, f_k) \right)^{1/p} \quad (9)$$

Y_{ij}^{int-de} = Intrinsic Y-parameters computed by direct embedding method.

$Y_{ij}^{int-cal}$ = Intrinsic Y-parameters computed by analytical reconstruction.

N_f = Number frequency points.

F_k = kth frequency.

N_{bias} = Number of S parameters used for optimization.

p can be either 2 or 1 that indicates norm-2 or norm-1 error form.

Figure 4 depicts the procedure for determining the error references, and Fig. 4b illustrates the flowchart for the optimization extraction. Improved artificial bee colony algorithm (Lee et al. 2001) can be used as the optimizer because of its outstanding global searching ability. Large sets of S-parameters under various operating voltages can be used in the optimization process. C_{pg}, C_{pd}, L_g, L_d, L_s, R_g, R_d and R_s are the optimized external parameters.

$$\% E_{xy}\Big|_{\substack{x=1,2 \\ y=1,2}} = 100x \sqrt{\frac{\sum_M \sum_N \left| S_{xy}^{cal} - S_{xy}^{meas} \right|^2}{\sum_M \sum_N \left| S_{xy}^{meas} \right|^2}} \quad (10)$$

Here, N and M represent the number of frequency and bias points, respectively.

4.1 Generation of Large Signal Model

The large signal model for a GaN HEMT is depicted in Fig. 5. The large signal model of a GaN HEMT has both linear and nonlinear parts. The linear parts include extrinsic parasitic components and the nonlinear part of the GaN HEMT large signal model consists of diodes D_{gd} and D_{gs}, temperature-dependent nonlinear charges Q_{gd}, Q_{ds}, and Q_{gs}, nonquasi-static resistances and drain current source I_{ds}. Transient nature of self-heating effects can be modelled by employing a thermal subcircuit. Moreover, two trapping subcircuits are utilized to depict the transient behaviours of the drain lag and gate lag, and in the harmonic balance simulation, they will be able to track the envelopes of the Vds/Vgs voltage–time waveforms. The design flow for this GaN HEMT large-signal model generation is displayed in Fig. 6. PIVs (pulsed I–V) together with a multilayer feedforward ANN is utilized to model the I_{ds} including dynamic trapping effects and electro-thermal effects.

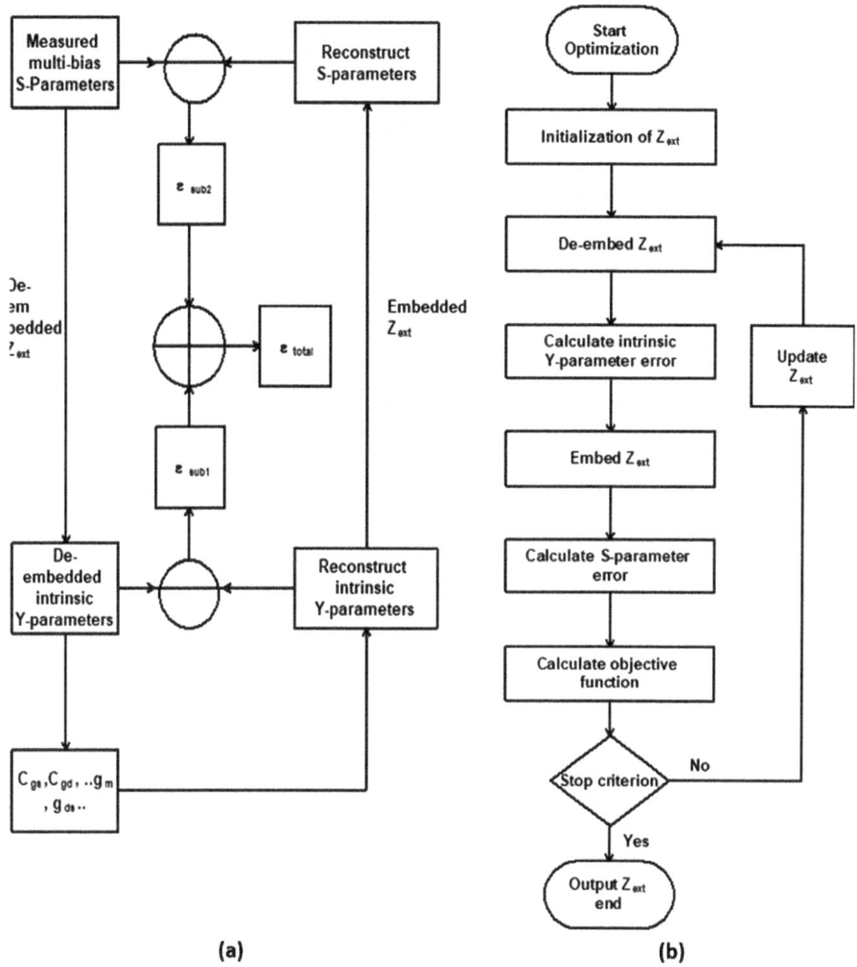

Fig. 4 **a** the procedure for determining the error references **b** Flowchart for the optimization extraction (Augustine Fletcher et al. 2019)

$$I_{ds\text{ - pulse}} = \mathrm{f_{ANN}}\left(V_{gs}, V_{ds}, T_j, \phi_G, \phi_D\right) \tag{11}$$

$$T_j = \mathrm{T_{amb}} + \mathrm{V_{dsq}} \times I_{dsq} \times R_{th} \tag{12}$$

$$\phi_G = \mathrm{Min}\left\{\mathrm{V_{gsq}}, \mathrm{V_{gs}}\right\} \tag{13}$$

$$\phi_D = \mathrm{Min}\left\{\mathrm{V_{dsq}}, \mathrm{V_{ds}}\right\} \tag{14}$$

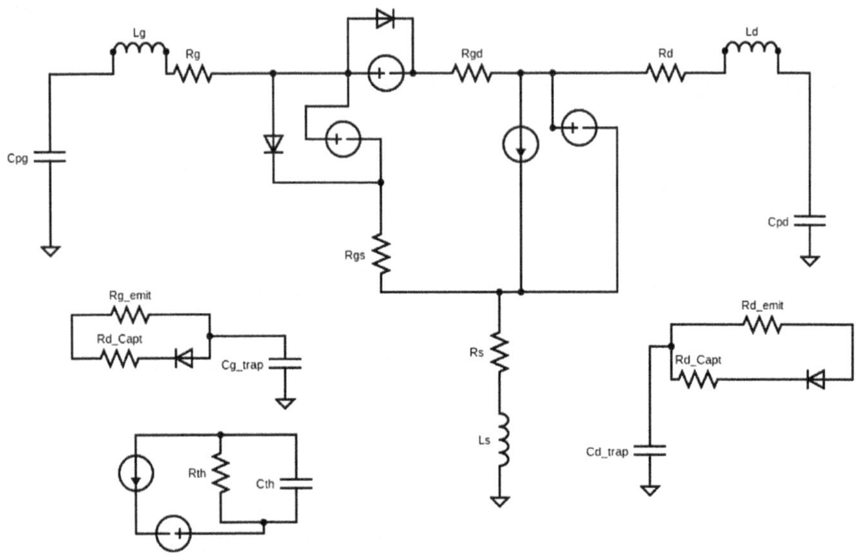

Fig. 5 Large signal model for a GaN HEMT (Augustine Fletcher et al. 2019)

Here, T_j = channel temperature, ϕ_D = drain trapping states, ϕ_G = gate trapping states.

For the DC current, the gate and drain trapping states are exactly modelled by the quiescent voltages as given below.

$$\phi_G = \mathrm{Min}\{V_{gsq}, V_{gsq}\} = V_{gsq} \tag{15}$$

$$\phi_D = \mathrm{Min}\{V_{dsq}, V_{dsq}\} = V_{dsq} \tag{16}$$

$$I_{ds\text{ - DC}} = f_{ANN}\left(V_{gsq}, V_{dsq}, T_j, V_{gsq}, V_{dsq}\right) \tag{17}$$

5 Machine Learning (ML) in GaN HEMT Modelling

In 2020, Tian-Li Wuetval. (Huang et al. Aug. 2016) demonstrated the use of a ML-based technique to study the process Dependencies on V_{TH} (Threshold Voltage) in AlGaN/GaN MIS-HEMTs (Fig. 7). The flowchart of ML-based technique to model the V_{TH} is depicted in Fig. 8.

To correlate with the V_{TH} (Output feature), 19 processing splits can be used as the input features in the machine learning model. It is a very difficult task to analyse the influence of each input feature on the V_{TH} using a traditional linear

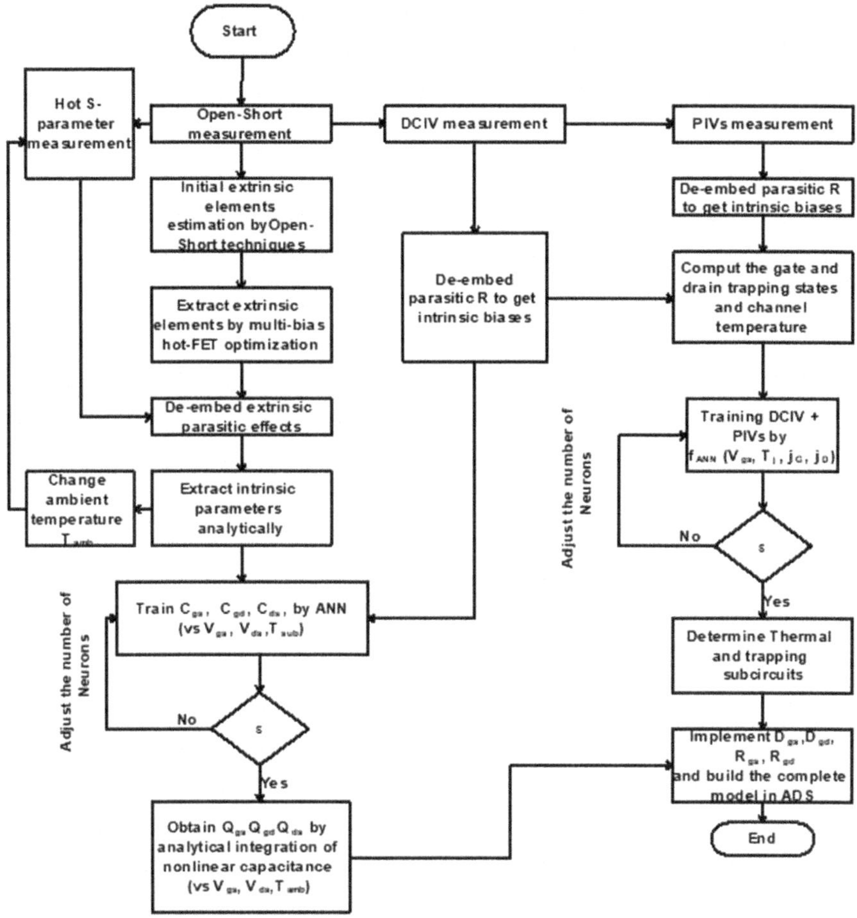

Fig. 6 The design flow for this GaN HEMT large-signal model generation (Augustine Fletcher et al. 2019)

regression-based machine learning due to the fact that the linear regression-based machine learning takes into account of all the input features to obtain the best fitting by modulating the weight (ω) and bias (b), that could result to an over fitting for all input features. Machine learning-based TR (Tikhonov regularization) technique offers the benefits of evaluating the input features by introducing a penalty term ($\lambda\omega^2$) in the cost function $(y_i - y_{i,predictions})^2$, which makes sure that the regularization of the least significant input features.

$$\sum_{i=1}^{k} \left(y_i - y_{i,\text{predictions}}\right)^2 = \sum_{i-1}^{k} \left(y_i - \left(\sum_{j=0}^{p} \omega_j \times x_{ij} + b_j \right) \right) \tag{18}$$

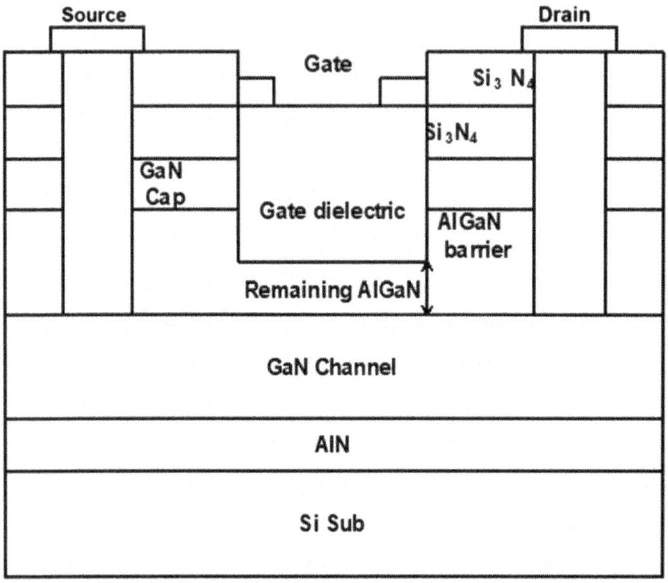

Fig. 7 MIS-HEMT with recessed gate (Huang et al. Aug. 2016)

$$\sum_{i=1}^{k}\left(y_i - \mathrm{y}_{\mathrm{i,predictions}}\right)^2 = \sum_{\mathrm{i-1}}^{\mathrm{k}}\left(y_i - \left(\sum_{\mathrm{j=0}}^{\mathrm{p}} \omega_j \times x_{ij} + \mathrm{b_j}\right)\right)^2 + \lambda \sum_{\mathrm{j=o}}^{\mathrm{p}} \omega_j^2 \tag{19}$$

The input features include the numerical input data, that is the remaining AlGaN thickness and text-based input data, such as gate dielectric, in-situ plasma and ex-situ wet clean. A text-based input needs to be modified into a numerical input with the help of one hot encoding vector (Gao et al. Jun. 2013). To obtain a good accuracy and predictability for the ML regularization algorithm, the Z-score (Wu and Kutub Dec. 2020) can be adopted to standardize the input features to middle around "zero" with a standard deviation (σ) of "one", as given below.

$$z = \frac{\mathrm{x} - \mu}{\sigma} \tag{20}$$

Here, μ = mean

The relationship obtained by using ML-based TR method between input process features and V_{TH} is given by

$$\text{Normalized } V_{TH} = \alpha \text{ (remaining AlGaN thickness)} + \beta \text{ (Ex} - \text{situ wet clean)} + \gamma \text{ (In} - \text{situ plasma)} + \eta \text{(gate dielectric)} + \xi_0 \tag{21}$$

Here, α, β, γ and η are coefficients and ξ_0 is a tuning parameter.

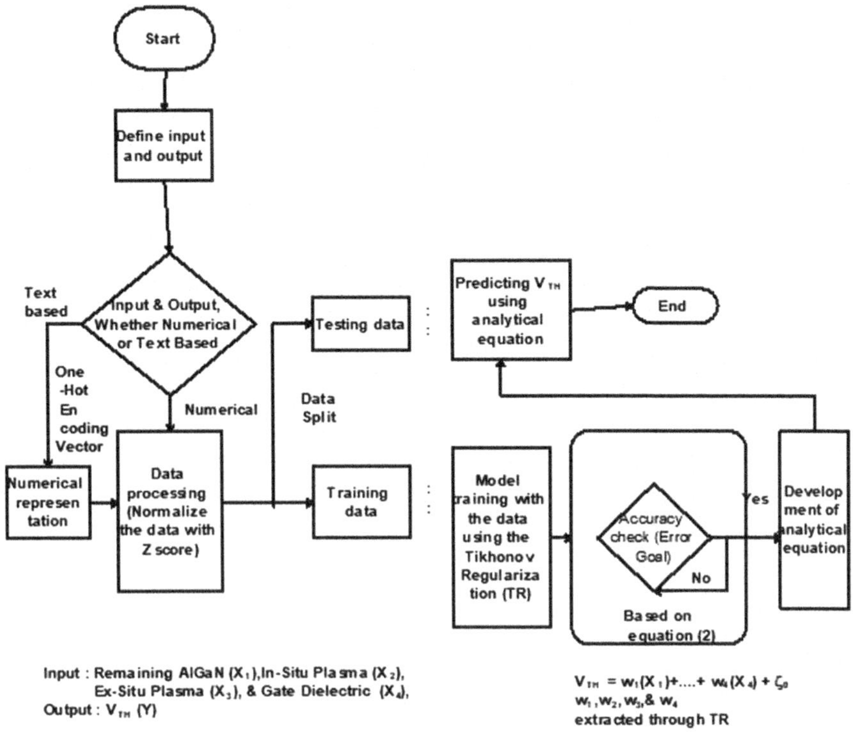

Fig. 8 The flowchart of ML-based technique to model the V_{TH} (Huang et al. Aug. 2016)

6 ANN-Based GaN HEMT Model with Buffer-Trapping Effects

In 2020, Xuekun Du et al. (Lantz 2015) developed an ANN-based LSM (large signal model) of AlGaN/GaN HEMTs with trapping effects in the buffer. A 22-component equivalent circuit (Fig. 9) is adopted for the small-signal modelling of GaN HEMTs.

ε_1 indicates the error between the modelled and measured S-parameters by using a normalized l_2-norm form in Colan Jan. (2013).

$$\varepsilon_1 = \frac{1}{4N_f}\sqrt{\sum_{k=1}^{N_f}\sum_{i,j=1}^{2}\frac{\left|\mathrm{Re}\left(S_{ij,k}^{error}\right)\right|^2+\left|\mathrm{Im}\left(S_{ij,k}^{error}\right)\right|^2}{\max_k\left\{\left|S_{ij}^{meas}(f_k)^2\right|\right\}}} \quad (22)$$

$$S_{ij,k}^{error} = S_{ij}^{meas}(f_k) - S_{ij}^{model}(f_k) \quad (23)$$

Here, f_k = kth frequency

N_f = number of frequency points.

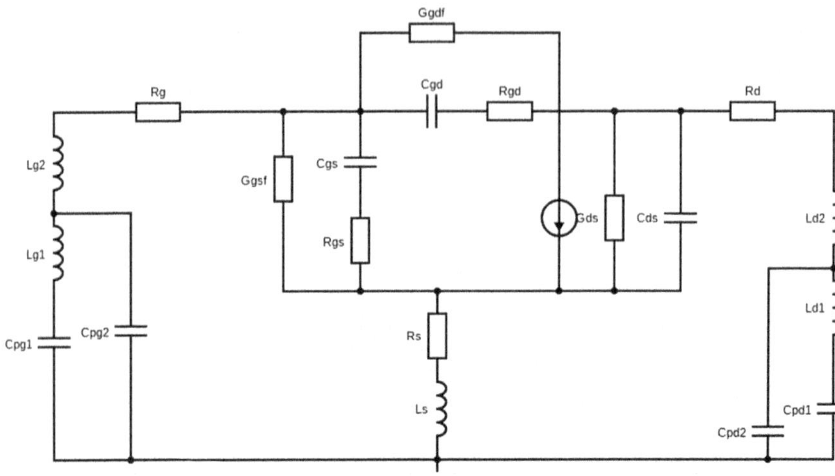

Fig. 9 Small-signal circuit for AlGaN/GaN HEMTs (Lantz 2015)

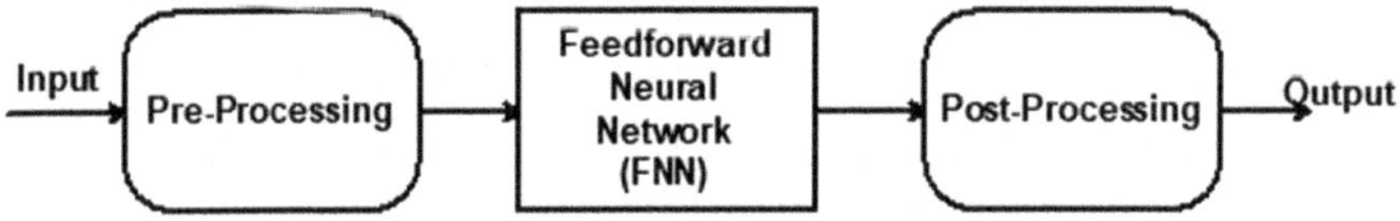

Fig. 10 Data postprocessing/preprocessing and FNN modelling adopted for characterizing the $I_{ds,iso}$ without considering the self-heating and trapping effects (Lantz 2015)

A more accurate error function (Du et al. July 2020) can be adopted to estimate the frequency-independent behaviour of intrinsic parasitic capacitances Cgd, Cds and Cgs, which is given by

$$\varepsilon_2 = \frac{1}{3}\left(\frac{\sigma^2_{C_{ds}}}{\mu^2_{C_{ds}}} + \frac{\sigma^2_{C_{gd}}}{\mu^2_{C_{gd}}} + \frac{\sigma^2_{C_{gs}}}{\mu^2_{C_{gs}}}\right) \tag{24}$$

Here, μ = mean

σ = variance.

Total error function is expressed as

$$\varepsilon_{total} = \frac{1}{\mathrm{N_{bias}}} \sum_{N_{bias}} [F_u(\varepsilon_1) + F_u(\varepsilon_2)] \tag{25}$$

$F_u(.)$ = feature scaling function.

To enhance the convergence speed of feedforward neural network (FNN), the feature scaling approach is needed to do the data postprocessing/preprocessing, such as Z-score normalization, mean normalization, and min–max normalization. The analytical expression for the implemented min–max scaling of [−1, 1] is provided in (26).

$$x\prime = \frac{2 \times [\mathrm{x} - \min(\mathrm{x})]}{\max(x) - \min(\mathrm{x})} - 1 \quad (26)$$

x = original value

x′ = normalized value.

The complete process of FNN modelling and data postprocessing/preprocessing is illustrated in Fig. 10.

The FNN model based $\mathrm{I}_{\mathrm{ds,iso}}$ is given by

$$\mathrm{I}_{\mathrm{ds,iso}} = \mathrm{f}_{\mathrm{ANN}}\left(V_{gs}, V_{ds}, \omega, b\right) \quad (27)$$

The buffer/surface related trapping effects can be characterized using $\mathrm{V}_{\mathrm{gs,eff}}$.

$$V_{gs,eff} = \mathrm{V}_{\mathrm{gs}} + \gamma_{\mathrm{subs}} + \gamma_{\mathrm{surf}} \quad (28)$$

$$\gamma_{surf} = \mathrm{k}_{\mathrm{g1}}\left(V_{gsq} - \mathrm{V}_{\mathrm{gsq0}}\right)\left(V_{gs} - \mathrm{V}_{\mathrm{gsq0}}\right) \quad (29)$$

Here, $\mathrm{V}_{\mathrm{gsq0}}$ and k_{g1} are fitting parameters.

It is proved that as V_{ds} is reduced from $\mathrm{V}_{\mathrm{dsq}}$, the CC (current collapse) is increasingly obvious. However, if V_{ds} is greater than $\mathrm{V}_{\mathrm{dsq}}$, a slight amount of CC still exists, and its quantity is unaffected beyond $\mathrm{V}_{\mathrm{dsq}}$. Therefore, buffer trapping effects can be considered as

$$\gamma_{subs} = \begin{cases} \tanh\left(\mathrm{k}_{\mathrm{g2}}\left(\mathrm{V}_{\mathrm{gs}} - \mathrm{V}_{\mathrm{gsq1}}\right)\left(V_{ds} - \mathrm{V}_{\mathrm{dsq}}\right)\right) + \mathrm{V}_{\mathrm{dsq}}\mathrm{k}_{\mathrm{g3}}\left(\mathrm{V}_{\mathrm{gs}} - \mathrm{V}_{\mathrm{gsq2}}\right), for V_{ds} \leq V_{dsq} \\ \mathrm{V}_{\mathrm{dsq}}\mathrm{k}_{\mathrm{g3}}\left(\mathrm{V}_{\mathrm{gs}} - \mathrm{V}_{\mathrm{gsq2}}\right), for V_{ds} > V_{dsq} \end{cases} \quad (30)$$

Here, $\mathrm{V}_{\mathrm{gsq1}}$, $\mathrm{V}_{\mathrm{gsq2}}$, k_{g2} and k_{g2} are fitting parameters.

R_{th}–C_{th} thermal subcircuit can be included to characterize the self-heating effects. The H(ω) (thermal frequency response) is given by

$$H(\omega) = \frac{\mathrm{j}\omega\tau_1}{1 + \mathrm{j}\omega\tau_1} \quad (31)$$

$$I_{ds,iso} = \mathrm{I}_{\mathrm{ds,iso}}\left(\mathrm{V}_{\mathrm{gs,eff}}, \mathrm{V}_{\mathrm{ds}}\right) \times [1 - \mathrm{K}_{\mathrm{T}}\mathrm{H}(\omega)P_{diss}] \quad (32)$$

$$K_T = \alpha_0 + \alpha_1[1 + \tanh[\alpha_2 + \alpha_3 Vd_s]] \quad (33)$$

Here, ω = operating frequency and

τ_1 = thermal time constant.

$\alpha0$, $\alpha1$, $\alpha2$, and $\alpha3$ are fitting parameters.

Under high operating power environment,

$$I_{ds}=\left(1+\mathrm{k}_{\mathrm{br2}}e^{k_b(V_{dg}-\mathrm{V_{tr}})}\right)\left(I_{ds,T}+\mathrm{k}_{\mathrm{br1}}e^{k_b(V_{dg}-\mathrm{V_{tr}})}\right) \tag{34}$$

$$i_{gs}=\mathrm{I_{gs0}}\left(e^{k_{gs1}V_{gs}}-1\right)e^{k_{gs2}V_{gd}} \tag{35}$$

$$i_{gd}=\mathrm{I_{gd0}}\left(e^{k_{gd1}V_{gd}}-1\right)e^{k_{gd2}V_{gd}} \tag{36}$$

Here, I_{gs0}, I_{gd0}, k_{gs1}, k_{gs2}, k_{gd1} and k_{gd2} are fitting parameters.

7 ANN-Based Gate Charge Model

In 2021, Wenrui Hu et al., (Lin and Kompa Jul. 1994) developed an accurate ANN-based gate charge model for GaN HEMTs. Various types of optimization problems can be solved effectively using a swarm intelligence technique called grey wolf optimizer (GWO) (Du et al. submitted for publication). α, β and δ are the three most suitable search agents in the GWO algorithm because these search agents have lower objective values compared with the other search agents.

$$X(iter+1)=(\mathrm{X}_1+\mathrm{X}_2+\mathrm{X}_3)/3 \tag{37}$$

$$X_1=\mathrm{X}_\alpha(iter)-\mathrm{A}\cdot|\mathrm{C}\cdot X_\alpha(iter)-\mathrm{X(iter)}| \tag{38}$$

$$X_2=\mathrm{X}_\beta(iter)-\mathrm{A}\cdot\left|\mathrm{C}\cdot X_\beta(iter)-\mathrm{X(iter)}\right| \tag{39}$$

$$X_{31}=\mathrm{X}_\delta(iter)-\mathrm{A}\cdot|\mathrm{C}\cdot X_\delta(iter)-\mathrm{X(iter)}| \tag{40}$$

$$A=2a.\mathrm{r}_1-a \tag{41}$$

$$C=2.\mathrm{r}_2 \tag{42}$$

Here, X = position vector of a search agent,

iter = current iteration.

X_α = position vectors of the alpha.

X_β = position vectors of the beta.

X_δ = position vector of delta.

r1 and r2 represent the random vectors in [0, 1].

The value of a is updated by a = 2(1 − iter/Iter).

Iter = maximum number of iterations.

Consider a 16-element small-signal equivalent circuit (Fig. 11) for GaN HEMT. Figure 12 demonstrates the MOGWO (multiobjective version of GWO)-based hybrid extraction. Direct methods can be used to extract the extrinsic parameters (bias-independent). Under low frequency and the pinch-off conditions, Cpd and Cpg (parasitic capacitances) can be extracted from the measured S-parameters (Hu et al. July 2021). The Rs, Rd and Rg (extrinsic resistances) and Ls, Ld and Lg (extrinsic inductances) can be extracted from the Z-parameters under the high frequency and unbiased conditions after capacitances are deembedded (Mirjalili et al. Mar. 2014). The extracted values can be adopted to initialize any one of the search agents, and the positions of the remaining search agents are initialized randomly. The objectives are given by Augustine Fletcher et al. (2019)

$$\varepsilon_Y = \frac{1}{4N_f N_p} \sum_{bias} \left(\sum_{i,j=1}^{2} \sum_{\text{freq}} \frac{\left|\text{Re}\left(Y_{ij}^d - \text{Y}_{\text{ij}}^{\text{c}}\right)\right| + \left|\text{Im}\left(Y_{ij}^d - \text{Y}_{\text{ij}}^{\text{c}}\right)\right|}{\max\left(\left|Y_{ij}^d\right|\right)} \right) \tag{43}$$

$$\varepsilon_S = \frac{1}{4N_f N_p} \sum_{bias} \left(\sum_{i,j=1}^{2} \sum_{\text{freq}} \frac{\left|\text{Re}\left(S_{ij}^m - \text{S}_{\text{ij}}^{\text{c}}\right)\right| + \left|\text{Im}\left(S_{ij}^m - \text{S}_{\text{ij}}^{\text{c}}\right)\right|}{\max\left(\left|S_{ij}^m\right|\right)} \right) \tag{44}$$

Here, N_f = number of the frequency points and Np = number of the bias points.

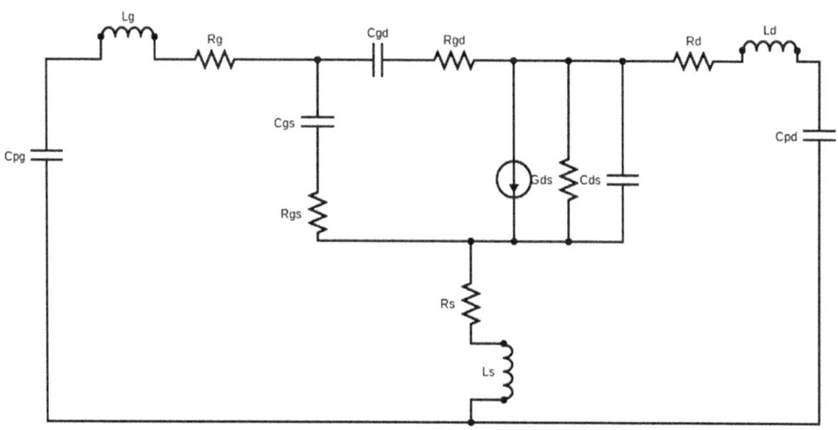

Fig. 11 16-element small-signal equivalent circuit (Lin and Kompa Jul. 1994)

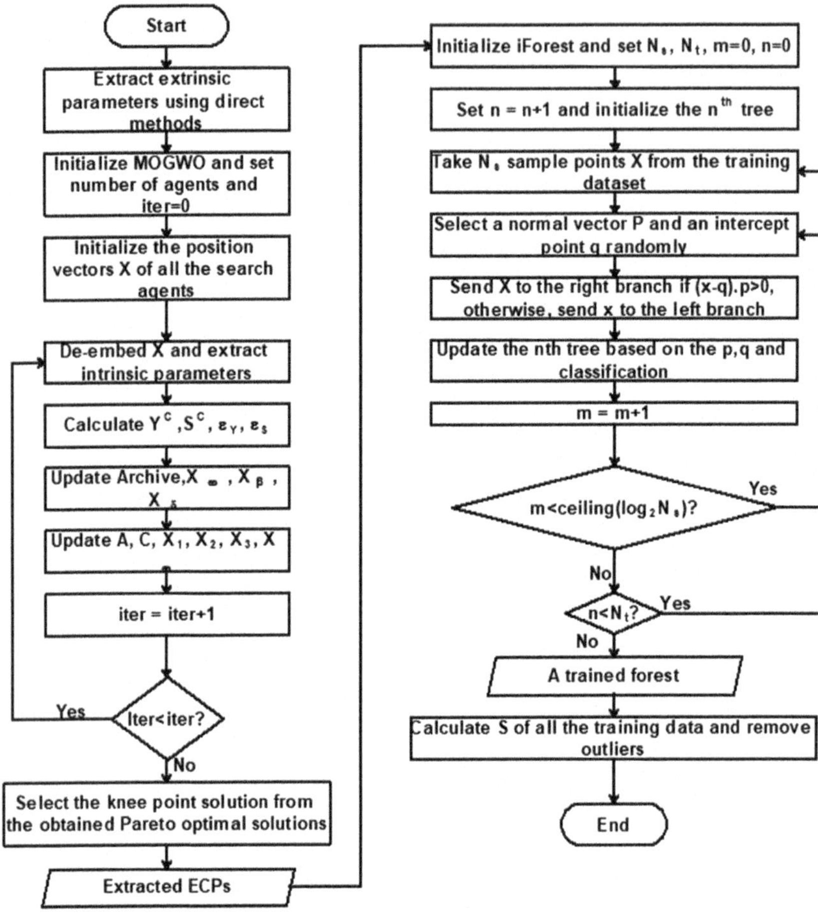

Fig. 12 MOGWO-based hybrid extraction (Lin and Kompa Jul. 1994)

The error between the computed and measured S-parameters is given by

$$\% \, E_{ij}\Big|_{\substack{i=1,2\\ j=1,2}} = 100 \times \sqrt{\frac{\sum N_p \sum N_f \left|S_{ij}^{c} - S_{ij}^{m}\right|^2}{\sum N_p \sum N_f \left|S_{ij}^{m}\right|^2}} \tag{45}$$

The ANN-based drain current model considering the trapping and self-heating effects is given by Lantz (2015); Brady et al. Jul. 2008)

$$I_{ds} = I_{ds,iso}\left(V_{gs,eff}, V_{ds}\right) \times (1 + K_T \Delta T) \tag{46}$$

$$I_{ds,iso} = f_{ANN}\left(V_{gs}, V_{ds}, \omega, b\right) \tag{47}$$

$$V_{gs,eff} = \gamma_{subs} + \gamma_{surf} + V_{gs} \tag{48}$$

$$\gamma_{surf} = k_{g1}\left(V_{gsq} - V_{gsq0}\right)\left(V_{gs} - V_{gsq0}\right) \tag{49}$$

$$\gamma_{subs} = \begin{cases} \tanh\left(k_{g2}\left(V_{gs} - V_{gsq1}\right)\left(V_{ds} - V_{dsq}\right)\right) + V_{dsq}k_{g3}\left(V_{gs} - V_{gsq2}\right), for V_{ds} \le V_{dsq} \\ V_{dsq}k_{g3}\left(V_{gs} - V_{gsq2}\right), for V_{ds} > V_{dsq} \end{cases} \tag{50}$$

$$K_T = \alpha_0 + \alpha_1[1 + \tanh[\alpha_2 + \alpha_3 V_{ds}]] \tag{51}$$

$$T = T_{ref} + \Delta T \tag{52}$$

$$\Delta T = P_{diss}R_{th} - T_{ref} + T_{amb} \tag{53}$$

$$\tau_{th} = C_{th}R_{th} \tag{54}$$

The gate current can be determined by differentiating gate charge (Q_g) with respect to time.

$$I_g = \frac{dQ_g}{dt} = \frac{\partial Q_g}{\partial V_{gs}}\frac{dV_{gs}}{dt} + \frac{\partial Q_g}{\partial V_{gd}}\frac{dV_{gd}}{dt} + \frac{\partial Q_g}{\partial T}\frac{dVT}{dt} \tag{55}$$

Here,

$$C_{gs} = {}^{\partial Q_g}/_{\partial V_{gs}} \tag{56}$$

$$C_{gd} = {}^{\partial Q_g}/_{\partial V_{gd}} \tag{57}$$

This model shows good match with experimental data.

8 Summary

In this chapter, we reviewed the popular ANN-based models of GaN HEMTs. The ANN-based AlGaN/GaN HEMT models can be made compact by using trapping states, ambient temperatures and the biases as the inputs of the ANN model (Jarndal et al. 2016; Yuk et al. Dec. 2009; Luo et al. Nov. 2022). The self-heating issue and trapping effects can also be taken into account in the ANN-based GaN HEMT models in addition to the dependence on ambient temperature to improve modelling accuracy.

References

Ajayan J, Nirmal D, Ramesh R, Bhattacharya S, Tayal S, Leo Joseph LMI, Thoutam LR, Ajitha D (2021) A critical review of AlGaN/GaN-heterostructure basedSchottky diode/HEMT hydrogen (H2) sensors for aerospace and industrial applications. Measurement 186:110100

Ajayan J, Nirmal D, Mohankumar P, Mounika B, Bhattacharya S, Tayal S, Augustine Fletcher AS (2022) Challenges in material processing and reliability issues inAlGaN/GaN HEMTs on silicon wafers for future RF power electronics & switching applications: a critical review. Mater Sci Semiconductor Process 151:106982

Arivazhagan L, Nirmal D, Reddy PPK et al (2021) A numerical investigation of heat suppression in HEMT for power electronics application. SILICON 13:3039–3046

Arivazhagan L, Nirmal D, Godfrey D, Ajayan J, Prajoon P, Augustine Fletcher AS, Amir Anton Jone A, Raj Kumar JS (2019) Improved RF and DC performance in AlGaN/GaN HEMT by P-type doping in GaN buffer for millimetre-wave applications. AEU Int J Electron Commun 108:189–194

Arivazhagan L, Nirmal D, Jarndal A, Huq HF, Chander S, Bhagyalakshmi S, Reddy PK, Ajayan J, Varghese A (2021) Applicability of double Channel Technique in AlGaN/GaN HEMT for future biosensing applications. Superlattices and Microstructures 160:107086

Augustine Fletcher AS, Nirmal D, Arivazhagan L, Ajayan J, Varghese A (2020) Enhancement of Johnson figure of merit in III–V HEMT combined with discrete field plate and AlGaN blocking layer. Int J RF Microw Comput Aided Eng 30:e22040. https://doi.org/10.1002/mmce.22040

Augustine Fletcher AS, Nirmal D, Ajayan J, Arivazhagan L (2019) Analysis of AlGaN/GaN HEMT using discrete field plate technique for high power and high frequency applications. AEU Int J Electron Commun 99:325–330

Brady RG, Oxley CH, Brazil TJ (2008) An improved small-signal parameter-extraction algorithm for GaN HEMT devices. IEEE Trans Microw Theory Techn 56(7):1535–1544

Colan SD (2013) The why and how of Z scores. J Amer Soc Echocardiograp 26(1):38–40. https://doi.org/10.1016/j.echo.2012.11.005

Du X et al (2020) ANN-based large-signal model of AlGaN/GaN HEMTs with accurate buffer-related trapping effects characterization. IEEE Trans Microw Theory Tech 68(7):3090–3099. https://doi.org/10.1109/TMTT.2020.2990171

Du X, Jarndal A, Helaoui M, Craig S, Noël J-P, Ghannouchi FM, Improved small-signal hybrid parameter-extraction method for AlGaN/GaN HEMTs. IEEEMicrow Wireless Compon Lett. submitted for publication

Fletcher ASA, Nirmal D, Arivazhagan L et al (2022) A 28-GHz low-loss AlGaN/GaN HEMT for T_X/R_X switches in 5G base stations. J Electron Mater 51:1215–1225

Gao W-F, Liu S-Y, Huang L-L (2013) A novel artificial bee colony algorithm based on modified search equation and orthogonal learning. IEEE Trans Cybern 43(3):1011–1024

Geng M, Crupi G, Cai J (2023) Accurate and effective nonlinear behavioral modeling of a 10-W GaN HEMT based on LSTM neural networks. IEEE Access 11:27267–27279. https://doi.org/10.1109/ACCESS.2023.3258691

Hu W, Luo H, Yan X, Guo Y-X (2021) An accurate neural network-based consistent gate charge model for GaN HEMTs by refining intrinsic capacitances. IEEE Trans Microw Theory Tech 69(7):3208–3218. https://doi.org/10.1109/TMTT.2021.3076064

Huang A-D, Zhong Z, Wu W, Guo Y-X (2016) An artificial neural network-based electrothermal model for GaN HEMTs with dynamic trapping effects consideration. IEEE Trans Microw Theory Tech 64(8):2519–2528. https://doi.org/10.1109/TMTT.2016.2586055

Husna Hamza K, Nirmal D, Augustine Fletcher AS, Ajayan J, Natarajan R (2022) Enhanced drain current and cut off frequency in AlGaN/GaN HEMT with BGaN back barrier. Mater Sci Eng B 284:115863

Husna Hamza K, Nirmal D, Augustine Fletcher AS, Arivazhagan L, Ajayan J, Natarajan R (2021) Highly scaled graded channel GaN HEMT with peak drain current of 2.48 A/mm. AEU Int J Electron Commun 136:153774

Jarndal A, Essaadali R, Kouki AB (2016) A reliable model parameter extraction method applied to AlGaN/GaN HEMTs. IEEE Trans Comput-Aided Design Integr Circuits Syst 35(2):211–219

Kumar JSR, Nirmal D, Ajayan J et al (2022a) Investigation on $L_G = 50$ nm tapered T-Gated AlGaN/GaN HEMT on silicon wafer with a f_T/f_{max} of 264/312 GHz for beyond 5G (B5G) applications. SILICON 14:11315–11322

Kumar JSR, Nirmal D, Hooda MK et al (2022b) Intensive study of field-plated AlGaN/GaN HEMT on silicon substrate for high power RF applications. SILICON 14:4277–4282

Lantz B (2015) Machine learning with R. Packet Publishing, Birmingham, U.K.

Lee SY et al. (2001) An X-band GaN HEMT power amplifier design using an artificial neural network modeling technique. IEEE Trans Electron Dev 48(3):495–501. https://doi.org/10.1109/16.906442

Li M et al., A scalable knowledge-based neural network model for GaN HEMTs with accurate trapping and self-heating effects characterization. In: IEEE transactions on microwave theory and techniques, https://doi.org/10.1109/TMTT.2023.3248225

Lin F, Kompa G (1994) FET model parameter extraction based on optimization with multiplane data-fitting and bidirectional search—a new concept. IEEE Trans Microw Theory Techn 42(7):1114–1121

Luo H, Yan X, Zhang J, Guo Y (2022) A neural network-based hybrid physical model for GaN HEMTs. IEEE Trans Microw Theory Tech 70(11):4816–4826. https://doi.org/10.1109/TMTT.2022.3206442

Mirjalili S, Mirjalili SM, Lewis A (2014) Grey wolf optimizer. Adv Eng Softw 69:46–61

Mounika B, Ajayan J (2023a) Sandip Bhattacharya, "Investigation on impact of AlxGa1-xN and InGaN back barriers and source-drain spacing on the DC/RF performance of Fe-doped recessed T-gated AlN/GaN HEMT on SiC wafer for future RF power applications." Micro Nanostructures 175:207504

Mounika B, Ajayan J (2023b) Sandip Bhattacharya, "An intensive study on effects of lateral scaling and gate metals on the RF/DC performance of recessed T-gated Fe-doped AlN/GaN/SiC HEMTs for future RF and microwave power applications." Microelectron Eng 271–272:111948

Mounika B, Ajayan J, Bhattacharya S, Nirmal D (2022) Recent developments in materials, architectures and processing of AlGaN/GaN HEMTs for future RF and power electronic applications: a critical review. Micro Nanostructures 168:207317

Mounika B, Ajayan J, Bhattacharya S, Nirmal D (2022) Investigation of back barrier material effects on the scalability of Fe-doped recess-gated AlN/GaN HEMTs for next generation RF power electronics. Micro Nanostructures 171:207431

Murugapandiyan P, Ravimaran S, William J, Ajayan J, Nirmal D (2017) DC and microwave characteristics of 20 nm T-gate InAlN/GaN high electron mobility transistor for high power RF applications. Superlattices Microstruct 109:725–734

Murugapandiyan P, Nirmal D, Tanvir Hasan Md, Varghese A, Ajayan J, Augustine Fletcher AS, Ramkumar N (2021) Influence of AlN passivation on thermal performance of AlGaN/GaN high-electron mobility transistors on sapphire substrate: a simulation study. Mater Sci Eng B 273:115449

Murugapandiyan P, Tanvir Hasan Md, Rajya Lakshmi V, Wasim M, Ajayan J, Ramkumar N, Nirmal D (2021) Breakdown voltage enhancement of gate field plate$Al_{0.295}Ga_{0.705}N$/GaN HEMTs. Int J Electron 108:1273–1287

Murugapandiyan P, Nirmal D, Ajayan J et al. (2022) Investigation of influence of SiN and SiO_2passivation in gate field plate double heterojunction $Al_{0.3}Ga_{0.7}N/GaN/Al_{0.04}Ga_{0.96}N$ high electron mobility transistors. Silicon 14:1421–1429

Nirmal D, Arivazhagan L, Augustine Fletcher AS, Ajayan J, Prajoon P (2018) Current collapse modeling in AlGaN/GaN HEMT using small signal equivalent circuit for high power application. Superlattices Microstr 113:810–820

S. Franklin A, Jebalin BK, Subhash Chander IV, Kumar R, Ajayan J, Nirmal D (2023) Investigation on Fe-Doped AlGaN/GaN HEMT at 148 GHz Using E-FPL technology for high-frequency communication systems 12:035006

Wu T-L, Kutub SB (2020) Machine learning-based statistical approach to analyze process dependencies on threshold voltage in recessed gate AlGaN/GaN MIS-HEMTs. IEEE Trans Electron Dev 67(12):5448–5453. https://doi.org/10.1109/TED.2020.3032634

Yuk KS, Branner GR, McQuate DJ (2009) A wideband multiharmonic empirical large-signal model for high-power GaN HEMTs with self-heating and charge-trapping effects. IEEE Trans Microw Theory Techn 57(12):3322–3332

Large Signal Modelling of GaN HEMTs

Devika Jena, Sanghamitra Das, Aruna Tripathy, and Taraprasanna Dash

Abstract In this chapter, the large signal response of AlGaN/GaN HEMT device is presented. Input signal of 10 GHz is considered to investigate the DC and RF performance of the device. The output power, power gain and amplification efficiency have been extracted from transient analysis. Response of the device to high-frequency signals is presented in detail by varying the signal frequency up to 90 GHz. Physics-based modelling has been chosen to perform the simulations using TCAD tool Silvaco. The results can be significantly useful as the GaN HEMT device is frequently utilized for high-frequency and high-power applications.

Keywords AlGaN/GaN · Empirical modelling · Power amplifiers · Power gain · High power applications

1 Introduction

A semiconductor material system known as AlGaN/GaN is made up of layers of aluminium gallium nitride (AlGaN) and gallium nitride (GaN). In the realm of power electronics, this material system is frequently utilized, especially for high-frequency

D. Jena · T. Dash
Department of ECE, Siksha 'O' Anusandhan (Deemed to be University), Bhubaneswar 751030, India
e-mail: tpdiitkgp@gmail.com

S. Das
Depatrment of ECE, Silicon Institute of Technology, Bhubaneswar 751025, India
e-mail: sanghamitra.das@silicon.ac.in

D. Jena (✉) · A. Tripathy
School of Electronic Sciences, Odisha University of Technology and Research, Bhubaneswar 751003, India
e-mail: jenadevika@gmail.com

A. Tripathy
e-mail: atripathy@outr.ac.in

D. Nirmal and J. Ajayan (eds.), *Modeling of AlGaN/GaN High Electron Mobility Transistors*, Springer Tracts in Electrical and Electronics Engineering,
https://doi.org/10.1007/978-981-97-7506-4_5

and high-power applications like radio frequency (RF) amplifiers and power amplifiers (Flack et al. 2016; Ma and Gu 2019; Zhang et al. 2021; Runton et al. 2013; Arivazhagan et al. 2019). It is crucial to take into account the nonlinear properties of AlGaN/GaN devices when analysing the large signal behaviour of these devices because they have a big impact on performance.

The term "large signal analysis" describes the investigation of how a device responds to a strong input signal (Jarndal and Kompa 2007; Aamir Ahsan et al. 2017; Xu et al. 2014). This analysis pertains to AlGaN/GaN devices and seeks to comprehend the device's behaviour under high-power and high-frequency signals. Large signal analysis takes into account the nonlinearities that result from the high signal amplitudes involved, in contrast to small signal analysis, which assumes linear behaviour around an operational point.

The nonlinear behaviour in AlGaN/GaN devices is caused by several causes (Bajaj et al. 2017; Moon et al. 2020a; Choi et al. 2017; Wang et al. 2022; Song et al. 2021; Kumar et al. 2011). The high electron mobility in the GaN channel, which promotes nonlinearities while enabling effective power handling, is one important factor. Traps and surface states can also have an impact on how well the device works. The nonlinearities can alter the electrical properties of the device and result in intermodulation products, signal distortion, and power-dependent fluctuations.

Several methodologies are used to examine the large signal behaviour of AlGaN/GaN devices. Among them are load-pull measurements, which entail adjusting the load impedance to see how the gadget responds to various circumstances. The ideal load impedance for maximizing power output or efficiency can be found through load-pull measurements. Another method is harmonic balance analysis, which is frequently applied to simulations of massive signal circuits. It is based on the iterative solution of a series of nonlinear equations. The input and output signals are represented as a combination of various sinusoidal components. The device's performance, including power, efficiency, and distortion under situations of strong signal, is shown by this method. Additionally, in circuit simulations, the nonlinear behaviour of AlGaN/GaN devices is faithfully modelled using large signal models such as the dynamic-IV and X-parameter models. These models incorporate the dependence of voltage and current on signal amplitude and frequency, enabling more precise predictions of device performance in the presence of strong signals.

Analysing AlGaN/GaN devices' behaviour under high-power working circumstances is known as large signal analysis. It seeks to comprehend how, in the presence of strong input signals, the device's electrical properties alter. The design of effective power amplifiers for a variety of applications and the assurance of dependable device functioning all depend on large signal analysis.

Usually, the analysis includes looking at several important factors, including:

1. Power input and output: large signal analysis looks into the connection between the strength of the input signal and the final output power. The efficiency and power handling capabilities of the gadget are determined by this characterization.
2. Gain and linearity: The gain of a device indicates how much the input signal is amplified at the output and is expressed as an amplification factor. Large signal

analysis evaluates the gain under high power settings and examines the device's linearity, or its capacity to accurately and distortion-free repeat the input signal.
3. Effects of nonlinearity: At high power levels, AlGaN/GaN devices frequently display nonlinear behaviour. The investigation of nonlinear effects through large signal analysis includes memory effects, harmonic production, and intermodulation distortion (IMD). For the signal to remain intact and undesirable distortions to be reduced, it is essential to comprehend and mitigate these effects.
4. Power-added efficiency (PAE): The efficiency with which a device converts DC power to RF power is measured by a term called power-added efficiency, or PAE. By reducing power usage and raising overall energy efficiency, large signal analysis aids in device design optimization to achieve greater PAE.

1.1 Device Modelling Approaches

Device modelling methods can indeed be classified into two approaches. The first one is the physical modelling approach that involves creating a mathematical model of the device based on the fundamental physical principles and parameters that describe its geometry and technology. This approach relies on understanding the underlying physics and engineering principles governing the device's operation. It typically requires knowledge of material properties, device structure and fabrication processes. The physical model attempts to simulate the behaviour of the device by solving equations derived from these principles. The second one is the empirical modelling approach, which relies on measured characteristics or data that describe the behaviour of the device. Instead of focusing on the underlying physical principles, empirical models aim to capture the device's behaviour based on experimental observations. This approach involves collecting data from actual devices or experiments and using statistical or mathematical techniques to create a model that approximates the device's response. Empirical models are often based on regression analysis, curve fitting, or machine learning algorithms. The advantage of empirical modelling is that it can capture complex device behaviours without the need for detailed knowledge of the underlying physics. However, it may lack the ability to extrapolate accurately beyond the range of measured data.

2 Physical Modelling

GaN compact physical models are mathematical representations that capture the large signal behaviour of GaN devices based on the underlying physical processes. The biggest advantage of this method is that it describes the device's operation in terms of device physics. These models aim to provide a balance between accuracy and computational efficiency. Some commonly used large signal compact physical models:

i. Charge-Based Transport Models: These models simulate charge transport and device physics in GaN devices using physics-based equations. They incorporate drift–diffusion and/or hydrodynamic transport equations to capture the carrier transport, and Poisson's equation to account for electric field distribution. These models include physical parameters such as carrier mobility, recombination, and trapping effects.
ii. Temperature-Dependent Models: GaN devices can exhibit temperature-dependent behaviour and accurate modelling of temperature effects is crucial. Temperature-dependent compact models incorporate thermal effects, such as self-heating, thermal resistance, and thermal capacitance, into the electrical model.
iii. Thermal Noise Models: In addition to the large signal behaviour, compact physical models may also incorporate thermal noise sources, such as shot noise and thermal noise from resistive elements. These models provide insights into the noise performance of GaN devices, which is important for applications involving low-noise amplifiers or communication systems.
iv. Electrothermal Models: Electrothermal models combine the electrical and thermal aspects of GaN devices. They consider the heat generation due to the electrical power dissipation and the resulting temperature rise, which in turn affects the electrical characteristics of the device. Electrothermal models are particularly useful for predicting the dynamic behaviour of GaN devices under large signal operation, where the heat generated can significantly impact the device's performance.

3 Empirical Modelling

To effectively depict the large signal behaviour, numerous empirical models have been constructed. Listed below are a few popular models:

i. Charge-Based Models: These models describe the dynamic behaviour of GaN devices based on charge transport mechanisms. They typically include a set of nonlinear differential equations that represent the device physics. Examples of charge-based models for GaN devices include the Chalmers model and the North Carolina State University (NCSU) model.
ii. HiCUM (High Current Model): HiCUM is a popular empirical model developed by Wolfspeed (a Cree company) for GaN high-electron-mobility transistors (HEMTs). It is a scalable, physics-based model that accurately captures the large-signal characteristics of GaN HEMTs under different biasing conditions and operating frequencies.
iii. X-parameters: X-parameters are a type of nonlinear behavioural model that describes the input–output relationship of a device using a set of power-dependent complex scattering parameters (S-parameters). They can be used to model both small and large signal behaviour of GaN devices, including amplifiers and mixers.

iv. Artificial Neural Networks (ANN): ANN models are trained using experimental data to capture the large signal behaviour of GaN devices. These models can be highly accurate and versatile but require a significant amount of training data and computational resources.

4 Literature Review

A table-based large-signal model for AlGaN/GaN HEMTs is presented in the paper that takes self-heating and trapping-induced current dispersion into account (Jarndal and Kompa 2007). IMD simulations are enhanced by using the B-spline-approximation technique to build model elements. The model is verified by comparisons with observed outputs from the device, demonstrating its effectiveness in foretelling output power, harmonics, and IMD under various operating situations. GaN HEMT surface-potential-based RF large-signal model is demonstrated in the paper (Aamir Ahsan et al. 2017), and it is validated using measurements from a commercial GaN device. When the model is fitted to the DC-IV characteristics, the intrinsic charges and current of the device demonstrate self-consistency, which enables the automatic construction of the small-signal equivalent circuit for RF simulation. However, standardized processes are needed to extract gate resistance and parasitic capacitance.

An electromagnetic (EM) simulation-based scaled GaN-HEMT large-signal model is presented in the work (Lee et al. 2020). An analogous circuit of the connecting circuit and a large-signal model of the two-finger GaN-HEMT make up the 10-finger GaN-HEMT. According to the EM simulation results, the equivalent circuit of the connector circuit was extracted. The two-finger device's large-signal model is based on the typical Angelov channel current model. Measurements of load-pull were used to confirm the 10-finger device's large-signal concept. Both load-pull tests and simulations showed that the 10-finger GaN-HEMT had an output power of roughly 20 W. An improved empirical large signal model for a 0.1 μm AlGaN/GaN HEMT process was proposed (Li et al. 2018). The proposed model takes into account the short channel effect, including the Drain Induced Barrier Lowering (DIBL) effect and channel length modulation, to accurately describe the DC characteristics of the device.

A nonlinear large-signal statistical model incorporates empirical equivalent circuit techniques and includes the effects of nonlinear electrothermal behaviour and trapping effects (Chen et al. 2016). The technique combines multivariate methods, including principal component analysis, factor analysis, multiple regression algorithms, and Monte Carlo simulation. The electrothermal model is constructed using a one-stage nonlinear R–C electrothermal network, which accurately represents the electrical and thermal characteristics of the device. Trapping effects are modelled by an effective gate bias (V_{gseff}) based on the Angelov model, allowing for the characterization of trapping phenomena in the GaN HEMTs. To validate the model and extract its parameters, measurements are conducted on 34 GaN HEMTs from 10 different batches. The statistical model is integrated into Agilent-ADS, a widely used electronic design automation software, to evaluate the large-signal performances such as output power (Pout) and power added efficiency (PAE). To demonstrate its effectiveness, two S-band GaN HEMT power amplifiers are designed using the statistical model. An enhanced thermal modelling strategy specifically designed for 0.25 μm for GaN HEMTs with an electrothermal large signal model in a foundry process is proposed by C. Wang et al. (Wang and Chen 2017) This model, based on the Angelov GaN model, offers the capability to predict various characteristics and behaviours, including DC I-V (current–voltage) characteristics, pulsed I–V characteristics, S-parameters, and large-signal characteristics. It also incorporates an extrinsic thermal network to improve the accuracy of circuit and system predictions, particularly in different heat transfer conditions. This advancement enables optimized power amplifier (PA) designs for various heat transfer environments. A large-signal model for AlGaN/GaN HEMTs for large-scaled devices was presented (Liu et al. 2012). This model introduces two important improvements: the inclusion of sub-threshold current based on a previous reference and the consideration of self-heating effects at high drain-source voltage (V_{ds}) and large drain current (I_{ds}) in the device. The authors optimize the AlGaN/GaN HEMT semi-empirical DC model and achieve a good fit for the I-V (current–voltage) curves. Based on this optimized DC model, they develop a large-signal model for the AlGaN/GaN HEMTs. The simulation results obtained from this model are compared with experimental data, and favourable agreements are observed, particularly for output power and gain at a bias of $V_{gs} = -2$ V and $V_{ds} = 25$ V, at an operational frequency of 5.4 GHz.

5 Device Parameters and Simulation Environment

In Fig. 1, the schematic of the device is shown. Silvaco TCAD tool is used for simulation. A thick GaN substrate with n-type Gaussian doping was taken. The doping concentration of the substrate is 6×10^{19} with a peak of 0.023. For the drain and source connection, *Ti/Al/Ni/Au* metal stack is considered (Bajaj et al. 2017; Sohel et al. 2018). This particular source/drain stack is preferred as it reduces the contact resistance to get good ohmic contact on the AlGaN/GaN structure (Yan et al. 2012). *Ti/Al* layers react with AlGaN and form thin layers of TiN and *Al–TiN* layers with a low work function and leaving a lot of *N* vacancies and *Ni/Au/Ni* stack is used for the Schottky gate contact (Bajaj et al. 2017; Sohel et al. 2018; Moon et al. 2020b). The $Al_xGa_{1-x}N$ layer acts as a carrier supply layer that is important for creating the 2DEG. In the proposed structure, a 23 nm $Al_xGa_{1-x}N$ channel layer with *x*-composition 28% is used. A passivation layer of SiO_2 is deposited above the $Al_xGa_{1-x}N$ layer. Shockley–Read–Hall model is used to include the carrier generation and recombination effect in the GaN channel. In the SRH model, carrier recombination occurs through a two-step process involving a trap level within the bandgap of the semiconductor material. The model assumes that there are trap states present in the semiconductor material that can capture and release charge carriers (electrons or holes). These trap states can be associated with impurities, defects, or interface states within the device structure. Field dependent mobility model is used to account for the effect of velocity saturation at higher drain voltages in the simulation profile. This model describes the changes in electron mobility as a function of the electric field strength within the device. In GaN, the crystal structure gives rise to a built-in polarization due to the difference in electronegativity between Gallium (Ga) and Nitrogen (N) atoms. This built-in polarization creates an internal electric field within the material, which affects the charge distribution and band structure. Additionally, when GaN is subjected to external mechanical stress or strain, the piezoelectric effect further contributes to the overall polarization. So, to model this effect, polarization model is used as GaN device. The work function of 5.15 eV is used for Schottky gate contact and 3.93 eV is used in source/drain for ohmic contact. The low field mobility as a function of doping and lattice temperature has been modelled by using the Albrecht low field mobility model. The acceptor traps in AlGaN/GaN HEMTs are typically associated with deep levels in the bandgap of the semiconductor material. These deep levels can arise from various sources, such as crystal defects, impurities, or interface states. The specific origin and characteristics of acceptor traps in AlGaN/GaN HEMTs can vary depending on the fabrication process and material quality. When holes are trapped by acceptor traps in the channel region, the effective carrier density and mobility are reduced, leading to decreased device performance. The captured holes introduce a positive fixed charge, altering the threshold voltage and increasing the access resistance of the device. This can result in reduced current drivability and increased power consumption. Moreover, acceptor traps can also cause time-dependent degradation in the device characteristics. For example, when the device is exposed to high electric fields or elevated temperatures, trapped holes can be released from the traps, leading to dynamic changes in the device parameters. This phenomenon is commonly referred to as "trapping and de-trapping" and can contribute to device instability and reliability concerns. To consider the effect of interface traps in our simulation, we have considered the acceptor traps in the GaN layer by taking reference from the literature (Ma et al. 2023) whose concentration can vary from 10^{15} cm^{-3} to 10^{18} cm^{-3} in actual processes. The trap energy level and capture cross section are Ec—0.36 eV and 1×10^{-15} cm^{-2}, respectively. A trap density of $1 \times 10^{18}/cm^2$ is considered in the proposed device.

The TCAD simulated structure and the electron concentration are shown in Fig. 2a, b. To analyse the device for large signal, 10 sinusoidal signals of amplitude 1.75 V–4 V with an increment of 0.25 V are given to the gate electrode. The frequency of the signal is kept constant at 10 GHz.

6 Results and Discussions

In this section, we have shown the DC and RF performance of the simulated GaN HEMT device of Lg = 250 nm. Then the amplifier performance has been investigated at an input frequency of 10 GHz.

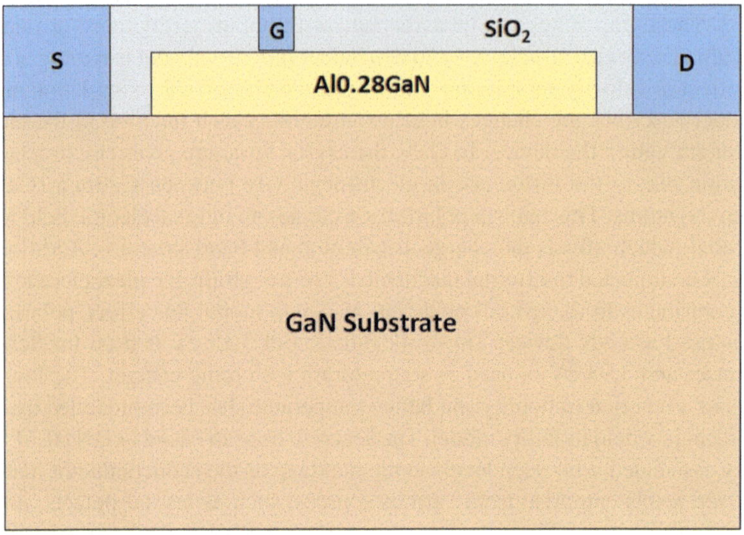

Fig. 1 Schematic diagram of the device structure considered for simulation

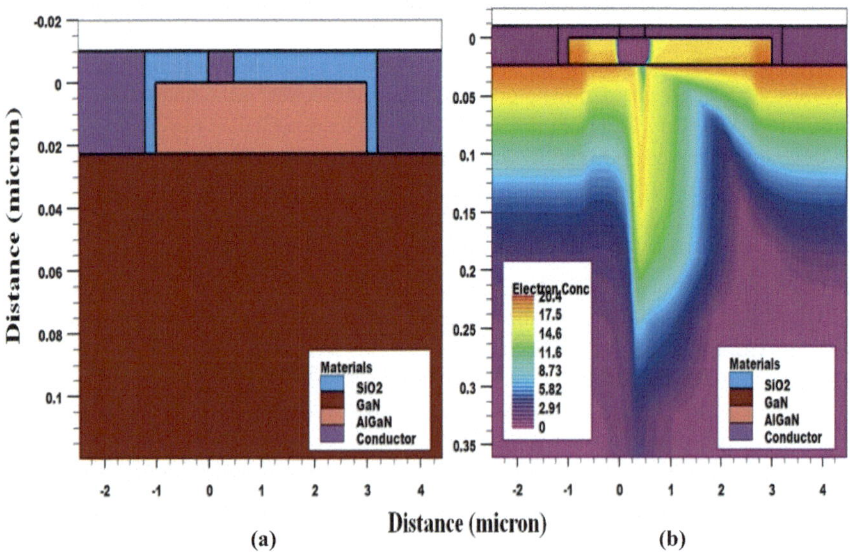

Fig. 2 Simulated structure and electron concentration

The Id-Vg and Id-Vd graphs are shown in Figs. 3 and 4, respectively. The Id-Vg graphs are simulated at drain voltages of 10 V and 30 V, respectively. The device shows normally ON behaviour due to the formation of 2DEG at the AlGaN/GaN interface. The threshold voltage reduces as the drain voltage is increased. The Id-Vd curves are shown for different gate voltage values of −7 V, −6 V, −5 V and −4 V.

As the drain current saturates beyond Vg = −4 V, hence, we have shown the device RF performance at Vd = 30 V and Vg = −4 V. The current gain and power gain versus frequency plots are shown in Figs. 5 and 6, respectively. From these gains, the cut-off frequency (f_T) and maximum frequency of oscillation (f_{max}) can be calculated when the gain is unity (0 dB). Here, the cut-off frequency is nearly 20 GHz and the f_{max} is 10 GHz for the simulated HEMT device of 250 nm gate length.

To investigate the large signal performance of the GaN HEMT device, simulations have been done by applying ten input signals at the gate with increasing amplitude levels with a frequency of 10 GHz. The graph of input and output signals versus time has been plotted. Also, the different parameters related to power output, such as power out versus power in, power gain and amplifier efficiency have been shown as the device is over-driven into high distortion.

Figure 7 shows the input and output signals versus time. The input signal is applied at the gate terminal and the output is measured at the drain terminal. The peak input and output voltages are extracted, followed by the exact times at which these peak voltages occurred. Then the input and output currents are extracted at the same times as the peak voltages occur. It is important to extract the currents at the same time as the peak voltages in order to take account of the phase between current and voltage, so that the correct power values can be calculated. The output power (P_{out}) and the power gain versus input power (P_{in}) are shown in Figs. 8 and 9, respectively. These are measured for a drain bias of 30 V.

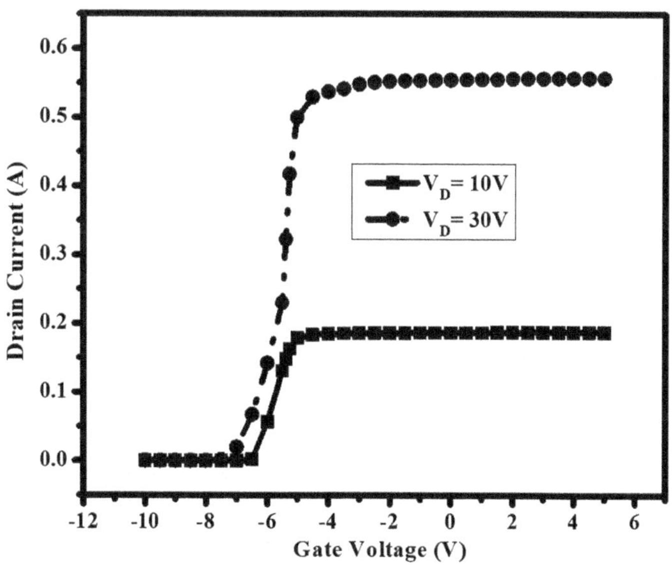

Fig. 3 Id-Vg characteristics for the GaN HEMT with gate length 250 nm

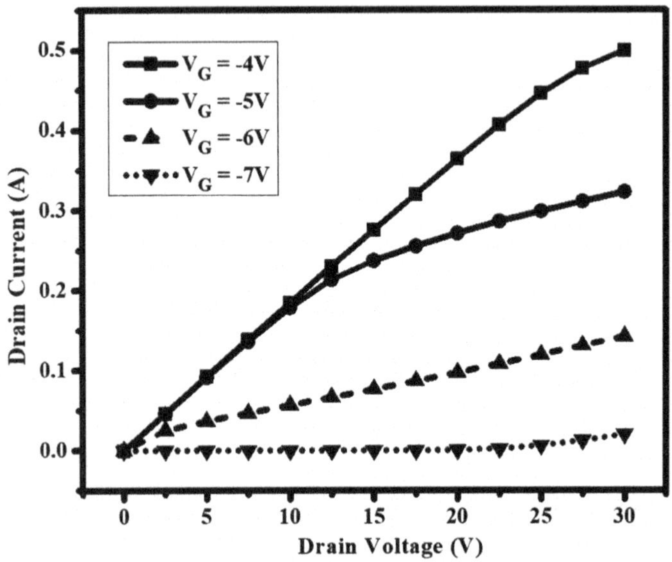

Fig. 4 Id-Vd characteristic for the GaN HEMT with gate length 250 nm

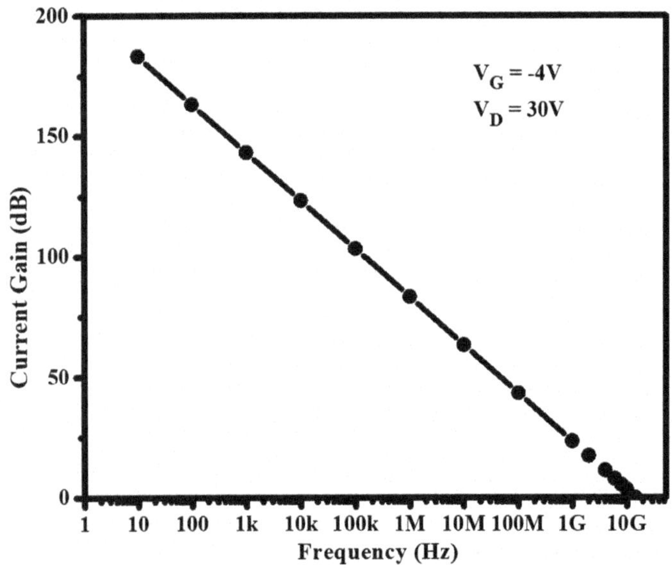

Fig. 5 Current gain versus frequency. The 0 dB cut-off frequency is at 20 GHz

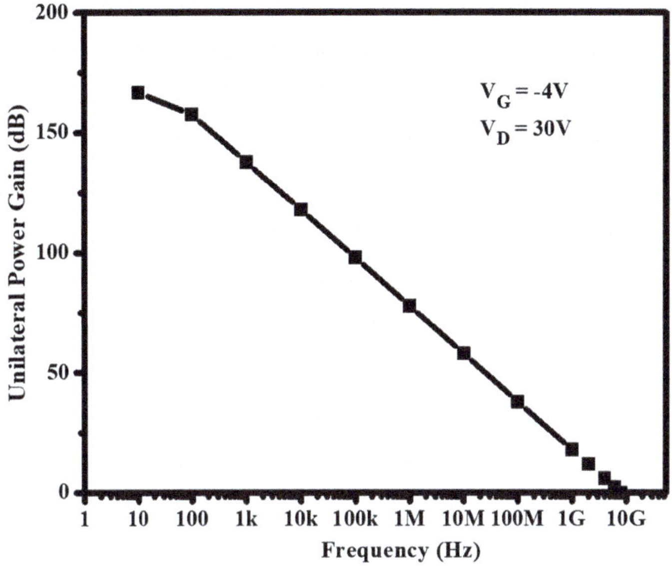

Fig. 6 Power gain versus frequency

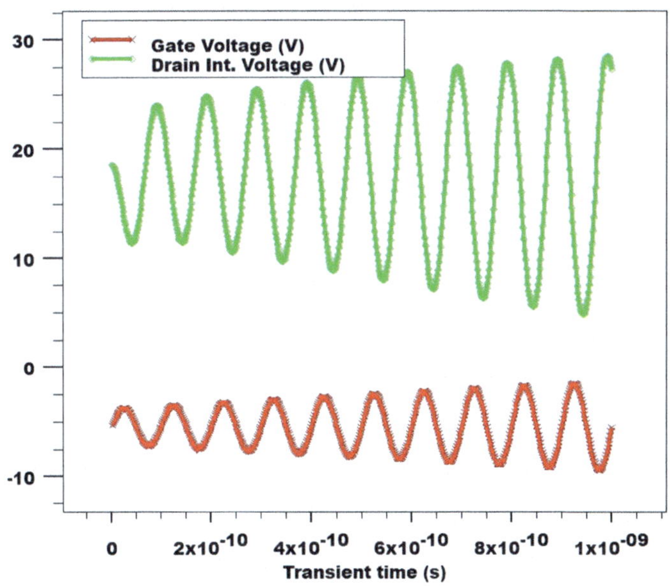

Fig. 7 Variation of input and output signal with time

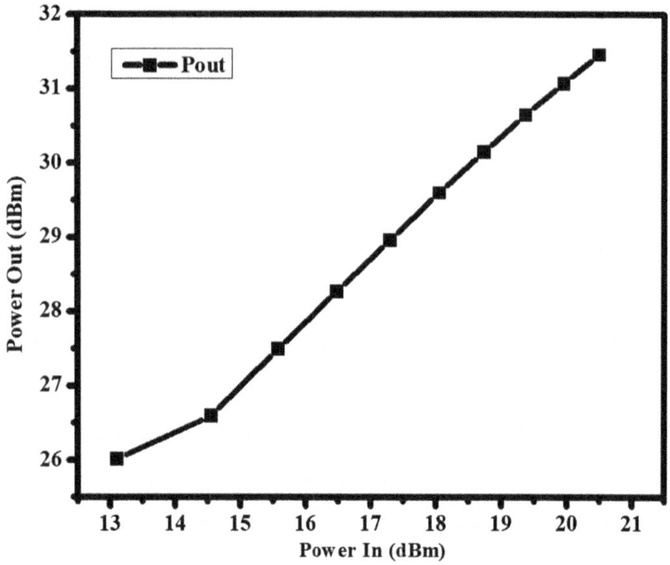

Fig. 8 Power out for 250 nm gate length GaN HEMT device

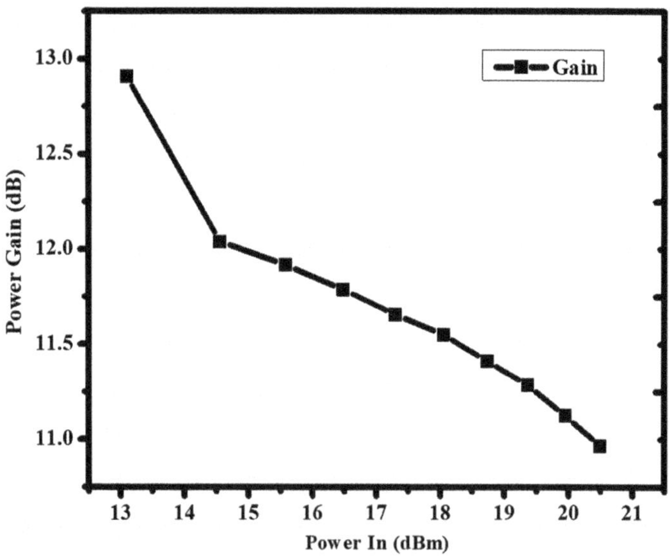

Fig. 9 Power gain for 250 nm gate length GaN HEMT device

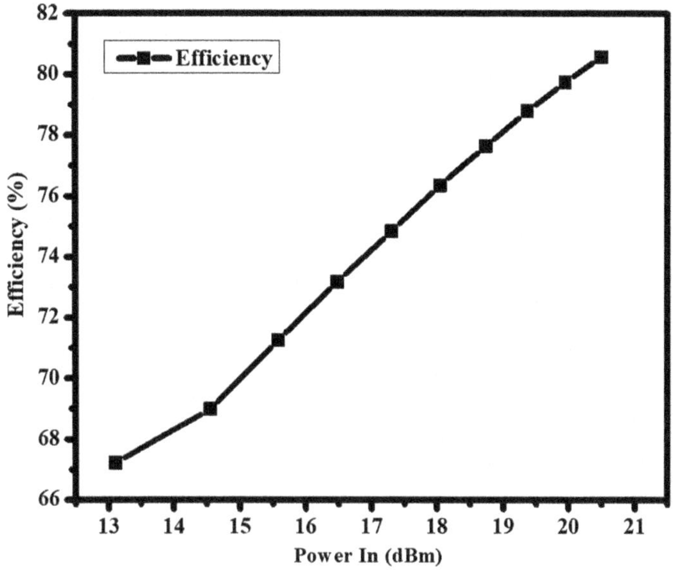

Fig. 10 Efficiency for 250 nm gate length GaN HEMT device

As it can be seen, the power gain reduces at higher input power levels. The efficiency which is measured as the ratio of output power to the supply voltage is shown in Fig. 10. As the output power increases with increase in input power at a constant supply voltage, the efficiency of the amplifier also increases.

To examine the effect of frequency on the HEMT device output power and power gain, we have calculated the output power and power gain for a single input wave at higher frequencies of 30 GHz, 50 GHz, 70 GHz and 90 GHz. The output power, gain and efficiency reduce as the input signal frequency increases. This can be attributed to the increased parasitic effects at higher frequencies (Figs. 11).

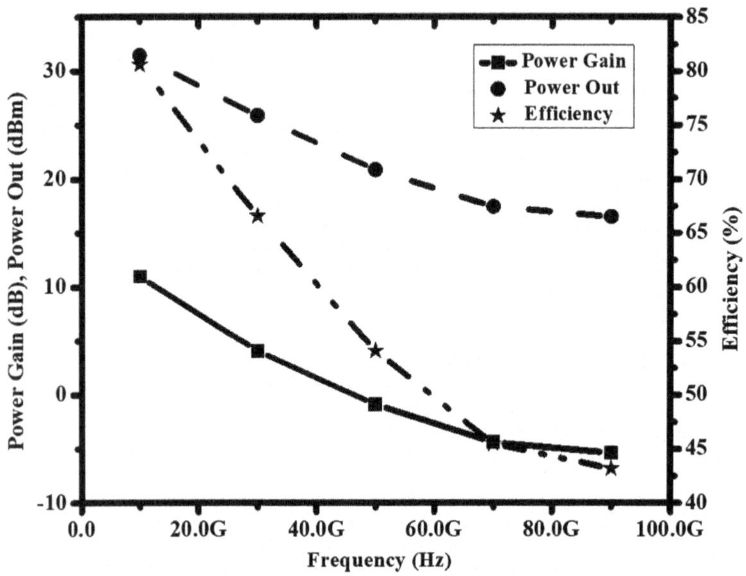

Fig. 11 Power gain, power output versus frequency

7 Conclusion

In this chapter, the large signal response of AlGaN/GaN HEMT device has been verified. The DC and RF performance of the device is shown at high drain bias of 30 V. The device cut-off frequency has been extracted to be 20 GHz at a gate voltage of −4 V. The output power, power gain and amplification efficiency have been extracted from transient analysis. Response of the device to high-frequency signals is presented in detail by varying the signal frequency up to 90 GHz. The power gain and efficiency decrease with increase in frequency due to the parasitic effects. The results can be significantly useful as the GaN HEMT device is frequently utilized for high-frequency and high-power applications.

References

Aamir Ahsan S, Ghosh S, Khandelwal S, Chauhan YS (2017) Physics-based multi-bias RF large-signal GaN HEMT modeling and parameter extraction flow. IEEE J Electron Devices Soc 5:310–319. https://doi.org/10.1109/JEDS.2017.2724839

Arivazhagan L, Nirmal D, Godfrey D, Ajayan J, Prajoon P, Augustine Fletcher AS, Amir Anton Jone A, Raj Kumar JS (2019) Improved RF and DC performance in AlGaN/GaN HEMT by P-type doping in GaN buffer for millimetre-wave applications. AEU Int J Electron Commun 108:189–194. https://doi.org/10.1016/J.AEUE.2019.06.015

Bajaj S, Yang Z, Akyol F, Park PS, Zhang Y, Price AL, Krishnamoorthy S, Meyer DJ, Rajan S (2017) Graded AlGaN channel transistors for improved current and power gain linearity. IEEE Trans Electron Devices 64:3114–3119. https://doi.org/10.1109/ted.2017.2713784

Chen Z, Xu Y, Zhang B, Chen T, Gao T, Xu R (2016) A GaN HEMTs nonlinear large-signal statistical model and its application in S-band power amplifier design. IEEE Microwave Wirel Compon Lett 26:128–130. https://doi.org/10.1109/LMWC.2016.2517444

Choi P, Radhakrishna U, Boon CC, Peh LS, Antoniadis D (2017) Linearity enhancement of a fully integrated 6-GHz GaN power amplifier. IEEE Microwave Wirel Compon Lett 27:927–929. https://doi.org/10.1109/LMWC.2017.2746673

Flack TJ, Pushpakaran BN, Bayne SB (2016) GaN technology for power electronic applications: a review. J Electron Mater 45:2673–2682. https://doi.org/10.1007/s11664-016-4435-3

Jarndal A, Kompa G (2007) Large-signal model for AlGaN/GaN HEMTs accurately predicts trapping- and self-heating-induced dispersion and intermodulation distortion. IEEE Trans Electron Devices 54:2830–2836. https://doi.org/10.1109/TED.2007.907143

Kumar SP, Agrawal A, Chaujar R, Gupta RS, Gupta M (2011) Device linearity and intermodulation distortion comparison of dual material gate and conventional AlGaN/GaN high electron mobility transistor. Microelectron Reliab 51:587–596. https://doi.org/10.1016/J.MICROREL.2010.09.033

Lee W, Kang H, Choi S, Lee S, Kwon H, Hwang KC, Lee KY, Yang Y (2020) Scaled GaN-HEMT large-signal model based on EM simulation. Electronics 9:632. https://doi.org/10.3390/ELECTRONICS9040632

Li J, Mao S, Xu Y, Zhao X, Wang W, Guo F, Zhang Q, Wu Y, Zhang B, Chen T, Yan B, Xu R, Li Y (2018) An improved large signal model for 0.1 μmAlGaN/GaN high electron mobility transistors (HEMTs) process and its applications in practical monolithic microwave integrated circuit (MMIC) design in W band. Micromachines 9:396. https://doi.org/10.3390/MI9080396

Liu D, Wang L, Chen X (2012) GaN HEMT large-signal model research. Proceedings of 2012 international workshop on microwave and millimeter wave circuits and system technology, MMWCST 2012. 319–322. https://doi.org/10.1109/MMWCST.2012.6238194

Ma CT, Gu ZH (2019) Review of GaN HEMT applications in power converters over 500 W. Electronics 8:1401. 8, 1401. https://doi.org/10.3390/ELECTRONICS8121401

Ma M, Cao Y, Lv H, Wang Z, Zhang X, Chen C, Wu L, Lv L, Zheng X, Tian W, Ma X, Hao Y (2022) Effect of acceptor traps in GaN buffer layer on breakdown performance of AlGaN/GaN HEMTs. Micromachines 14:79. https://doi.org/10.3390/MI14010079

Moon JS, Grabar R, Wong J, Antcliffe M, Chen P, Arkun E, Khalaf I, Corrion A, Chappell J, Venkatesan N, Fay P (2020b) High-speed graded-channel AlGaN/GaN HEMTs with power added efficiency >70% at 30 GHz. Electron Lett 56:678–680. https://doi.org/10.1049/EL.2020.0281

Moon JS, Grabar B, Antcliffe M, Wong J, Dao C, Chen P, Arkun E, Khalaf I, Corrion A, Chappell J, Venkatesan N, Fay P (2020) High-speed graded-channel GaN HEMTs with linearity and efficiency. IEEE MTT-S international microwave symposium digest. 573–575 (2020). https://doi.org/10.1109/IMS30576.2020.9223775

Runton DW, Trabert B, Shealy J, Vetury R (2013) History of GaN: High-power RF gallium nitride (GaN) from infancy to manufacturable process and beyond. IEEE Microw Mag 14:82–93. https://doi.org/10.1109/MMM.2013.2240853

Sohel SH, Xie A, Beam E, Xue H, Roussos JA, Razzak T, Bajaj S, Cao Y, Meyer DJ, Lu W, Rajan S (2018) X-band power and linearity performance of compositionally graded AlGaN channel transistors. IEEE Electron Device Lett 39:1884–1887. https://doi.org/10.1109/LED.2018.2874443

Song W, Zheng Z, Chen T, Wei J, Yuan L, Chen KJ (2021) RF linearity enhancement of GaN-on-Si HEMTs with a closely coupled double-channel structure. IEEE Electron Device Lett 42:1116–1119. https://doi.org/10.1109/LED.2021.3087785

Wang(王鹏飞) P, Mi(宓珉瀚) M, Zhang(濛) M, Zhu(祝杰杰) J, Zhou(周雨威) Y, Liu(刘捷龙) J, Liu(刘思佳) S, Yang(杨凌) L, Hou(侯斌) B, Ma(马晓华) X, Hao(郝跃) Y, Wang(王鹏飞)

P, Mi(宓珉瀚) M, Zhang(张濛) M, Zhu(祝杰杰) J, Zhou(周雨威) Y, Liu(刘捷龙) J, Liu(刘思佳) S, Yang(杨凌) L, Hou(侯斌) B, Ma(马晓华) X, Hao(郝跃) Y (2022) High linearity AlGaN/GaN HEMT with double-Vth coupling for millimeter-wave applications. Chinese Phys B 31:27103–027103. https://doi.org/10.1088/1674-1056/AC2B21

Wang C, Chen Y (2018) An electrothermal large signal model of GaN HEMTs foundry process. 2017 IEEE electrical design of advanced packaging and systems symposium, EDAPS 2017. 2018-January, 1–3. https://doi.org/10.1109/EDAPS.2017.8276956

Xu Y, Fu W, Wang C, Ren C, Lu H, Zheng W, Yu X, Yan B, Xu R (2014) A scalable GaN HEMT large-signal model for high-efficiency RF power amplifier design 28, 1888–1895. https://doi.org/10.1080/09205071.2014.947440

Yan W, Zhang R, Du Y, Han W, Yang F (2012) Analysis of the ohmic contacts of Ti/Al/Ni/Au to AlGaN}/{GaN {HEMTs} by the multi-step annealing process. J Semicond 33:64005. https://doi.org/10.1088/1674-4926/33/6/064005

Zhang Y, Zubair A, Liu Z, Xiao M, Perozek J, Ma Y, Palacios T (2021) GaN FinFETs and trigate devices for power and RF applications: review and perspective. Semicond Sci Technol 36:054001. https://doi.org/10.1088/1361-6641/ABDE17

Thermal Resistance Modeling and its Different Aspects on AlGaN/GaN HEMTs: A Comprehensive Review

Abdul Naim Khan, K. Jena, Gaurav Chatterjee, Meenakshi Chauhan, and S. Routray

Abstract Thermal heat management profoundly affects AlGaN/GaN high-electron-mobility transistors (HEMTs) overall performance and reliability. Understanding and improving the thermal performance of these devices depends on thermal resistance modeling. This work extensively reviews the methods and tools used for AlGaN/GaN HEMT thermal resistance modeling. Different fundamentals of thermal resistance modeling with their characteristics are also incorporated. The impact of the heat transfer mechanism in AlGaN/GaN HEMT is discussed in this review. We review the various simulation methods, material characteristics, and experimental procedures used to determine the various simulation methods, material characteristics, and testing approaches used in determining thermal resistance. We also investigate how various material characteristics, device designs, key parameters influencing thermal resistance, and packaging strategies affect thermal performance. The overview of steady-state and transient measurement techniques are analyzed to examine the thermal behavior of the AlGaN/GaN HEMT. Different analytical models are evaluated with their performance for accurate thermal resistance estimation. Numerical simulation techniques FEA, CFD, FDM, and Monte Carlo Method are briefly discussed with their geometry and boundary conditions. Material properties and their characterization for AlGaN/GaN HEMT thermal resistance modeling are observed.

A. N. Khan (✉) · K. Jena · G. Chatterjee · M. Chauhan
Department of Electronics and Communication Engineering, The LNM Institute of Information Technology, Jaipur, India
e-mail: abdulnaimkhan.y19@lnmiit.ac.in

K. Jena
e-mail: kanjalochan.jena@lnmiit.ac.in

G. Chatterjee
e-mail: gaurav.chatterjee@lnmiit.ac.in

M. Chauhan
e-mail: meenakshichauhan.y20phd@lnmiit.ac.in

S. Routray
Department of Electronics and Communication Engineering, SRM Institute of Science and Technology, Chennai, India
e-mail: s.r.routray@ieee.org

D. Nirmal and J. Ajayan (eds.), *Modeling of AlGaN/GaN High Electron Mobility Transistors*, Springer Tracts in Electrical and Electronics Engineering,
https://doi.org/10.1007/978-981-97-7506-4_6

We have also reviewed different design strategies to mitigate the thermal resistance in AlGaN/GaN HEMT. A Unify packaging technique significantly improved heat dissipation and thermal resistance management. Advanced approaches and future trends of AlGaN/GaN HEMT with thermal resistance modeling are discussed at last. To enable the creation of improved and more reliable AlGaN/GaN HEMT devices, we summarize present challenges and potential future paths in thermal resistance modeling. To provide opportunities for researchers and engineers to improve their design and thermal management methodologies, an exhaustive investigation of the thermal behavior of AlGaN/GaN HEMTs is needed.

Keywords AlGaN/GaN HEMT · Thermal resistance · Thermal model · Thermal management · and Trapping effects

1 Introduction

1.1 Background and Significance of Thermal Resistance Modeling in AlGaN/GaN HEMTs

High-electron-mobility transistors (HEMTs) made of AlGaN/GaN have exceptional high-frequency (Dubey et al. 2022; Eastman et al. 2002; Fletcher et al. 2019; Haziq et al. 2022; Palacios et al. 2006; Saadat et al. 2009; Khan et al. 2023) and high-power performance (Wu et al. 2001; Gangwani et al. 2007; Wu et al. 1997; Khan et al. 2023), which makes them advantageous in radar systems (Palacios 2009; Mounika et al. 2022; Jeong et al. 2014), wireless communications (Mizuno et al. 2012; Quay et al. 2018), and power amplifiers (Ohki et al. 2009; Chung et al. 2002; Quay et al. 2008). The reality is that these devices produce much heat at higher junction temperatures and negatively impact the results, which deteriorates their performance, dependability, and durability. Understanding and controlling AlGaN/GaN HEMTs' thermal behavior depends heavily on thermal resistance modeling (Arivazhagan et al. 2020; Alim et al. 2020; Sommet et al. 2012). It entails forecasting and examining heat flow within the device design, including heat creation, conduction, and dissipation. Researchers and engineers can effectively handle the significant challenges by precisely characterizing the thermal Resistance of AlGaN/GaN HEMTs. AlGaN/GaN HEMTs' performance and dependability can suffer from an excessive temperature rise (Chou et al. 2004; Huque et al. 2009); reducing can also reduce their lifespan. Potential hotspots are identified and made possible through thermal resistance modeling, which also offers information into the device's capacity to dissipate heat. This information aids in developing effective thermal management plans to ensure the durability and dependability of the equipment.

The electrical characteristics of AlGaN/GaN HEMTs, such as electron mobility and carrier concentration, can be significantly affected by excessive heat (Heller and Crespo 2008; Wang et al. 2012). Researchers may enhance the design and layout of the device to improve heat dissipation, reduce temperature rise, and optimize electrical performance by precisely modeling the thermal behavior. Planning efficient cooling techniques requires understanding the thermal Resistance of AlGaN/GaN HEMTs. By estimating the thermal resistance, researchers may evaluate the effectiveness of various cooling techniques, including flip-chip bonding, heat sinks, microchannel, and thermoelectric cooling (Quay 2008; Wang et al. 2012; Wijesundara et al. 2011). This information helps choose and optimize the best cooling mechanism to maintain the device's operating temperature within reasonable bounds. Thermal resistance modeling shows how AlGaN/GaN HEMTs' overall thermal performance is impacted by packaging approaches such as material choice, thermal interface materials, and bonding processes. It assists in enhancing heat dissipation and minimizing thermal resistance by optimizing package design and integration. As AlGaN/GaN HEMT performance and power handling capabilities improve, efficient thermal management becomes increasingly important. To overcome thermal constraints and propel future developments in AlGaN/GaN HEMT technology, thermal resistance modeling directs the development of new materials, device architectures, and cooling techniques.

1.2 Overview of Thermal Management Challenges and Implications on Device Performance

Certainly, AlGaN/GaN high-electron-mobility transistors (HEMTs) require precise thermal management due to the direct impacts on device performance. The difficulties with heat management and their effects on the functionality of AlGaN/GaN HEMTs are summarized as follows:

Temperature Rise: When operating, AlGaN/GaN HEMTs produce much heat. The device's inside temperature may increase due to needing more thermal management. High temperatures can harm the functionality and dependability of a device. Higher temperatures can result in worsened device performance and reduced lifespan due to lower carrier mobility, increased leakage current, and faster material degradation (Choi et al. 2012; Cai et al. 2007).

Performance Degradation: High temperatures may influence the AlGaN/GaN HEMTs' electrical characteristics and performance. Increased temperatures can impact the threshold voltage, decrease carrier mobility, improve contact resistances, and lower maximum power output. These consequences can result in decreased device performance, reduced power handling capacity, and weakened signal integrity (Chou et al. 2004).

Reliability issues: Poor heat management may accelerate mechanisms that cause interconnects and solder connections to wear out, like electromigration and thermal fatigue. These dependability problems may cause connection problems, more resistance, and shorter device life. Thermal overstress can affect a device, making it useless adversely (Ajayan et al. 2022).

Thermal Hotspots: The device could acquire localized hotspots due to ineffective thermal management. These "hotspots" concentrate heat in those places, accelerating the rise in temperature and raising the thermal load on components close by. Hotspots can result in localized material deterioration, more intense performance, and higher failure susceptibility (Dundar et al. 2019).

Packaging and Integration: Packaging and incorporating devices face challenges due to effective thermal control. Thermal resistance must be decreased through proper heat sinking, the choice of thermal interface materials, and adequate heat transmission mechanisms. However, these factors could result in design limitations, such as a more significant, substantial, and expensive final product (Das et al. 2006).

These thermal management issues must be overcome for maximum device performance, dependability, and robustness. To mitigate the impact of temperature on device performance and improve overall thermal management capabilities, approaches can be used, including effective heat dissipation, advanced cooling techniques (such as heat sinks and liquid cooling), package optimization, and thermal modeling. By adequately regulating heat, designers may improve the dependability and realize the full potential of AlGaN/GaN HEMTs in various high-power applications.

2 Fundamentals of Thermal Resistance Modeling

2.1 Overview of Thermal Resistance Concepts and Definitions

The heat dissipation characteristics in AlGaN/GaN high-electron-mobility transistors (HEMTs) involve understanding thermal resistance concepts and terminologies. The junction-to-case thermal Resistance (Rjc) represents the resistance to heat transfer from the Junction, which is the active part of the HEMT, to the device's outer case or heat sink. It evaluates the efficiency of heat dissipation through the device's package and thermal interaction materials. Better heat dissipation capacities are indicated by a lower Rjc value, which allows the device to maintain lower junction temperatures (Cappello et al. 2017). Junction-to-Ambient Thermal Resistance (Rja) measures the impedance to heat transfer from the Junction, the operational part of the HEMT, to the ambient environment. It considers the device's overall capacity to dissipate heat and the effects of the package, thermal interface materials, and any additional cooling systems. The working temperature of the device declines through more excellent heat dissipation to the surroundings, as indicated by a lower Rja value (Won et al. 2015). Various thermal resistances inside the device and its package are considered a network in thermal resistance modeling. Mentioned in these resistances are the junction-to-case and junction-to-ambient resistances, along with any other resistances connected to the materials used as thermal interfaces, heat spreaders, or other cooling elements. The combination of these resistances impacts the overall thermal performance of the

HEMT. Another measure commonly used in thermal analysis is thermal impedance. It provides a more thorough evaluation of the device's thermal behavior by analyzing the impacts of thermal capacitance and resistance. Dynamic thermal analysis is made possible by thermal impedance, which explains how the device interacts with changes in heat dissipation over an extended period. Heat distribution within the HEMT and its surrounding components is called heat spreading (Wang et al. 2013). It is affected by the device's design, the materials used, and how effectively the thermal interfaces work. Adequate heat distribution helps to prevent localized hotspots and ensures a more even distribution of compassion throughout the device.

The study of Arivazhagan et al. (2020) also aims to demonstrate the Variable thermal resistance model of GaN on SiC with the substrate scalability and develop the drain current, transfer, and transconductance model with the impact of thermal resistance. Device structure and different types of modeling and simulation results are illustrated in Fig. 1). Optimizing the performance, dependability, and lifespan of AlGaN/GaN HEMTs requires understanding and managing these thermal resistance concepts. Designers can reduce temperature-related challenges and improve these devices' overall thermal management capabilities by reducing thermal resistance and providing efficient heat dispersion.

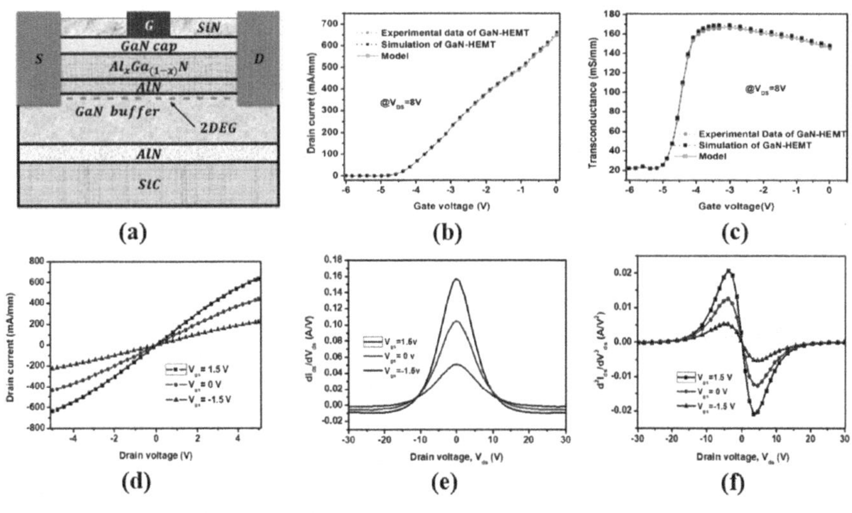

Fig. 1 **a** Device structure **b** Drain current versus the gate voltage characteristics **c** Transconductance characteristics **d** Drain current versus the drain voltage (with Vds swept in positive and negative directions) **e** The first-order derivative of the drain current with respect to the drain voltage **f** Second-order derivative of the drain current with respect to the drain voltage. Reprinted from Ref. (Arivazhagan et al. 2020), L. Arivazhagan et al., "Variable thermal resistance model of GaN-on-SiC with substrate scalability," J Comput Electron 19, 1546–1554 (2020), https://doi.org/10.1007/s10825-020-01561-y. Copyright (2020), with permission from Springer

2.2 Description of Heat Transfer Mechanisms in AlGaN/GaN HEMTs

Several heat transfer processes are essential in the heat dissipation process in AlGaN/GaN high-electron-mobility transistors (HEMTs). The typical heat transfer methods seen in AlGaN/GaN HEMTs are described as follows:

Conduction: In AlGaN/GaN HEMTs, conduction is the primary heat transfer mechanism. Conduction happens between semiconductor materials and other solid components like the substrate and packaging, which is how heat created inside the device spreads. How well heat is transferred away from the heat-generating regions depends on the thermal conductivity of these materials (Hirama 2011; Aubry et al. 2007).

Convection: Convection is the process of transferring heat which is transferred via a fluid medium, usually air or a refrigerant. Convective heat transfer happens in AlGaN/GaN HEMTs when a cooling mechanism, like a heat sink or a liquid coolant, removes heat from the component. Heat dissipation is assisted by fluid movement, which eliminates heat and exchanges it with more excellent fluid (Zhang et al. 2011; Jia et al. 2018).

Radiation: Heat is transferred through radiation by way of electromagnetic waves. Radiation can still slightly affect heat transfer in AlGaN/GaN HEMTs, even though it is not the primary mechanism when the device operates at high temperatures. Radiation enables heat to be released from the equipment into the surrounding area (Vitusevich et al. 2003; Polyakov et al. 2012).

Joule Heating: The heat produced by electrical resistance inside a device is called Joule heating. The movement of current through the channel and other parts of AlGaN/GaN HEMTs causes resistive losses, which cause heat to be produced. It is essential to manage Joule heating because it adds to the device's overall heat burden (Heller and Crespo 2008; Asubar et al. 2014).

The work of Yingshuo Qin et al. (2022) also aims to demonstrate the self-heating effects and high-power microwave that can cause device heating, and impacts on the various parameters viz. mobility, velocity, and thermal conductivity are analyzed. Different types of experimental results like Electrical Characteristics, Low-field mobility, and thermal conductivity are generated. This work also shows that self-heating problems also reduced by proper material selection, device layout as well as heat transfer, and heat sinking is important. Device structure and different types of modeling and simulation results are shown in Fig. 2.

For AlGaN/GaN HEMTs to avoid excessive temperature rise, maintain device performance, and improve reliability, efficient heat transmission is essential. Practical thermal management approaches are used to improve the heat transfer mechanisms and ensure efficient heat dissipation from the device, such as optimized device design, suitable material selection, efficient cooling strategies, and thermal interface materials.

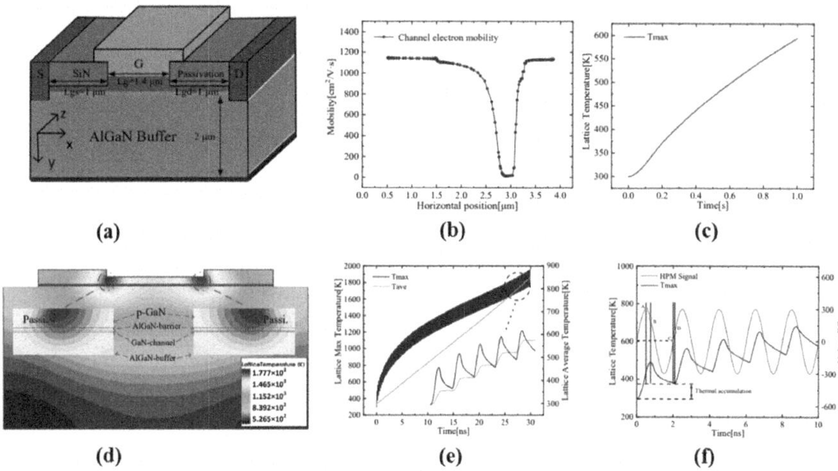

Fig. 2 **a** Device structure **b** Variation of electron mobility **c** Tmax curve with self-heating effect over time **d** Internal thermal distribution under HPM effect **e** Variation of Tmax **f** Tmax and HPM signal change within 10 ns (Qin et al. 2022)

2.3 Key Parameters Influencing Thermal Resistance

AlGaN/GaN high-electron-mobility transistors (HEMTs) are one example of an electronic device whose thermal resistance is influenced by several important factors. For effective thermal management, it is essential to comprehend these parameters. The following three crucial variables have a substantial impact on thermal resistance:

Thermal Conductivity: Thermal conductivity is a fundamental characteristic of materials that describes how well they transport heat. It illustrates how quickly heat can move through a substance. Low thermal resistance is achieved by allowing heat to easily penetrate materials with high thermal conductivity. The overall thermal Resistance of AlGaN/GaN HEMTs is significantly influenced by the thermal conductivity of the substrate, epitaxial layers, and any additional thermal interface materials or heat spreaders. Materials with higher thermal conductivity aid in heat dissipation and lower temperature rise (Chakraborty et al. 2022; Li et al. 2022).

The work of Surajit Chakraborty et al. (Chakraborty et al. 2022) is on an explicit thermal resistance model of AlGaN/GaN HEMT. This model worked for lincar and non-linear thermal conductivity profiles without time delay as compared to other numerical simulations. This work also demonstrated the different modeling equations related to channel temperature and dissipation power, thermal conductivity, and accurate channel temperature. Thermal resistance behavior with different temperatures is also observed with the incorporation of this explicit thermal model. Device structure and different types of experimental results are demonstrated in Fig. 3.

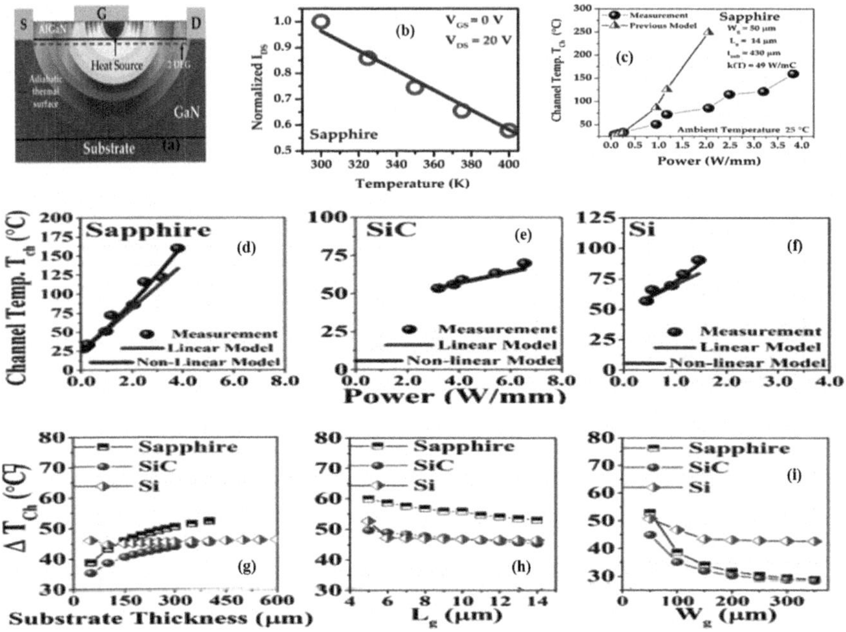

Fig. 3 **a** Cross-section diagram of AlGaN/GaN HEMT **b** Temperature dependence of drain current at high VDS value **c** Channel Temperature versus Power **d**Thermal resistance measurement and the linear and non-linear model of sapphire **e** SiC **f** Si **g** Thermal resistance dependence of substrate thickness **h** gate length **i** gate width (Chakraborty et al. 2022)

Heat Capacity: The quantity of heat energy necessary to raise a material's temperature by a particular amount is known as heat capacity, sometimes referred to as specific heat. It establishes a material's capacity to take in and hold thermal energy. More heat can be absorbed by materials with greater heat capacity before their temperature noticeably rises. The heat capacity of the device layers and surrounding materials impact heat transfer efficiency in AlGaN/GaN HEMTs. By absorbing and storing more heat energy, materials with higher heat capacities can reduce temperature rise (Darwish et al. 2004; Riedel et al. 2008).

Thermal Boundary Resistance: At the boundaries between various materials or layers, thermal boundary resistance develops. It is a representation of the heat transfer resistance at these interfaces and has a substantial effect on the overall thermal resistance. Thermal boundary resistance can develop in AlGaN/GaN HEMTs at the interfaces between the device's active region and the substrate and between the device and the packaging components. To reduce thermal boundary resistance and increase heat transfer efficiency, choosing thermal interface materials carefully and optimizing their interaction quality is crucial (Manoi et al. 2010; García et al. 2016).

It is possible to significantly lower the thermal Resistance in AlGaN/GaN HEMTs by considering and optimizing these parameters. Critical tactics for effective thermal management include raising heat capacity, lowering thermal boundary resistance, and improving the thermal conductivity of materials. These efforts aid in maintaining lower device temperatures, enhancing performance, and improving reliability.

3 Thermal Resistance Measurement Techniques

3.1 Experimental Methods for Thermal Resistance Characterization

Several techniques are frequently used for the experimental characterization of heat resistance in AlGaN/GaN high-electron-mobility transistors (HEMTs). These techniques enable researchers to assess and measure the gadgets' thermal behavior up close. Here are various experimental methods for characterizing thermal resistance in AlGaN/GaN HEMTs:

Electromotive Force Thermometry: This technique entails inserting temperature sensors, like thermocouples or resistance temperature detectors (RTDs), inside the device or close by. The thermal resistance can be calculated by measuring the temperature rise under different operation situations. This approach makes it possible to characterize thermal gradients and hotspots by providing localized temperature readings (Gucmann et al. 2021).

Infrared Thermography: Using infrared cameras, infrared (IR) thermography records the temperature distribution of a surface. Researchers can locate hotter spots and assess the efficiency of heat dissipation by analyzing thermal pictures. A non-contact, in-the-moment visualization of the thermal behavior is provided by IR thermography, allowing for qualitative evaluations of thermal resistance (Killat et al. 2010).

Thermoreflectance Imaging: This method uses variations in light reflectance to map the temperature distribution within a device. Thermal pictures can be obtained using a modulated light source and examining the reflected light. Using the method's high-resolution temperature maps makes it possible to analyze thermal resistance in great detail and locate hotspots (El Helou et al. 2020).

Micro-Raman Spectroscopy: This technique can assess the device's internal phonon modes' temperature-dependent frequency shift. Researchers can determine temperature and evaluate thermal behavior by analyzing the Raman spectra. High spatial resolution localized temperature measurements are provided by this method (Kuball et al. 2002; Baczkowski et al. 2015).

Transient Thermal Analysis: In transient thermal analysis, the device is exposed to a predetermined heat pulse, and the temperature response is tracked over time. Thermal time constants can be calculated using this method, and heat dissipation abilities can be assessed. The transient thermal analysis sheds light on the device's dynamic thermal behavior (Stoffels et al. 2015).

These experimental techniques aid in understanding and characterizing the thermal resistance in AlGaN/GaN HEMTs, together with numerical simulations and analytical models. These methods allow for evaluating the efficacy of thermal management measures such as packaging, heat sinks, and interface materials, leading to better thermal design and improved device performance.

3.2 Overview of Steady-State and Transient Measurement Techniques

Both steady-state and transient measuring approaches are frequently used to characterize AlGaN/GaN high-electron-mobility transistors (HEMTs) to examine their thermal behavior. These methods offer insight into how the devices react thermally under various operating circumstances. An overview of the steady-state and transient measuring methods applied in AlGaN/GaN HEMTs is given below:

Stable-State Measurement Methods: Steady-state measurements supply the device with a constant power or current while monitoring the resulting temperature in an equilibrium environment. Several regularly employed steady-state methods are as follows:

- Electrical Resistance Thermometry: This method embeds temperature sensors within or near the apparatus, such as resistance temperature detectors (RTDs) or thermocouples. The device's temperature is measured as it continues to run at a steady power or current. This offers readings of the steady-state temperature that make it possible to calculate the thermal resistance (Pavlidis et al. 2018).
- Infrared Thermography: Using infrared cameras, infrared (IR) thermography records the temperature distribution of a surface. By examining the thermal images, researchers can assess steady-state temperature profiles and locate hotter regions. The thermal behavior can be seen in real time and without contact, thanks to IR thermography (Zhao et al. 2013).

Techniques for Transient Measurement: Transient measurements entail abruptly changing the device's power or current while tracking the device's temperature response over time. These methods shed light on the device's dynamic thermal behavior. Transient methods that are frequently employed include the following:

- Transient Thermoreflectance: During a transient event, this technique maps the temperature evolution in the device by observing changes in the reflectance of light. Using a modulated light source and examining the reflected light allows for creating high-resolution temporal temperature maps (Maize et al. 2014).
- Transient Electrical Thermometry: In this method, the temperature rise of the device caused by a sudden shift in power or current is measured over time using temperature sensors like RTDs or thermocouples. This makes it possible to characterize thermal time constants and assess the device's capacity to dissipate heat (Cho et al. 2012).
- Transient Raman Spectroscopy: During a transient event, transient Raman spectroscopy tracks the device's temperature-dependent frequency shift of phonon

modes. Researchers can determine temperature and evaluate transient thermal behavior by extracting temperature information from the temporal evolution of the Raman spectra (Zhang et al. 2011).

These complementary steady-state and transient measuring methods offer a thorough understanding of the thermal properties of AlGaN/GaN HEMTs. They support the assessment of hotspot detection techniques, the enhancement of device performance and reliability, and the evaluation of thermal management techniques. These methods enable researchers to enhance the overall thermal management capabilities of AlGaN/GaN HEMTs and optimize thermal design parameters.

3.3 Challenges and Limitations in Accurate Thermal Resistance Determination

Various difficulties and restrictions exist in measuring thermal resistance accurately in AlGaN/GaN high-electron-mobility transistors (HEMTs). Several variables may impact the accuracy and precision of the thermal resistance measurements. Here are some typical issues and restrictions related to precise thermal resistance measurement in AlGaN/GaN HEMTs.

Due to variances in the device's structure, the material's characteristics, and the heat sink design, AlGaN/GaN HEMTs frequently experience non-uniform heat dissipation (Heller and Crespo 2008; Riccio et al. 2011). This non-uniformity makes it harder to measure the thermal resistance precisely. The actual thermal resistance value may need to be clarified due to temperature differences across the device. As AlGaN/GaN HEMTs are continuously scaled down, the size of the active zone shrinks, leading to better power densities (Chu et al. 2004). Due to the temperature monitoring techniques' low spatial resolution and high power density, measuring the temperature rise and thermal resistance is difficult. Smaller device dimensions also make interfacial thermal resistance more significant, which makes it more difficult to calculate total thermal resistance accurately (Guo et al. 2017). Measurements taken in steady-state or quasi-steady-state conditions are frequently used to determine thermal resistance. However, transient effects can strongly impact the thermal behavior of AlGaN/GaN HEMTs. The precise characterization of the thermal resistance is complicated by transient events such as self-heating (Chakraborty et al. 2022), temperature overshoots (Jacquet et al. 2004), and thermal time constants (Yang et al. 2015). Manufacturing processes or material heterogeneity can cause variations in a material's attributes, such as thermal conductivity and heat capacity (Chakraborty et al. 2022). As a result, the assumed material properties may not accurately reflect the actual qualities of the device being tested, which impacts the accuracy of the thermal resistance calculation. When calculating thermal resistance, it's essential to consider how the device's thermal interface is with its surroundings or heat sinks. Uncertainties can be introduced into the observed thermal resistance by imperfect contact, interface layers, and interfacial thermal resistance. It can be difficult to

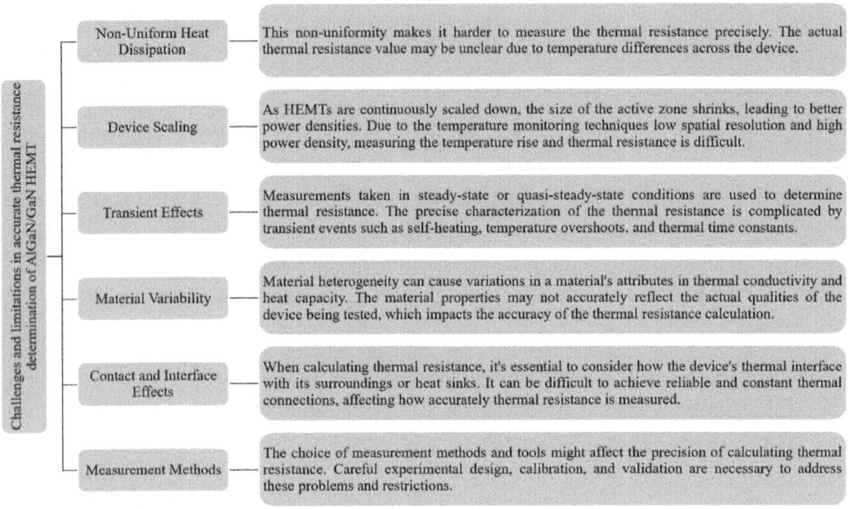

Fig. 4 Challenges and Limitations in accurate thermal resistance determination of AlGaN/GaN HEMT

achieve reliable and constant thermal connections, affecting how accurately thermal resistance is measured. The choice of measurement methods and tools might affect the precision of calculating thermal resistance. Each method's drawbacks include accuracy, sensitivity, and spatial resolution, which might cause inaccurate results.

Careful experimental design, calibration, and validation are necessary to address these problems and restrictions. To increase the precision of the thermal resistance measurement in AlGaN/GaN HEMTs, it is essential to consider the unique device characteristics, reduce measurement errors, and apply thorough modeling and simulation approaches. The above-discussed challenges and limitations in accurate thermal resistance determination of AlGaN/GaN HEMT are shown in Fig. 4.

4 Analytical Models for Thermal Resistance

4.1 Overview of Analytical Models for Thermal Resistance Estimation

Analytical models mostly predict the thermal Resistance of AlGaN/GaN high-electron-mobility transistors (HEMTs). These models enable quick evaluations and preliminary design exploration by offering simplified yet correct predictions of the device's thermal behavior. Here is a summary of various popular analytical models for estimating thermal resistance in AlGaN/GaN HEMTs (Jia et al. 2018; Chakraborty et al. 2022; Darwish et al. 2004).

The One-Dimensional (1D) Heat Conduction Model: The 1D heat conduction model assumes that heat flows in just one dimension, simplifying the thermal behavior of the apparatus. This model considers the thermal resistances connected to various layers, such as the substrate, buffer, and epitaxial layers, but ignores the effects of lateral spreading. The separate resistances are added up to determine the overall thermal resistance. While this model is relatively simple, it quickly estimates the overall thermal resistance (Azarifar and Donmezer 2016, 2017).

Cauer Network Model: The Cauer network model uses an electrical analogy to depict how thermal resistance is distributed across the device. Different layers' thermal resistances are viewed as resistors connected in series and parallel, much like an electrical circuit. Estimating the overall thermal resistance by resolving the network equations is possible. By considering the consequences of lateral heat dispersion, the Cauer network model produces more precise results (Zhang et al. 2011; Baczkowski et al. 2015).

Semi-Empirical Models: Semi-empirical models blend analytical equations with empirical data to calculate thermal resistance. These models consider the material's temperature-dependent thermal conductivity and elements like device geometry and heat transfer processes. Usually, experimental data fitting is used to determine the equations' parameters. Semi-empirical models provide a good compromise between accuracy and computing efficiency(Khandelwal et al. 2012; Liu et al. 2019).

Compact Models: Compact models thoroughly represent the gadget, including its mechanical, thermal, and electrical properties. These models combine analytical equations, numerical techniques, and actual data to calculate thermal resistance; these models combine analytical equations, numerical techniques, and actual data. Compact models offer more accurate and detailed predictions but require more accurate and detailed ones that necessitate more complicated calculations (Jia et al. 2018; Wu et al. 2017).

The research work of Shuman Mao et al. (2018) investigated the impact of the I–V kink effect on the large signal capabilities of AlGaN/GaN high-electron-mobility transistors (HEMTs). To accurately depict the I–V kink effect, an additional accurate compact model was presented. The I–V kink effect, according to the results, will describe output power, gain, and power-added efficiency to reduce at the saturation zone. This model describes the characterization of the I–V Kink effect. Different types of investigations and their simulation results like Validation of the Large Signal Model, Influence of I–V Kink Effect with different Input Power, and Influence of I–V Kink Effect under different Bias Points are observed and demonstrated in Fig. 5.

It's crucial to remember in mind that analytical models only offer estimations and simplifications, and their correctness can change based on the precise device structure and operating circumstances. They are valuable resources for preliminary design investigation, fast evaluations, and as a roadmap for additional examination. Numerical simulations or experimental observations are often needed to make more accurate predictions. However, analytical models are indispensable for estimating the thermal resistance of AlGaN/GaN HEMTs and supporting the creation of effective thermal management techniques.

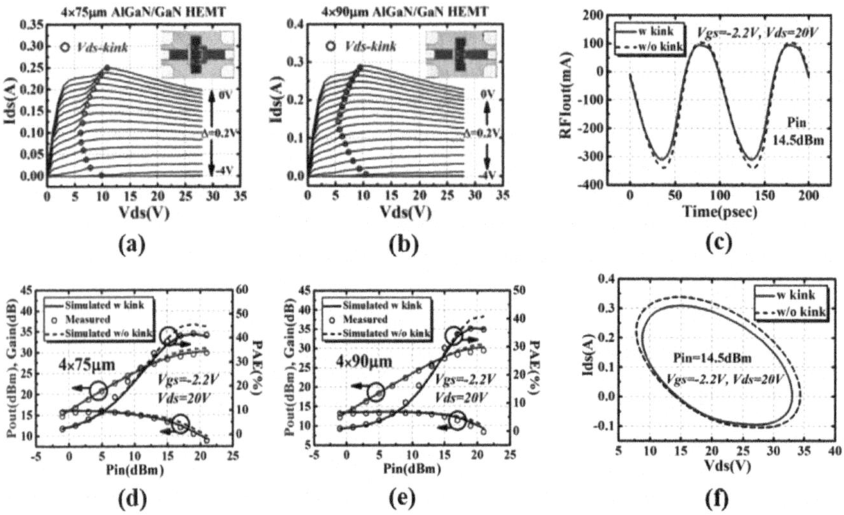

Fig. 5 **a** I–V kink effect in 0.25 μm AlGaN/GaN HEMTs with 4 × 75 μm gate width **b** 4 × 90 μm gate width **c** Time-domain waveform **d** I–V kink effect on large signal performance 4 × 75 μm gate width **e** 4 × 90 μm gate width **f** AC dynamic load line (Mao and Xu 2018)

4.2 Discussion of Critical Assumptions and Simplifications

Certain crucial assumptions and simplifications are used to analyze and model AlGaN/GaN high-electron-mobility transistors (HEMTs) to make understanding and predicting device behavior easier. Although these presumptions and simplifications speed up the analytical process, they might need to fully account for how complex the device operates. Here, we go through some crucial suppositions and simplifications that are frequently used in AlGaN/GaN HEMTs:

Two-Dimensional (2D) Modeling: Consideration of a two-dimensional model, where the device behavior is considered to vary solely in the device's plane, is a typical assumption in AlGaN/GaN HEMTs. This simplification overlooks the lateral changes and three-dimensional structural influences that may affect the calculation of thermal resistance and heat dissipation. However, the 2D assumption lowers the computing complexity and offers a good approximation in many situations (Azarifar and Donmezer 2017; Asgari et al. 2005).

Homogeneous Material Properties: Analytical and numerical models frequently presuppose that the device's materials are homogeneous throughout. Any spatial changes or non-uniformity in a material's attributes, such as thermal conductivity and heat capacity, are ignored by this assumption. In practice, composition gradients, flaws, or development processes cause material characteristics to fluctuate throughout the device. Estimates of thermal resistance may be more precise if regional differences in material characteristics are considered (Gamarra et al. 2013; Kakanakova-Georgieva et al. 2007).

Ignoring Phonon Transport: In solid-state devices, phonon transport, which involves the movement of lattice vibrations, is essential for heat conduction. However, some models ignore or oversimplify the phonon transport effects. This presumption is based on the idea that heat dissipation in AlGaN/GaN HEMTs is dominated by electron transport. Although this simplification is sometimes justified, ignoring phonon transmission might reduce the precision of thermal resistance predictions, particularly at lower temperatures or in devices made of particular materials (Chatterjee et al. 2020; Hao et al. 2017).

Steady-State Analysis: Many estimates of thermal resistance are made under steady-state circumstances, assuming a continuous supply of power or current. The initial warm-up phase or dynamic changes in heat dissipation during device operation are left out by this simplification of the situation. Transient effects can significantly impact thermal resistance, particularly in devices with high power densities or under rapidly changing operating conditions. Incorporating transient analysis can give you a more in-depth insight into how the device will behave thermally (Russo et al. 2012; Sridharan et al. 2008).

Ideal Interfaces: A few models rely on flawless thermal interfaces between the various layers of the device or between the gadget and its container (Zhang et al. 2000).

This presumption ignores any potential interfacial heat resistance brought on by surface abrasion, oxide deposits, or other interface flaws. Interfacial thermal resistance can obstruct heat transmission and influence the device's overall thermal resistance. The accuracy of thermal resistance estimation can be increased by considering more realistic contact conditions. When evaluating the findings of thermal resistance models in AlGaN/GaN HEMTs, it's critical to be aware of these presumptions and simplifications. While these simplifications make analysis and calculation quicker, they could also limit the models' accuracy and scope of use. As a result, confirming the findings with experimental data and considering more intricate models or simulations can offer a more thorough knowledge of the thermal behavior of AlGaN/GaN HEMTs.

5 Numerical Simulation Techniques

Numerical simulation techniques are advantageous for modeling heat resistance in AlGaN/GaN HEMTs because they can consider complex geometries, material changes, and boundary conditions. They explain the devices' temperature distributions, heat transfer processes, and thermal resistance features. To get accurate findings, it is necessary to ensure correct modeling of device architectures, suitable boundary conditions, and precise material properties. Combined with experimental validation, numerical simulations contribute substantially to the comprehension and optimization of thermal management strategies in AlGaN/GaN HEMTs.

We introduce numerical simulation techniques for simulating thermal resistance in AlGaN/GaN HEMTs in this context:

FEA: Finite Element Analysis FEA is a numerical technique that divides the device into tiny, connected pieces to tackle complex engineering challenges. For AlGaN/GaN HEMTs, FEA can thoroughly understand the temperature distribution inside the component, enabling precise thermal resistance calculation. To solve the heat conduction equation for each element, FEA considers the material qualities, heat sources, and boundary conditions. This makes it possible to calculate the device's internal temperature gradients, heat flow patterns, and localized hotspots. For a thorough examination of thermal behavior, FEA is especially helpful for analyzing complex device geometries, non-uniform material characteristics, and heterogeneous architectures (Jones et al. 2015; Heller et al. 2011).

Huaixin Guo et al. (2020) presented a methodology shown in Fig. 6 that demonstrated the three-dimensional simulation with finite element method of the transient thermal response of an AlGaN/GaN high-electron-mobility transistor (HEMT) used in high-power amplifiers. In this work, Different types of simulation results of Transient Channel Temperature in Pulsed Operation, The Pulse Width Effect, The Gate Structures Effect, and The Power Density Effect are demonstrated in the AlGaN/GaN HEMT.

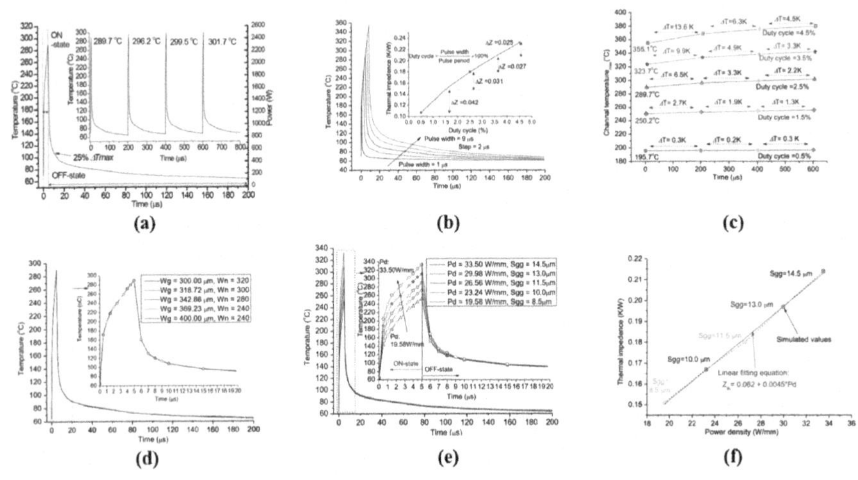

Fig. 6 **a** Transient channel temperature and periodic transient pulse temperature versus Time **b** Dependence of transient channel temperature versus Time and thermal impedance versus Duty Cycle **c** Impact of pulse width on maximal channel temperature during **d** Relationships between transient channel temperature **e** Relationships between transient channel temperature and the changes of power density and gate pitch spacing **f** The impact of power density on the thermal impedance of device (Guo et al. 2020)

CFD: Computational fluid dynamics CFD is a method used to study heat transfer and fluid flow. CFD is used in AlGaN/GaN HEMTs to examine convective heat transmission from the device to the environment. CFD simulations consider fluid flow patterns, heat convection, and temperature distributions to estimate the thermal resistance precisely. CFD offers helpful insights into the convective heat dissipation and cooling mechanisms by considering the suitable boundary conditions, fluid characteristics, and device shape. This knowledge aids in the improvement of the device's thermal management techniques, such as the layout of heat sinks or cooling systems (Chen et al. 2014; Vanko et al. 2011).

To thoroughly investigate the thermal behavior of AlGaN/GaN HEMTs, FEA and CFD are frequently used in conjunction. Compared to CFD, which permits the investigation of convective heat flow from the device surface, FEA gives detailed temperature distribution throughout the device. This integrated method makes it possible to identify any thermal problems affecting the device's performance and reliability and estimate thermal resistance more precisely. The optimization of device architectures, materials, and cooling techniques can be guided by the findings from FEA and CFD simulations, improving thermal management and enhancing device performance. It is crucial to remember that accurate input parameters, such as material properties, boundary conditions, and device geometries, are necessary for FEA and CFD calculations. To guarantee the precision and dependability of the predictions, it is essential to validate the simulation results using experimental measurements. Engineers and researchers may better understand and optimize the thermal behavior of these devices under various operating situations thanks to the inclusion of FEA and CFD methodologies in the analysis of AlGaN/GaN HEMTs.

FDM: Finite Difference Method: By discretizing the device into a grid of points, the finite difference method approximates the heat conduction equation. Based on the heat conduction equation and the temperature values of nearby sites, it determines the temperature at each location. FDM offers a clear-cut and somewhat easy method for resolving heat transfer issues. FDM may be used to analyze the temperature distribution inside the device and calculate the thermal resistance in AlGaN/GaN HEMTs. FDM simulations make it possible to assess temperature gradients and heat flow patterns by considering the material's qualities, heat sources, and boundary conditions. FDM can shed light on the overall thermal behavior of devices and is particularly helpful for analyzing those with simpler shapes (Kandasamy and NB 2019; Ludwig et al. 2019).

Monte Carlo Method: The Monte Carlo Method is a statistical simulation method that uses statistical analysis and random sampling to estimate thermal behavior. The Monte Carlo Method can be applied to AlGaN/GaN HEMTs to trace the interactions and energy transfers between phonons and electrons as they move through the device. Numerous random simulations are run to provide statistical averages that reveal data on temperature distributions, thermal resistance, and other thermal properties. The study of nanoscale devices and materials, where stochastic effects are significant, benefits significantly from the Monte Carlo Method. It makes it possible to analyze intricate heat transfer processes and accurately depict thermal behavior's statistical basis (Yamakawa et al. 2004; Linkohr et al. 2012).

When simulating the thermal behavior of AlGaN/GaN HEMTs, the Finite Difference Method and the Monte Carlo Method have advantages and disadvantages. The Finite Difference Method provides an easy-to-understand methodology ideal for quickly investigating temperature distributions in devices with uncomplicated geometries. However, the intrinsic stochastic nature of heat transmission in nanoscale devices can be captured statistically using the Monte Carlo Method. It enables a thorough examination of intricate heat transfer events but can be computationally demanding. It is significant to highlight that both methods require precise input parameters, such as material properties, boundary conditions, and device geometries, to produce findings that can be trusted. Furthermore, experimental validation is necessary to guarantee the precision and dependability of the simulation results. When used in conjunction with other computational and experimental techniques, the Finite Difference Method and the Monte Carlo Method help to provide a thorough knowledge of the thermal behavior in AlGaN/GaN HEMTs.

6 Material Properties and Characterization

6.1 Overview of Relevant Material Properties Influencing Thermal Resistance

Several material characteristics significantly influence the thermal resistance of AlGaN/GaN high-electron-mobility transistors (HEMTs). These characteristics substantially impact the device's thermal behavior and internal heat transmission processes. The following is a summary of the pertinent material characteristics that affect the thermal resistance in AlGaN/GaN HEMTs:

Temperature Conductivity: A material's capacity to conduct heat is characterized by its thermal conductivity, which is a fundamental attribute. The thermal conductivity of components like AlGaN, GaN, and substrates impacts how well heat is transferred across the device in AlGaN/GaN HEMTs. More excellent thermal conductivity makes better heat dissipation and lower thermal resistance possible. However, defects, contaminants, or interfaces can lower thermal conductivity and raise thermal resistance (Mitterhuber et al. 2020; Vitanov et al. 2010).

Heat Capacity: The quantity of heat energy needed to raise the temperature of a substance by a given amount is referred to as heat capacity, also known as specific heat. The ability of various materials in the device stack to store and release heat is influenced by their heat capacity, including the substrate, buffer layer, and semiconductor layers. Materials with greater heat capacity are better able to absorb and disperse heat, which lowers temperature rise and thermal resistance (Chatterjee et al. 2020; Nigam et al. 2017).

Thermal Expansion Coefficient: The thermal expansion coefficient describes how a material's dimensions alter due to temperature changes. Materials with various thermal expansion coefficients are frequently layered together in AlGaN/GaN HEMTs, which causes mechanical stress and strain during thermal cycling. This stress may affect the device's thermal resistance by altering the thermal contact and causing deformations that change the heat transfer pathways (Fletcher et al. 2021; Cordier et al. 2008).

Interface and Contact Resistance: Interfaces and contacts may increase thermal resistance within the device. The thermal contact resistance at material surfaces impacts heat transfer effectiveness between several layers. Increased thermal resistance and decreased heat dissipation efficiency are the results of higher contact resistance. Optimizing the thermal performance of AlGaN/GaN HEMTs requires understanding and reducing interface and contact resistance (Chen et al. 2019, 2015).

Thermal Boundary Resistance: In a device, thermal boundary resistance appears at the intersections of various materials or layers. It develops due to mismatches in phonon transmission and phonon scattering between materials. The transfer of heat across surfaces is impeded by higher thermal boundary resistance, which raises thermal resistance. Enhancing material selections and interface engineering strategies can decrease thermal boundary resistance (García et al. 2016; Sarua et al. 2007).

It is significant to note that combining these material characteristics affects the thermal resistance of AlGaN/GaN HEMTs. Thermal performance can be significantly enhanced using materials to design devices with good thermal conductivity, the right heat capacity, suitable thermal expansion coefficients, and low interface and contact resistances. It is crucial to accurately characterize and comprehend these material features to improve the reliability and performance of AlGaN/GaN HEMTs and create effective thermal management strategies.

6.2 Thermal Conductivity Measurement Techniques for GaN-based Materials

Understanding the thermal characteristics of these devices depends on the measurement of thermal conductivity in materials based on GaN, which are frequently utilized in AlGaN/GaN HEMTs. The thermal conductivity of GaN-based materials is measured using various techniques, providing essential details about their heat transfer properties. Here, we give a summary of the typical measuring methods for thermal conductivity in materials based on GaN:

Laser Flash Analysis (LFA): (LFA) Laser Flash Analysis LFA is a method frequently used to assess the thermal conductivity of solid materials, especially those based on GaN. In LFA, a quick laser pulse is fired into the sample's surface, instantly heating it. A sensor, usually a thermocouple, measures the temperature increase that results. The thermal diffusivity, obtained by examining the temperature rise and the material's dimensions, combines density and specific heat information to get the thermal conductivity (Faqir et al. 2011; Schlenker 2018).

Transient Thermoreflectance (TTR): TTR is a non-contact optical method for determining the thermal conductivity of bulk materials and thin films. In TTR, a small spot on the sample's surface is heated using a pulsed laser, and the ensuing temperature change is gauged using a probe laser. By examining the time-dependent thermal response, the temperature-dependent reflectivity of the material is used to calculate the thermal conductivity (Maize et al. 2014; Zhou et al. 2017).

Three-Technique: The three-technique is frequently used to assess the thermal conductivity of multilayer structures and thin films, particularly GaN. The voltage response at the third harmonic frequency (3) is then measured after applying an alternating current (AC) to a thin metallic layer coated on the sample's surface. The voltage signal's amplitude and phase can be examined to estimate the thermal conductivity (Mion et al. 2006; Killat et al. 2012).

Time-Domain Thermoreflectance (TDTR): The thermal conductivity of thin films and nanoscale structures is measured using TDTR, a pump-probe technique. Two pulsed lasers are employed in TDTR; one is used to heat the sample, and the other is utilized to measure the temperature response. The temperature response's temporal evolution is examined to determine the thermal conductivity. When describing materials with limited thermal conductivity or intricate layered structures, TDTR is especially helpful (Mitterhuber et al. 2020; Siddique et al. 2019).

Scanning Thermal Microscopy (SThM): The nanoscale thermal conductivity can be measured using SThM, a scanning probe microscopy technique. It entails scanning a heated probe tip over the sample's surface while keeping an eye on the heat transmission. The thermal conductivity can be calculated by comparing the thermal resistance of the probe–sample interface with that of a reference material (Raj and Bindra 2013; Fotsing et al. 2005).

These testing methods provide insightful data on the thermal conductivity of GaN-based materials, which is crucial knowledge for AlGaN/GaN HEMT thermal management strategies. To achieve accurate and trustworthy results, it is crucial to consider aspects like sample preparation, calibration, and any potential impacts of anisotropy or flaws on the measurements. These methods can enhance the thermal performance of AlGaN/GaN HEMTs and provide researchers with a better understanding of the thermal characteristics of GaN-based materials.

7 Design Considerations for Thermal Management

The design consideration for the thermal management of AlGaN/GaN HEMT and its subsections with different approaches are shown in Fig. 7.

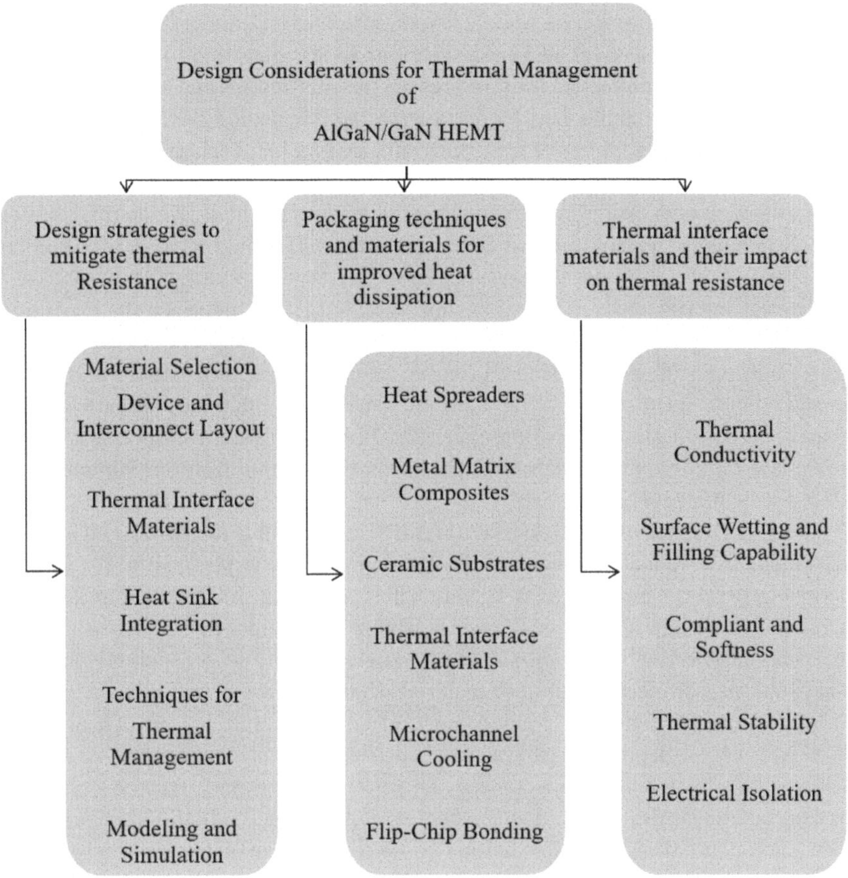

Fig. 7 Design considerations for Thermal management of AlGaN/GaN HEMT

7.1 Review of Design Strategies to Mitigate Thermal Resistance in AlGaN/GaN HEMTs

AlGaN/GaN high-electron-mobility transistors (HEMTs) thermal performance and overall device reliability must be improved through design tactics to reduce thermal resistance. Several design approaches have been used to overcome thermal resistance in AlGaN/GaN HEMTs.

Material Selection: One of the primary methods for reducing thermal resistance is carefully choosing materials with good thermal conductivity. By selecting materials such as superior GaN substrates and epitaxial layers with improved thermal conductivity, heat can be dissipated effectively, lowering the overall thermal resistance. Improved thermal management in AlGaN/GaN HEMTs can be achieved by optimizing the composition of AlGaN layers, which can also increase their thermal conductivity (Mounika et al. 2022; Reddy et al. 2018).

Device and Interconnect Layout: Improving heat dissipation and lowering thermal resistance can be accomplished by optimizing the device and connection arrangement. AlGaN/GaN HEMTs' thermal performance can be enhanced by minimizing device size, narrowing the space between devices, and strategically placing heat-generating elements. Adopting effective heat-spreading methods like thermal vias or metal heat sinks can also decrease thermal resistance (Stoffels et al. 2014; Lambert et al. 2012).

Thermal Interface Materials: Choosing suitable thermal interface materials and using them effectively are essential for reducing thermal resistance at material surfaces. High-performance TIMs that promote thermal conduction across various layers and components and reduce the thermal resistance at interfaces include thermal greases, pads, and phase transition materials. The overall heat transport within the device can be improved by careful TIM selection and optimization (Narang et al. 2022; Ejeckam et aol. 2014).

Heat Sink Integration: Heat dissipation can be considerably improved, and thermal resistance can be reduced by including heat sinks in the device packaging. To promote thermal conduction, heat sinks can be built with features like fins or heat pipes, which increase their surface area for heat transfer. Thermal resistance at the device-package interface must be minimized by effectively connecting the device and the heat sink using soldering or thermal interface materials (Sang 2022; Chernykh et al. 2022).

Techniques for Thermal Management: Thermal Resistance in AlGaN/GaN HEMTs can be effectively reduced using active thermal management strategies. Device performance can be enhanced, and thermal resistance can be decreased by thermal annealing, which involves heating components to high temperatures to lower fault density and increase thermal conductivity. Heat dissipation and thermal resistance-reducing on-chip or off-chip cooling techniques such as microchannel, liquid, or thermoelectric can be beneficial (Mohanty et al. 2019; Minoura et al. 2020).

Modeling and Simulation: Simulation modeling the design of devices can be improved, and thermal behavior can be predicted using sophisticated thermal modeling and simulation techniques. Simulations using computational fluid dynamics (CFD) and finite element analysis (FEA) can reveal information about temperature distributions, hotspots, and the effects of different design changes on thermal resistance. This enables the determination of the best heat management techniques and iterative design enhancements (Vitanov et al. 2010; Belkacemi and Hocine 2018).

Researchers and engineers can successfully reduce thermal resistance in AlGaN/GaN HEMTs by applying these design ideas. These devices can be used in various high-power electronic applications by improving their thermal management because doing so increases their reliability, power handling capacity, and overall performance.

7.2 *Packaging Techniques and Materials for Improved Heat Dissipation*

AlGaN/GaN high-electron-mobility transistors (HEMTs) benefit significantly from improved heat dissipation and thermal resistance management due to packaging strategies and materials. The thermal performance of these devices can be improved using efficient packaging techniques, ensuring their dependable operation at high power levels. Different packaging methods and materials used to enhance heat dissipation in AlGaN/GaN HEMTs are as follows:

Heat Spreaders: Heat spreaders are frequently employed in AlGaN/GaN HEMT packaging to improve thermal conductivity and effectively distribute heat. These heat spreaders are in direct contact with the device to help spread heat over a greater surface area. They are often composed of materials with excellent thermal conductivity, like copper (Cu) or diamond. Heat spreaders efficiently move heat away from the device's active areas, improving heat dissipation and lowering thermal resistance (Zhou et al. 2017; Sang 2022).

For high-power electronic applications, L. Arivazhagan et al. (Arivazhagan et al. 2021) proposed a GaN HEMT on a Si substrate with a diamond heat spreader. In order to mitigate the self-heating effect, a GaN-on-Si HEMT with a diamond (Dia) heat spreader is proposed. GaN-on-Si with a diamond heat spreader prevents the device from excessive heating and provides a higher saturation drain current than GaN-on-Si without one. Furthermore, compared to GaN-on-Si, GaN-on-Si with a Diamond heat spreader generates a greater transconductance and cut-off frequency. With this improved structure, a low-cost device with improved thermal properties will be conceivable for higher-powered applications. This work also proposed a Theoretical model of heat conduction and also generates different thermal analysis simulation results. Different types of thermal analysis and AC characteristics are shown in Fig. 8.

Metal Matrix Composites (MMCs): MMCs are composite materials that combine ceramic particles such as silicon carbide (SiC) or aluminum oxide (Al2O3) with a metal matrix such as aluminum (Al) or copper. MMCs are excellent for thermal management in AlGaN/GaN HEMTs because they have more mechanical strength and thermal conductivity than pure metals. These materials can be heat spreaders or packaging substrates to improve heat dissipation and lower thermal resistance (Faqir et al. 2012; Guggenheim and Rodes 2017).

Ceramic Substrates: Because of their superior thermal conductivity, ceramic substrates like aluminum nitride (AlN) or beryllium oxide (BeO) are frequently utilized in AlGaN/GaN HEMT packaging. These substrates offer a surface for mounting and dispersing heat from thermally efficient devices. Additionally, ceramic substrates have strong electrical insulating qualities that provide correct electrical separation and reduce the chance of short circuits (Zehetner et al. 2016; Kisiel et al. 2021).

Thermal Interface Materials (TIMs): Thermal Interface Materials are essential for maximizing heat transmission at interfaces between various package components. To improve thermal conduction, these materials are positioned between the device

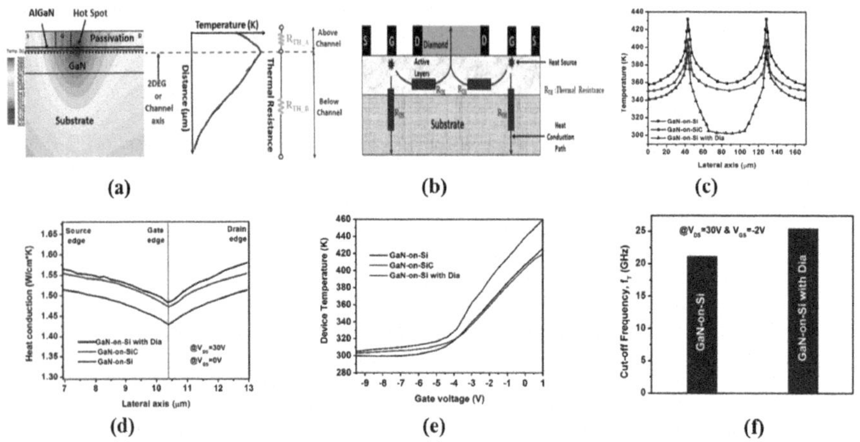

Fig. 8 **a** Device structure **b** Drain current versus the gate voltage characteristics **c** Transconductance characteristics **d** Drain current versus the drain voltage (with Vds swept in positive and negative directions) **e** The first-order derivative of the drain current with respect to the drain voltage **f** Second-order derivative of the drain current with respect to the drain voltage. Reprinted from Ref. (Arivazhagan et al. 2021), L. Arivazhagan et al., "GaN HEMT on Si substrate with diamond heat spreader for high power applications," J Comput Electron 20, 873–882 (2021), https://doi.org/10.1007/s10825-020-01646-8. Copyright (2021), with permission from Springer

and the heat sink or between layers of packaging. Thermal greases, thermal pads, and phase change materials are examples of frequently used TIMs. TIMs reduce interface heat resistance by bridging gaps, removing air holes, and enhancing thermal contact between surfaces (Mitterhuber et al. 2020; Narang et al. 2022).

Microchannel Cooling: A cutting-edge packaging method, microchannel cooling uses microfluidic or microfluidic channels to improve heat dissipation. In this method, the AlGaN/GaN HEMT device is directly underneath a coolant, such as liquid or gas, which efficiently dissipates heat. High heat transfer coefficients and precise temperature control provided by microchannel cooling lead to improved thermal management and decreased thermal resistance (Pu et al. 2016; Blevins et al. 2016).

Flip-Chip Bonding: Using conductive adhesives or solder bumps, flip-chip bonding is a packaging technology that directly bonds the AlGaN/GaN HEMT device to the substrate or package. With a more direct thermal channel between the device and substrate made possible by this method, heat may dissipate effectively and decrease thermal resistance. Improved electrical connections are also provided through flip-chip bonding, which reduces parasitic inductance and capacitance (Das et al. 2006; Wu et al. 1998).

AlGaN/GaN HEMTs can achieve enhanced heat dissipation and decreased thermal resistance using these packaging methods and materials with high thermal conductivity. These devices can run at higher power levels while still delivering dependable performance thanks to effective heat dissipation, improved thermal interfaces, and novel cooling techniques. The total thermal management of AlGaN/GaN HEMTs is improved by ongoing developments in packaging methods and materials, fueling the creation of next-generation high-power electronics.

7.3 Thermal Interface Materials and Their Impact on Thermal Resistance

AlGaN/GaN high-electron-mobility transistors (HEMTs) require the effective management of thermal resistance, which is achieved by using thermal interface materials (TIMs). These substances enhance heat transmission at interfaces within the device package between various components, such as between the HEMT device and the heat sink or between various packaging layers. AlGaN/GaN HEMTs' thermal performance can be significantly improved, and their thermal resistance can be decreased with the appropriate selection and application of TIMs. Here, we go through the significance of TIMs and how they affect these devices' thermal resistance:

Thermal Conductivity: First, thermal conductivity Enhancing thermal conduction at surfaces is a primary function of TIMs. Effective heat transfer between the device and the heat sink or other packaging layers is made possible by TIMs with high thermal conductivity, such as thermal greases, thermal pads, or phase transition materials. Substantial thermal conductivity TIMs reduce thermal resistance and improve heat dissipation by filling in cracks and defects between mating surfaces (Mitterhuber et al. 2020; Mion et al. 2006).

Surface Wetting and Filling Capability: TIMs must have good wetting capabilities for Surface Wetting and Filling to cling to the mating surfaces effectively. TIMs with good wetting properties can fill minute air gaps or spaces by conforming to surface imperfections and roughness. There is less thermal resistance with better thermal contact, lower interfacial air resistance, and improved heat transmission (Zhu et al. 2020; Pal et al. 2022).

Compliant and Softness: To account for mechanical changes and mismatches between various components, TIMs should have a particular level of compliance and softness. Due to the TIM's ability to adapt to surface flaws, thanks to its compliance, contact is improved, and thermal resistance is decreased. The device is less likely to be stressed or damaged due to thermal cycling or temperature changes thanks to soft and compliant TIMs (Motala et al. 2020; Zhang et al. 2022).

Thermal Stability: TIMs should have solid thermal stability throughout the AlGaN/GaN HEMTs' operational temperature range. They should not significantly deteriorate or undergo phase changes while retaining their thermal qualities, such as conductivity and interface integrity. The device's thermal performance is consistent and reliable over the long term (Hatano et al. 2014; Arulkumaran et al. 2016).

Electrical Isolation: TIMs should have good electrical insulating qualities in AlGaN/GaN HEMTs, where electrical isolation is essential. The gadget and the heat sink or other conducting elements in the package shouldn't be able to have any electrical short circuits or leakage currents. The risk of destroying the gadget is eliminated by proper electrical separation, which also preserves the device's electrical integrity (Kim et al. 2017; Werquin et al. 2005).

The application requirements, operating temperatures, interface materials, and requirements for thermal conductivity are only a few of the variables that affect the choice of TIMs. To achieve the necessary thermal performance, it is critical to carefully choose TIMs compatible with the materials and surfaces involved and optimize their thickness and application techniques. The best materials for lowering thermal resistance in AlGaN/GaN HEMTs can be found by experimental characterization and testing several TIMs. In conclusion, AlGaN/GaN HEMTs' thermal resistance is strongly affected by the appropriate choice and use of high-performance TIMs. TIMs help AlGaN/GaN HEMTs transport heat effectively, reduce interfacial resistance, and improve overall thermal management by promoting thermal conductivity, wetting and filling capabilities, compliance, assuring thermal stability, and providing electrical isolation.

8 Advanced Approaches and Future Trends

The advanced approaches and future trends of AlGaN/GaN HEMT with thermal resistance modeling are shown in Fig. 9.

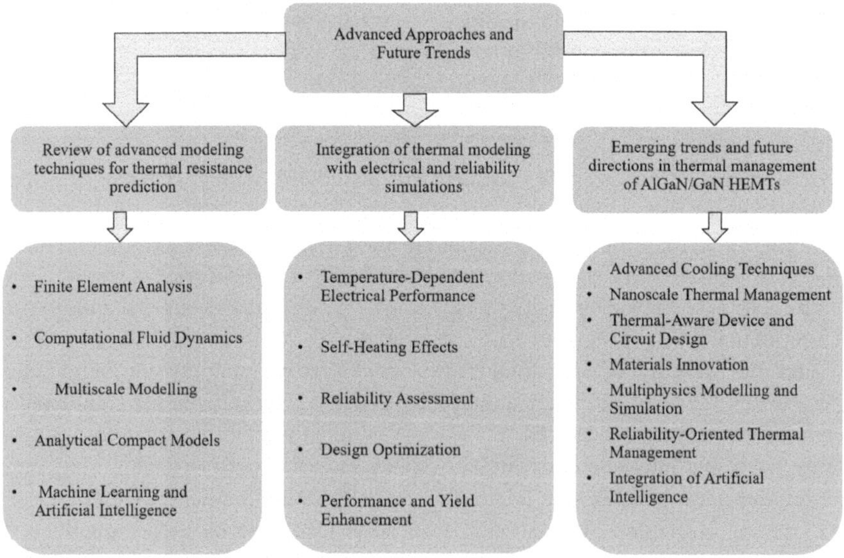

Fig. 9 Advanced approaches and future trends for thermal modeling of AlGaN/GaN HEMT

8.1 *Review of Advanced Modeling Techniques for Thermal Resistance Prediction*

Advanced modeling techniques have become practical tools for forecasting and analyzing heat resistance in AlGaN/GaN high-electron-mobility transistors (HEMTs). These modeling methods help design and optimize the devices' thermal management strategies by providing valuable insights into their thermal behavior. In this literature, we go through a few of the cutting-edge modeling approaches used to forecast thermal resistance in AlGaN/GaN HEMTs:

Finite Element Analysis (FEA) is a numerical modeling method frequently used to simulate and examine the thermal behavior of complicated structures, such as AlGaN/GaN HEMTs. FEA breaks the device into small finite elements to estimate the temperature distribution and thermal resistance and solves the heat transfer equations. The exact forecast of thermal performance is made possible by FEA's detailed modeling of various parameters, including device shape, material properties, and boundary conditions (Jones et al. 2015; Heller et al. 2011).

Computational Fluid Dynamics (CFD): CFD is a modeling approach for researching fluid flow and heat transfer in complex systems. CFD can simulate the fluid flow and heat transfer within the device packaging or cooling systems in AlGaN/GaN HEMTs. It shows how various cooling methods, like liquid or microchannel, affect thermal resistance. CFD simulations aid in optimizing cooling plans and detecting possible hotspots within the apparatus (Chen et al. 2014; Vanko et al. 2011).

Multiscale Modeling: Multiscale modeling incorporates several modeling approaches to accurately represent the thermal behavior of AlGaN/GaN HEMTs at various length scales. To account for events happening at various scales entails merging models at the macroscopic level (such as FEA or CFD) with models at the microscopic level (such as molecular dynamics simulations). Multiscale modeling, which considers microstructural impacts, material surfaces, and heat transmission mechanisms at various length scales, provides a thorough knowledge of thermal resistance (Heller et al. 2013).

Analytical Compact Models: Compact analytical models: The heat resistance of AlGaN/GaN HEMTs can be estimated using compact analytical models, which offer streamlined mathematical formulations or empirical connections. To anticipate thermal behavior, these models consider important factors, including device geometry, material properties, and operating circumstances. Analytical compact models are helpful for initial design exploration and parameter sensitivity research since they are computationally fast and enable quick estimates of thermal resistance (Jia et al. 2018).

Machine Learning and Artificial Intelligence: Machine Learning and Artificial Intelligence Techniques in AlGaN/GaN HEMTs have shown promise in forecasting and optimizing thermal resistance. With the help of massive datasets and algorithmic models, these techniques can find intricate connections between input factors and thermal behavior. To improve device design and thermal management tactics, machine learning algorithms can learn from historical data and produce precise predictions of thermal resistance (Wu and Kutub 2020; Husain et al. 2020).

The ability to estimate thermal resistance in AlGaN/GaN HEMTs using advanced modeling tools provides substantial advantages that allow researchers and engineers to investigate design alternatives, optimize thermal management techniques, and increase device reliability. These modeling techniques offer insightful information on the thermal behavior of the devices, assisting in the creation of effective cooling plans and improving overall device performance. The development of thermal management in AlGaN/GaN HEMTs will continue to be aided by improvements in modeling approaches.

8.2 Integration of Thermal Modeling with Electrical and Reliability Simulations

Integrating thermal modeling with electrical and reliability simulations has become necessary in researching and designing AlGaN/GaN high-electron-mobility transistors (HEMTs). This integration makes a thorough understanding of device performance possible, including how thermal behavior, electrical properties, and long-term dependability interact. By combining these models, engineers and researchers can learn much about how heat impacts affect device function and reliability. In this section, we go through the importance of combining electrical and reliability simulations with thermal modeling in AlGaN/GaN HEMTs:

Temperature-Dependent Electrical Performance: AlGaN/GaN HEMTs are significantly influenced by temperature in terms of their electrical behavior. Thermal and electrical simulations can be combined to analyze the device's performance as a function of temperature. Electrical simulations can use the temperature distribution provided by thermal modeling as input. This integrated method enables a more precise evaluation of electrical parameters under various operating settings, including current–voltage characteristics, threshold voltage shift, and output power (Jeon et al. 2021).

Self-Heating Effects: Due to their high power dissipation and low thermal conductivity, AlGaN/GaN HEMTs may experience self-heating effects. These consequences may impact device performance and dependability. Combining thermal modeling with electrical simulations can assess the degree of self-heating and its effect on the device's electrical properties. This makes it possible to identify any potential thermally induced performance decline and to analyze the device's operational limits more accurately (Nigam et al. 2017).

Reliability Assessment: Thermal stresses from temperature gradients can impact how long-lastingly reliable AlGaN/GaN HEMTs are. Analysis of the effects of temperature cycling, thermal cycling fatigue, and thermal-induced stress on the device's lifetime and failure causes is made possible by merging thermal modeling with reliability simulations. By providing insights into probable failure modes, this integrated approach enables the creation of reliable device designs and optimized temperature management techniques (Kim et al. 2001).

Design Optimization: AlGaN/GaN HEMT design optimization is facilitated by integrating thermal modeling with electrical and reliability simulations. Engineers can investigate numerous design factors and evaluate their effects on thermal behavior, electrical performance, and dependability. These factors include device shape, material qualities, and cooling techniques. Identifying the best design configurations that balance thermal, electrical, and reliability factors is made possible by this integrated approach, which enhances device performance and durability (Guo et al. 2020).

Performance and Yield Enhancement: Device performance and yield can be enhanced by considering thermal effects and electrical and reliability calculations. Thermal hotspots, which can reduce device performance and cause reliability problems, can be identified and mitigated using the integrated method. Additionally, it makes it possible to compare various cooling methods and packaging options to improve thermal management and guarantee uniform device performance across a batch or wafer (El-Helou et al. 2020).

In conclusion, combining thermal modeling with electrical and reliability simulations provides a comprehensive understanding of the performance and reliability of AlGaN/GaN HEMTs. This method permits the analysis of self-heating effects, the assessment of temperature-dependent electrical behavior, the optimization of design, and the improvement of performance. Engineers can create reliable and effective AlGaN/GaN HEMTs that satisfy the requirements of high-power applications by considering the interaction between thermal behavior, electrical properties, and reliability.

8.3 Emerging Trends and Future Directions in Thermal Management of AlGaN/GaN HEMTs

High-electron-mobility transistors (HEMTs) made on AlGaN/GaN continue to be the subject of extensive study and development. For these devices to operate reliably and efficiently, they must overcome thermal issues as they approach power density limits and operating temperatures. Several new trends and future directions are shaping thermal management in AlGaN/GaN HEMTs. Here, we talk about a few of these developments and their possible effects:

Advanced Cooling Techniques: The limitations of conventional cooling methods, like air cooling or heat sinks, may be reached when dealing with the heat dissipation demands of next-generation AlGaN/GaN HEMTs. New methods like phase change, jet impingement, and microchannel are becoming more popular. These methods provide better thermal uniformity, increased heat transfer efficiency, and decreased thermal resistance. To address the thermal requirements of the next AlGaN/GaN HEMTs, it will be essential to investigate and optimize innovative cooling strategies (Ajayan et al. 2022).

Nanoscale Thermal Management: Nanoscale thermal management becomes more significant as device dimensions get smaller. To improve heat dissipation and lower thermal resistance at the micro- and nanoscales, researchers are looking into nanoscale thermal interface materials, nanofluids, and nanoscale heat spreaders. An emerging trend that can potentially enhance thermal management in AlGaN/GaN HEMTs is the development of innovative nanostructured materials and processes (Belkacemi and Hocine 2018).

Thermal-Aware Device and Circuit Design: Incorporating thermal considerations into the device and circuit design process is gaining popularity. Design approaches considering thermal effects on device performance, such as self-heating and reliability difficulties caused by temperature, are being investigated. This method makes optimizing the design of the device's internal structures, choice of materials, and layout arrangements to reduce thermal resistance and increase overall performance possible (Ramella et al. 2019).

Materials Innovation: Material science improvements significantly impacting material science have a significant impact on AlGaN/GaN HEMT heat control. Researchers are looking at new materials that have improved device compatibility, reduced thermal resistance, and increased thermal conductivity. To improve heat dissipation and thermal conductivity in these devices, cutting-edge material systems, including graphene, carbon nanotubes, and high thermal conductivity substrates, are being studied (Feng et al. 2009; Chaturvedi et al. 2020).

Multiphysics Modeling and Simulation: Multiphysics modeling and simulation, which integrates mechanical, electrical, and thermal elements, will continue to advance. Integrated simulations make possible a thorough understanding of the relationships between thermal behavior, electrical performance, and mechanical reliability. AlGaN/GaN HEMT, thermal management methods, will be accurately predicted and optimized thanks to the development of multiphysics modeling tools (Jones et al. 2015).

Reliability-Oriented Thermal Management: AlGaN/GaN HEMTs' thermal management directly affects their reliability. The main focus of future research is to develop thermal management systems that consider long-term reliability issues in addition to heat dispersion. Promoting reliable and long-lasting device operation involves reducing thermally induced stress, improving temperature cycling performance, and mitigating thermomechanical failures (Menozzi et al. 2015).

Integration of Artificial Intelligence: Incorporating artificial intelligence (AI) methods, including deep learning and machine learning, can potentially enhance thermal control in AlGaN/GaN HEMTs. AI algorithms may analyze large datasets, discovering trends and revealing insights for improving thermal management tactics. AI-based methods can automate thermal design, optimize cooling, and forecast device behavior (Zhu et al. 2019).

In conclusion, AlGaN/GaN HEMT thermal management is a dynamic field with several new developments and potential future directions. Advanced cooling methods, nanoscale thermal management, thermal-aware design, materials innovation, multiphysics modeling, reliability-oriented approaches, and the integration of artificial intelligence are shaping the future of thermal management.

9 Conclusion

9.1 Summary of the Essential Findings and Insights From the Review

Analyzing AlGaN/GaN high-electron-mobility transistors (HEMTs) offers insightful information on thermal management-related topics. Here is a summary of the review's key conclusions and learnings:

Thermal Resistance Modeling: To calculate the thermal resistance in AlGaN/GaN HEMTs, analytical models, numerical simulations, and experimental methods are used. Thermal conductivity, heat capacity, and thermal boundary resistance are all factors these models consider. Device geometry, material characteristics, and heat dissipation systems influence the models' accuracy.

Heat Transfer Mechanisms: Conduction, convection, and radiation affect heat transfer in AlGaN/GaN HEMTs. Low thermal conductivity and high power density make efficient heat dissipation difficult for these devices. Maintaining device performance and reliability requires designing effective cooling techniques and optimizing heat transfer systems.

Material Properties: The thermal resistance of AlGaN/GaN HEMTs is greatly influenced by the thermal conductivity and heat capacity of the materials used in their construction. Materials with higher thermal conductivity and heat capacity are preferred for better heat dissipation. The thermal boundary resistance at material boundaries further impacts the device's overall thermal resistance.

Experimental Characterization: Experimental techniques characterize the thermal resistance of AlGaN/GaN HEMTs, including steady-state and transient studies. These methods assist in validating the correctness of thermal models and offer valuable insights into device behavior under various operating situations.

Design Techniques: Different design techniques reduce thermal resistance in AlGaN/GaN HEMTs. These include refining packaging methods, enhancing thermal interface materials, adding heat spreaders, and optimizing device shape. An effective thermal management design is essential to prevent performance deterioration and guarantee device reliability.

Advanced Modeling Techniques: AlGaN/GaN HEMT thermal behavior is predicted using advanced modeling techniques such as finite element analysis (FEA), computational fluid dynamics (CFD), finite difference method (FDM), and Monte Carlo method. These methods optimize device performance and reliability by offering thorough insights into temperature distributions, hotspots, and thermal gradients.

Integration of Thermal Modeling: For a thorough understanding of device behavior, thermal modeling must be combined with electrical and reliability simulations. The analysis of temperature-dependent electrical performance, evaluation of self-heating effects, evaluation of long-term dependability, optimization of design parameters, and improvement of device performance and yield are all aided by this integration.

Future Directions: New developments in AlGaN/GaN HEMT thermal management include sophisticated cooling methods, nanoscale thermal management, thermally conscious design, material advancements, multiphysics simulations, reliability-oriented strategies, and the incorporation of artificial intelligence. The issues of increased power densities, temperature cycling, and long-term reliability are being addressed by these trends.

Overall, the review offers insightful information on AlGaN/GaN HEMT thermal management, emphasizing the significance of precise thermal resistance modeling, efficient heat transfer mechanisms, optimized design approaches, and integrating thermal modeling with electrical and reliability simulations. These discoveries aid in the creation of AlGaN/GaN HEMTs for high-power applications that are more dependable and efficient.

9.2 Importance of Thermal Resistance Modeling in Optimizing AlGaN/GaN HEMT Performance

AlGaN/GaN high-electron-mobility transistors (HEMTs) performance must be optimized via thermal resistance modeling. It is crucial to comprehend and correctly estimate the thermal resistance in these devices for several reasons:

Device Reliability: In AlGaN/GaN HEMTs, excessive heat generation can cause device degradation, higher failure rates, and shorter lifespans. Using thermal resistance modeling, designers can adopt efficient cooling solutions and avoid early device failure by identifying probable hotspots and regions of high-temperature gradients.

Performance Degradation: High temperatures can adversely affect the electrical performance of AlGaN/GaN HEMTs. The overall functionality of a device may be harmed by elements such as increased electron scattering, decreased carrier mobility, and altered device properties. By modeling thermal resistance, designers can optimize device operation settings to reduce temperature-induced performance degradation and precisely guarantee dependable and consistent performance.

Power Handling Capability: AlGaN/GaN HEMTs are renowned for their capacity to handle high power densities. However, excessive heat buildup may reduce their ability to handle power. Designers might propose techniques to improve the device's capacity for handling power by pinpointing important regions where an understanding of thermal resistance constrains heat dissipation. This may entail enhancing heat-sinking strategies, refining device architectures, or investigating cutting-edge cooling approaches.

Efficiency Enhancement: The efficiency of AlGaN/GaN HEMTs is directly impacted by thermal resistance. Overheating can result in energy losses and a reduction in overall effectiveness. Designers may pinpoint inefficient locations and apply thermal management strategies to increase heat dissipation, reduce thermal gradients, and increase device efficiency by correctly simulating thermal resistance.

Design Optimization: Thermal resistance modeling shows how AlGaN/GaN HEMTs react thermally under various operating circumstances. This knowledge is crucial for device designers since it enables them to optimize device designs, material selections, and thermal management techniques. Designers can create AlGaN/GaN HEMTs that are more effective and reliable and meet the performance specifications for specific applications by taking thermal resistance into account throughout the design phase.

System-Level Considerations: AlGaN/GaN HEMTs are incorporated into more extensive electronic systems in many real-world applications. The thermal behavior of various components may significantly impact the system's overall thermal management. Designers may guarantee correct thermal integration and maximize the overall system performance and reliability by accurately modeling the thermal resistance of HEMTs.

In summary, thermal resistance modeling enhances AlGaN/GaN HEMT performance. It enables designers to pinpoint and resolve thermal-related issues, enhance power handling capabilities, boost efficiency, prevent performance degradation, improve device dependability, and optimize device design and integration into more significant systems. AlGaN/GaN HEMTs can be effectively managed to realize their full potential in high-power and high-temperature applications by considering thermal resistance in the design and operational stages.

9.3 Future Research Directions and Potential Advancements in the Field

High-electron-mobility transistors (HEMTs) made of AlGaN/GaN are a lively topic of study with many potential future possibilities. The following are some crucial study topics that could advance AlGaN/GaN HEMTs:

Material Innovations: Future research aims to discover new materials or novel material combinations that can improve the thermal performance of AlGaN/GaN HEMTs. Increasing heat dissipation involves producing high thermal conductivity epitaxial layers, exploring enhanced thermal interface materials, and looking into alternate substrate materials.

Nanoscale Thermal Management: As electronics continue to get smaller, nanoscale thermal management becomes increasingly important. Future research will examine nanoscale heat transfer methods such as phonon engineering and thermal conductivity modulation to improve heat dissipation and thermal performance at the nanoscale level.

Advanced Cooling Techniques: To overcome the thermal problems of AlGaN/GaN HEMTs, novel methods such as microfluidics, heat pipes, and thermoelectric cooling will be examined. These methods can effectively remove heat from the system and regulate temperature, allowing for better power densities and increased device dependability.

Thermal-Aware Design: Designing with thermal impacts in mind will develop thermal-aware design techniques. Future studies will concentrate on heat-aware device layouts, heat dissipation structure optimization, and early design integration of thermal management techniques.

Multiphysics Simulations: Multiphysics simulations will fully comprehend how AlGaN/GaN HEMTs behave by combining electrical, thermal, and mechanical factors. These simulations will illuminate the interactions between mechanical stresses, thermal behavior, and electrical performance, assisting in creating reliable and effective device designs.

Reliability-Oriented Approaches: Future research will focus on reliability-oriented strategies to overcome the difficulties of AlGaN/GaN HEMTs' long-term operation and dependability. Evaluating the device's reliability under varied operating situations involves researching failure sources, comprehending degradation processes, and designing expedited testing techniques.

Artificial Intelligence (AI) Integration: Incorporating artificial intelligence (AI) approaches like machine learning and data-driven modeling will enable advanced thermal management techniques in AlGaN/GaN HEMTs. AI algorithms can analyze massive datasets, forecast device behavior, optimize thermal designs, and create automated thermal management systems for increased performance and dependability.

System-Level Integration: Upcoming research will concentrate on AlGaN/GaN HEMT integration into larger electronic systems and system-level thermal management optimization. This entails researching cutting-edge packaging methods, considering thermal interactions between components, and creating effective thermal routes across the entire system.

These new research areas hope to improve AlGaN/GaN HEMT performance, efficiency, reliability, and integration by addressing the persisting issues with thermal management. AlGaN/GaN HEMTs will continue progressing and opening up new opportunities for high-power and high-temperature applications by utilizing cutting-edge cooling methods, thermally conscious design procedures, and emerging technologies like AI and multiphysics simulations.

References

Ajayan J, Nirmal D, Mohankumar P, Mounika B, Bhattacharya S, Tayal S, Fletcher AA (2022) Challenges in material processing and reliability issues in algan/gan hemts on silicon wafers for future rf power electronics & switching applications: A critical review. Mater Sci Semiconductor Process 151:106982

Alim MA, Afrin S, Rezazadeh A, Gaquière C (2020) Thermal response and correlation between mobility and kink effect in gan hemts. Microelectron Eng 219:111148

Arivazhagan L, Jarndal A, Nirmal D (2021) Gan hemt on si substrate with diamond heat spreader for high power applications. J Computat Electron 20:873–882

Arivazhagan L, Nirmal D, Chander S, Ajayan J, Godfrey D, Rajkumar J, Bhagya Lakshmi S (2020) Variable thermal resistance model of gan-on-sic with substrate scalability. J. Comput. Electron. 19(4):1546–1554
Arulkumaran S, Ranjan K, Ng GI, Kennedy J, Murmu PP, Bhat TN, Tripathy S (2016) Thermally stable device isolation by inert gas heavy ion implantation in algan/gan hemts on si. J Vacuum Sci Technol B 34(4)
Asgari A, Kalafi M, Faraone L (2005) A quasi-two-dimensional charge transport model of algan/gan high electron mobility transistors (hemts). Physica E: Low-Dimensional Syst Nanostr 28(4):491–499
Asubar JT, Yatabe Z, Hashizume T (2014) Reduced thermal resistance in algan/gan multi-mesa-channel high electron mobility transistors. Appl Phys Lett 105(5)
Aubry R, Jacquet J-C, Weaver J, Durand O, Dobson P, Mills G, di Forte-Poisson M-A, Cassette S, Delage S-L (2007) Sthm temperature mapping and nonlinear thermal resistance evolution with bias on algan/gan hemt devices. IEEE Trans Electron Dev 54(3):385–390
Azarifar M, Donmezer N (2017) A multiscale analytical correction technique for two-dimensional thermal models of algan/gan hemts. Microelectron Reliab 74:82–87
Azarifar M, Donmezer N (2016) A roadmap for building thermal models for algan/gan hemts: simplifications and beyond. In: Heat transfer summer conference, vol 50336. American Society of Mechanical Engineers, p V002T11A006
Baczkowski L, Jacquet J-C, Jardel O, Gaquière C, Moreau M, Carisetti D, Brunel L, Vouzelaud F, Mancuso Y (2015) Thermal characterization using optical methods of algan/gan hemts on sic substrate in rf operating conditions. IEEE Trans Electron Dev 62(12):3992–3998
Belkacemi K, Hocine R (2018) Efficient 3d-tlm modeling and simulation for the thermal management of microwave algan/gan hemt used in high power amplifiers sspa. J Low Power Electron Appl 8(3):23
Blevins J, Via G, Bar-Cohen A, Sivananthan A (2016) Developing a new thermal paradigm for gallium nitride (gan) device technology. In: Proceedings compound semiconductor mantechnical conference, pp 141–146
Cai Y, Cheng Z, Yang Z, Tang CW, Lau KM, Chen KJ (2007) High-temperature operation of algan/gan hemts direct-coupled fet logic (dcfl) integrated circuits. IEEE Electron Dev Lett 28(5):328–331
Cappello T, Santarelli A, Florian C (2017) Dynamic ron characterization technique for the evaluation of thermal and off-state voltage stress of gan switches. IEEE Trans Power Electron 33(4):3386–3398
Chakraborty S, Amir W, Shin J-W, Shin K-Y, Cho C-Y, Kim J-M, Hoshi T, Tsutsumi T, Sugiyama H, Matsuzaki H et al (2022) Explicit thermal resistance model of self-heating effects of algan/gan hemts with linear and non-linear thermal conductivity. Materials 15(23):8415
Chatterjee B, Shoemaker D, Song Y, Shi T, Huang H-L, Keum D, Krishnan A, Foley BM, Jovanovic I, Hwang J et al (2020) Cumulative impacts of proton irradiation on the self-heating of algan/gan hemts. ACS Appl Electron Mater 2(4):980–991
Chatterjee B, Dundar C, Beechem TE, Heller E, Kendig D, Kim H, Donmezer N, Choi S (2020) Nanoscale electro-thermal interactions in algan/gan high electron mobility transistors. J Appl Phys 127(4)
Chaturvedi N, Singh K, Kachhawa P, Lossy R, Mishra S, Chauhan A, Kharbanda DK, Jain AK, Thakur RR, Saxena D et al (2020) Algan/gan hemt based sensor and system for polar liquid detection. Sensors Actuators A Phys 302:111799
Chen J-T, Persson I, Nilsson D, Hsu C-W, Palisaitis J, Forsberg U, Persson PO, Janzén E (2015) Room-temperature mobility above 2200 cm2/v · s of two-dimensional electron gas in a sharp-interface algan/gan heterostructure. Appl Phys Lett 106(25)
Chen X, Donmezer FN, Kumar S, Graham S (2014) A numerical study on comparing the active and passive cooling of algan/gan hemts. IEEE Trans Electron Dev 61(12):4056–4061
Chen D, Wan L, Li J, Liu Z, Li G (2019) Ohmic contact to algan/gan hemt with electrodes in contact with heterostructure interface. Solid-State Electron 151:60–64

Chernykh M, Andreev A, Ezubchenko I, Chernykh I, Mayboroda I, Kolobkova E, Khrapovitskaya YV, Grishchenko J, Perminov P, Sedov V et al (2022) Gan-based heterostructures with cvd diamond heat sinks: A new fabrication approach towards efficient electronic devices. Appl Mater Today 26:101338

Cho J, Li Z, Bozorg-Grayeli E, Kodama T, Francis D, Ejeckam F, Faili F, Asheghi M, Goodson KE (2012) Improved thermal interfaces of gan-diamond composite substrates for hemt applications. IEEE Trans Components Packaging Manuf Technol 3(1):79–85

Choi S, Heller ER, Dorsey D, Vetury R, Graham S (2012) The impact of bias conditions on self-heating in algan/gan hemts. IEEE Trans Electron Dev 60(1):159–162

Chou Y-C, Leung D, Smorchkova I, Wojtowicz M, Grundbacher R, Callejo L, Kan Q, Lai R, Liu P, Eng D et al (2004) Degradation of algan/gan hemts under elevated temperature lifetesting. Microelectron Reliab 44(7):1033–1038

Chu K, Chao P, Pizzella M, Actis R, Meharry D, Nichols K, Vaudo R, Xu X, Flynn J, Dion J et al (2004) 9.4-w/mm power density algan-gan hemts on free-standing gan substrates. IEEE Electron Dev Lett 25(9):596–598

Chung Y, Hang C, Cai S, Chen Y, Lee W, Wen C, Wang K, Itoh T (2002) Effects of output harmonic termination on pae and output power of algan/gan hemt power amplifier. IEEE Microwave and Wirel Components Lett 12(11):421–423

Cordier Y, Portail M, Chenot S, Tottereau O, Zielinski M, Chassagne T (2008) Realization of algan/gan hemts on 3c-sic/si (111) substrates. Physica Status Solidi C 5(6):1983–1985

Darwish AM, Bayba AJ, Hung HA (2004) Thermal resistance calculation of algan-gan devices. IEEE Trans Microwave Theory Techn 52(11):2611–2620

Das J, Oprins H, Ji H, Sarua A, Ruythooren W, Derluyn J, Kuball M, Germain M, Borghs G (2006) Improved thermal performance of algan/gan hemts by an optimized flip-chip design. IEEE Trans Electron Dev 53(11):2696–2702

Dubey SK, Mishra M, Islam A (2022) Characterization of algan/gan based hemt for low noise and high frequency application. Int J Numer Model Electron Netw Dev Fields 35(1):e2932

Dundar C, Kara D, Donmezer N (2019) The effects of gate-connected field plates on hotspot temperatures of algan/gan hemts. IEEE Trans Electron Dev 67(1):57–62

Eastman LF, Tilak V, Kaper V, Smart J, Thompson R, Green B, Shealy J, Prunty T (2002) Progress in high-power, high frequency algan/gan hemts. Physica Status Solidi (A) 194(2):433–438

Ejeckam F, Francis D, Faili F, Dodson J, Twitchen DJ, Bolliger B, Babic D (2014) Diamond for enhanced gan device performance. In: Fourteenth intersociety conference on thermal and thermomechanical phenomena in electronic systems (ITherm). IEEE, pp 1206–1209

El Helou A, Komarov P, Tadjer MJ, Anderson TJ, Francis DA, Feygelson T, Pate BB, Hobart KD, Raad PE (2020) High-resolution thermoreflectance imaging investigation of self-heating in algan/gan hemts on si, sic, and diamond substrates. IEEE Trans Electron Dev 67(12):5415–5420

El-Helou A, Cui Y, Tadjer M, Anderson T, Francis D, Feygelson T, Pate B, Hobart K, Raad P (2020) Full thermal characterization of algan/gan high electron mobility transistors on silicon, silicon carbide, and diamond substrates using a reverse modeling approach. Semiconductor Sci Technol 36(1):014008

Faqir M, Manoi A, Mrotzek T, Knippscheer S, Massiot M, Buchta M, Blanck H, Rochette S, Vendier O, Kuball M (2011) New gan power-electronics packaging solutions: a thermal analysis using raman thermography. J Microelectron Electron Packaging 8(3):110–113

Faqir M, Batten T, Mrotzek T, Knippscheer S, Massiot M, Buchta M, Blanck H, Rochette S, Vendier O, Kuball M (2012) Improved thermal management for gan power electronics: Silver diamond composite packages. Microelectron Reliab 52(12):3022–3025

Feng Q, Li L-M, Hao Y, Ni J-Y, Zhang J-C (2009) The improvement of ohmic contact of ti/al/ni/au to algan/gan hemt by multi-step annealing method. Solid-State Electron 53(9):955–958

Fletcher AA, Nirmal D, Ajayan J, Arivazhagan L (2019) Analysis of algan/gan hemt using discrete field plate technique for high power and high frequency applications. AEU-Int J Electron Commun 99:325–330

Fletcher AA, Nirmal D, Ajayan J, Arivazhagan L (2021) An intensive study on assorted substrates suitable for high jfom algan/gan hemt. Silicon 13:1591–1598

Fotsing JN, Dietzel D, Chotikaprakhan S, Meckenstock R, Pelzl J, Cassette S (2005) Evaluation of active semiconductor structures by combined scanning thermo-elastic microscopy and finite element simulations. Journal de Physique IV (Proceedings) 125. EDP sciences, pp 117–120

Gamarra P, Lacam C, Tordjman M, di Forte-Poisson M-A (2013) Impact of the substrate and of the nucleation layer on the properties of algan/gan hemts on sic. J Crystal Growth 370:282–287

Gangwani P, Pandey S, Haldar S, Gupta M, Gupta R (2007) Polarization dependent analysis of algan/gan hemt for high power applications. Solid-State Electron 51(1):130–135

García S, Íñiguez-de-la Torre I, Mateos J, González T, Pérez S (2016) Impact of substrate and thermal boundary resistance on the performance of algan/gan hemts analyzed by means of electro-thermal monte carlo simulations. Semiconductor Sci Technol 31(6):065005

Gucmann F, Pomeroy JW, Kuball M (2021) Scanning thermal microscopy for accurate nanoscale device thermography. Nano Today 39:101206

Guggenheim R, Rodes L (2017) Roadmap review for cooling high-power gan hemt devices. In: 2017 IEEE International conference on microwaves, antennas, communications and electronic systems (COMCAS). IEEE, pp 1–6

Guo H, Kong Y, Chen T (2017) Thermal simulation of high power gan-on-diamond substrates for hemt applications. Diamond Related Mater 73:260–266

Guo H, Chen T, Shi S (2020) Transient simulation for the thermal design optimization of pulse operated algan/gan hemts. Micromachines 11(1):76

Hao Q, Zhao H, Xiao Y (2017) A hybrid simulation technique for electrothermal studies of two-dimensional gan-on-sic high electron mobility transistors. J Appl Phys 121(20)

Hatano M, Taniguchi Y, Kodama S, Tokuda H, Kuzuhara M (2014) Reduced gate leakage and high thermal stability of algan/gan mis-hemts using zro2/al2o3 gate dielectric stack. Appl Phys Express 7(4):044101

Haziq M, Falina S, Manaf AA, Kawarada H, Syamsul M (2022) Challenges and opportunities for high-power and high-frequency algan/gan high-electron-mobility transistor (hemt) applications: A review. Micromachines 13(12):2133

Heller ER, Vetury R, Green DS, AFRLW-PAOMM, DIRECTORATE/SURVIVABILITY, DIV SM (2011) Development of a versatile physics-based finite-element model of an algan/gan hemt capable of accommodating process and epitaxy variations and calibrated using multiple dc parameters (postprint)

Heller E, Crespo A (2008) Electro-thermal modeling of multifinger algan/gan hemt device operation including thermal substrate effects. Microelectron Reliab 48(1):45–50

Heller E, Choi S, Dorsey D, Vetury R, Graham S (2013) Electrical and structural dependence of operating temperature of algan/gan hemts. Microelectron Reliab 53(6):872–877

Hirama K, Taniyasu Y, Kasu M (2011) Algan/gan high-electron mobility transistors with low thermal resistance grown on single-crystal diamond (111) substrates by metalorganic vapor-phase epitaxy. Appl Phys Lett 98(16)

Huque M, Eliza S, Rahman T, Huq H, Islam S (2009) Temperature dependent analytical model for current-voltage characteristics of algan/gan power hemt. Solid-State Electron 53(3):341–348

Husain S, Hamadeh A, Zaatar O, Al Gumaei T, Jarndal A (2020) Temperature dependent svr and ann based iv models for gan hemts. In: 2020 International conference on communications, computing, cybersecurity, and informatics (CCCI). IEEE, pp 1–4

Jacquet J, Aubry R, Gérard H, Delos E, Rolland N, Cordier Y, Bussutil A, Rousseau , M, Delage S (2004) Analytical transport model of algan/gan hemt based on electrical and thermal measurement

Jeon D-Y, Koh Y, Cho C-Y, Park K-H (2021) Impact of temperature-dependent series resistance on the operation of algan/gan high electron mobility transistors. AIP Adv 11(11)

Jeong J-C, Jang D-P, Han B-G, Yom I-B (2014) A compact c-band 50 w algan/gan high-power mmic amplifier for radar applications. ETRI J 36(3):498–501

Jia Y, Xu Y, Guo Y (2018) A universal scalable thermal resistance model for compact large-signal model of algan/gan hemts. IEEE Trans Microwave Theory Techn 66(10):4419–4429

Jones JP, Heller E, Dorsey D, Graham S (2015) Transient stress characterization of algan/gan hemts due to electrical and thermal effects. Microelectron Reliab 55(12):2634–2639
Kakanakova-Georgieva A, Forsberg U, Ivanov IG, Janzén E (2007) Uniform hot-wall mocvd epitaxial growth of 2 inch algan/gan hemt structures. J Crystal Growth 300(1):100–103
Kandasamy S, NB B (2019) Modeling and simulation of dual-material-gate algan/gan high-electron-mobility transistor using finite difference method. Int J Numer Model Electron Netw Dev Fields 32(3):e2546
Khan AN, Mishra S, Routray S, Chatterjee G, Jena K (2023) Analytical modeling and simulation of lattice-matched ferro pzt algan/gan moshemt for high-power and rf/microwave applications. J Comput Electron 22(3):827–838
Khan AN, Bhat AM, Jena K, Lenka TR, Chatterjee G (2023) Improved breakdown voltage mechanism in algan/gan hemt for rf/microwave applications: Design and physical insights of dual field plate. Microelectron Reliab 147:115036
Khandelwal S, Fjeldly TA (2012) A physics based compact model for drain current in algan, gan hemt devices. 24th International symposium on power semiconductor devices and ICs. IEEE, pp 241–244
Killat N, Montes M, Pomeroy J, Paskova T, Evans K, Leach J, Li X, Ozgur Ü, Morkoc H, Chabak K et al (2012) Thermal properties of algan/gan hfets on bulk gan substrates. IEEE Electron Dev Lett 33(3):366–368
Killat N, Kuball M, Chou T-M, Chowdhury U, Jimenez J (2010) Temperature assessment of algan, gan hemts: a comparative study by raman, electrical and ir thermography. IEEE international reliability physics symposium. IEEE, 528–531
Kim D-S, Lee J-H, Yeo S, Lee J-H (2017) Proton irradiation effects on algan/gan hemts with different isolation methods. IEEE Tran Nuclear Sci 65(1):579–582
Kim H, Tilak V, Green B, Smart J, Schaff W, Shealy J, Eastman L (2001) Reliability evaluation of high power algan/gan hemts on sic substrate. Physica Status Solidi (a) 188(1):203–206
Kisiel R, Guziewicz M, Taube A, Kaminski M, Sochacki M (2021) Development of assembly techniques for connection of algan/gan/si chips to dbc substrate. Circuit World 47(2):146–152
Kuball M, Hayes J, Uren M, Martin I, Birbeck J, Balmer R, Hughes B (2002) Measurement of temperature in active high-power algan/gan hfets using raman spectroscopy. IEEE Electron Dev Lett 23(1):7–9
Lambert B, Labat N, Carisetti D, Karboyan S, Tartarin J-G, Thorpe J, Brunel L, Curutchet A, Malbert N, Latu-Romain E et al (2012) Evidence of relationship between mechanical stress and leakage current in algan/gan transistor after storage test. Microelectron Reliab 52(9–10):2184–2187
Li Y, Chen Y, Zhang C, Wang X, Pu H (2022) Accurate estimation of thermal properties for algan/gan hemts on diamond. AIP Adv 12(7)
Linkohr S, Pletschen W, Polyakov V, Himmerlich M, Lorenz P, Krischok S, Kirste L, Müller S, Ambacher O, Cimalla V (2012) Influence of plasma treatments on the properties of gan/algan/gan hemt structures. Physica Status Solidi c 9(3–4):1096–1098
Liu C, Wang X, Huang S, Ma X, Wang Y, Zhang S, Zhao R, Shi W, He Q, Yin H et al (2019) A large-signal pspice modeling of gan-based mis-hemts. Superlattices Microstr 130:499–511
Ludwig F, Bauer M, Lisauskas A, Roskos HG (2019) Circuit-based hydrodynamic modeling of algan/gan hemts. In: ESSDERC 2019-49th European solid-state device research conference (ESSDERC). IEEE, pp 270–273
Maize K, Pavlidis G, Heller E, Yates L, Kendig D, Graham S, Shakouri A (2014) High resolution thermal characterization and simulation of power algan, gan hemts using micro-raman thermography and 800 picosecond transient thermoreflectance imaging. IEEE compound semiconductor integrated circuit symposium (CSICS). IEEE, pp 1–8
Manoi A, Pomeroy JW, Killat N, Kuball M (2010) Benchmarking of thermal boundary resistance in algan/gan hemts on sic substrates: Implications of the nucleation layer microstructure. IEEE Electron Dev Lett 31(12):1395–1397
Mao S, Xu Y (2018) Investigation on the i–v kink effect in large signal modeling of algan/gan hemts. Micromachines 9(11):571

Menozzi R, Cova P, Delmonte N, Giuliani F, Sozzi G (2015) Thermal and electro-thermal modeling of components and systems: A review of the research at the university of parma. Facta Universitatis, Series: Electron Energetics 28(3):325–344

Minoura Y, Ohki T, Okamoto N, Yamada A, Makiyama K, Kotani J, Ozaki S, Sato M, Nakamura N (2020) Surface activated bonding of sic/diamond for thermal management of high-output power gan hemts. Japanese J Appl Phys 59(SG):SGGD03

Mion C, Muth J, Preble EA, Hanser D (2006) Thermal conductivity, dislocation density and gan device design. Superlattices Microstr 40(4–6):338–342

Mitterhuber L, Hammer R, Dengg T, Spitaler J (2020) Thermal characterization and modelling of algan-gan multilayer structures for hemt applications. Energies 13(9):2363

Mizuno S, Yamada F, Yamamoto H, Nishihara M, Yamamoto T, Sano S (2012) Development of gan hemt for microwave wireless communications. SEI Techn Rev 74:71–74

Mohanty SK, Chen Y-Y, Yeh P-H, Horng R-H (2019) Thermal management of gan-on-si high electron mobility transistor by copper filled micro-trench structure. Scientific Reports 9(1):19691

Motala MJ, Blanton , EW, Hilton A, Heller C, Muratore C, Burzynski K, Brown JL, Chabak K, Durstock M, Snure M et al. (2020) Transferrable algan/gan high-electron mobility transistors to arbitrary substrates via a two-dimensional boron nitride release layer. ACS Appl Mater Interfaces 12(19):21 837–21 844

Mounika B, Ajayan J, Bhattacharya S, Nirmal D (2022) Recent developments in materials, architectures and processing of algan/gan hemts for future rf and power electronic applications: A critical review. Micro Nanostr 168:207317

Narang K, Singh VK, Pandey A, Khan R, Bag RK, Rawal D, Padmavati M, Tyagi R, Singh R (2022) Suitability of thin-gan for algan/gan hemt material and device. J Mater Sci 57(10):5913–5923

Nigam A, Bhat TN, Rajamani S, Dolmanan SB, Tripathy S, Kumar M (2017) Effect of self-heating on electrical characteristics of algan/gan hemt on si (111) substrate. AIP Adv 7(8)

Ohki T, Kikkawa T, Kanamura M, Imanishi K, Makiyama N, Okamoto N, Joshin K, Hara N (2009) An over 100 w algan/gan enhancement-mode hemt power amplifier with piezoelectric-induced cap structure. Physica Status Solidi C 6(6):1365–1368

Palacios T (2009) Beyond the algan/gan hemt: new concepts for high-speed transistors. Physica Status Solidi (A) 206(6):1145–1148

Palacios T, Dora Y, Chakraborty A, Sanabria C, Keller S, DenBaars S, Mishra U (2006) Optimization of algan/gan hemts for high frequency operation, pp 1845–1850

Pal P, Pratap Y, Kabra S (2022) T-zno/algan/gan hemt uric acid sensor-sensitivity analysis and effect of surface wettability for improved performance. IEEE Sensors J 22(12):11 819–11 826

Pavlidis G, Kendig D, Heller ER, Graham S (2018) Transient thermal characterization of algan/gan hemts under pulsed biasing. IEEE Trans Electron Dev 65(5):1753–1758

Polyakov AY, Smirnov N, Govorkov A, Kozhukhova E, Pearton SJ, Ren F, Liu L, Johnson J, Lim W, Kolin N et al. (2012) Comparison of neutron irradiation effects in algan/aln/gan, algan/gan, and inaln/gan heterojunctions. J Vacuum Sci Technol B 30(6)

Pu S, Luo W, Shuai Y, Wu C, Zhang W (2016) Simulations of diamond heat spreader for the thermal management of gan hemt. In: 2016 5th International conference on measurement, instrumentation and automation (ICMIA 2016). Atlantis Press, pp 184–188

Qin Y, Chai C, Li F, Liang Q, Wu H, Yang Y (2022) Study of self-heating and high-power microwave effects for enhancement-mode p-gate gan hemt. Micromachines 13(1):106

Quay R (2008) Integration, thermal management, and packaging. Gallium Nitride Electron, 337–366

Quay R, Schwantuschke D, Ture E, van Raay F, Friesicke C, Krause S, Müller S, Breuer S, Godejohann B, Brückner P (2018) High-power microwave gan/algan hemts and mmics on sic and silicon substrates for modern radio communication. Physica Status Solidi (A) 215(9): 1700655

Quay R, van Raay F, Kuhn J, Kiefer R, Waltereit P, Zorcic M, Musser M, Bronner W, Dammann M, Seelmann-Eggebert M (2008) Efficient algan, gan hemt power amplifiers et al (2008) European microwave integrated circuit conference. IEEE, 87–90

Raj B, Bindra S (2013) Thermal analysis of algan/gan hemt: Measurement and analytical modeling techniques. Int J Comput Appl 75(18)

Ramella C, Pirola M, Reale A, Ramundo M, Colantonio P, Der Maur MA, Camarchia V, Piacibello A, Giofrè R (2019) Thermal-aware gan/si mmic design for space applications. In: 2019 IEEE international conference on microwaves, antennas, communications and electronic systems (COMCAS). IEEE, pp 1–6
Reddy BPK, Teja KBR, Kandpal K (2018) Investigation on high-κ dielectric for low leakage algan/gan mis-hemt device, using material selection methodologies. Semiconductors 52:420–430
Riccio M, Pantellini A, Irace A, Breglio G, Nanni A, Lanzieri C (2011) Electro-thermal characterization of algan/gan hemt on silicon microstrip technology. Microelectron Reliab 51(9–11):1725–1729
Riedel G, Pomeroy J, Hilton K, Maclean J, Wallis D, Uren M, Martin T, Kuball M (2008) Nanosecond timescale thermal dynamics of algan/gan electronic devices. IEEE Electron Dev Lett 29(5):416–418
Russo S, d'Alessandro V, Costagliola M, Sasso G, Rinaldi N (2012) Analysis of the thermal behavior of algan/gan hemts. Mater Sci Eng B 177(15):1343–1351
Saadat OI, Chung JW, Piner EL, Palacios T (2009) Gate-first algan/gan hemt technology for high-frequency applications. IEEE Electron Dev Lett 30(12):1254–1256
Sang L (2022) Diamond as the heat spreader for the thermal dissipation of gan-based electronic devices. Funct Diamond 1(1):174–188
Sarua A, Ji H, Hilton K, Wallis D, Uren MJ, Martin T, Kuball M (2007) Thermal boundary resistance between gan and substrate in algan/gan electronic devices. IEEE Trans Electron Dev 54(12):3152–3158
Schlenker EL (2018) Modeling and characterization of high-power electronic devices: System analysis of laser diodes with flash boiling and gan hemt reliability modeling
Siddique A, Ahmed R, Anderson J, Nazari M, Yates L, Graham S, Holtz M, Piner EL (2019) Structure and interface analysis of diamond on an algan/gan hemt utilizing an in situ sin x interlayer grown by mocvd. ACS Appl Electron Mater 1(8):1387–1399
Sommet R, Mouginot G, Quéré R, Ouarch Z, Camiade M (2012) Thermal modeling and measurements of algan/gan hemts including thermal boundary resistance. Microelectron J 43(9):611–617
Sridharan S, Venkatachalam A, Yoder P (2008) Electrothermal analysis of algan/gan high electron mobility transistors. J Comput Electron 7:236–239
Stoffels S, Oprins H, Baelmans M, Decoutere S, De Wolf I (2015) A modeling and experimental method for accurate thermal analysis of algan, gan powerbars. IEEE 27th international symposium on power semiconductor devices & IC's (ISPSD). IEEE, pp 377–380
Stoffels S, Ronchi N, Venegas R, De Jaeger B, Marcon D, Decoutere S (2014) Comparison of algan/gan mishemt powerbar designs. Physica Status Solidi (C) 11(3–4):906–910
Vanko G, Držík M, Vallo M, Lalinskỳ T, Kutiš V, Stančík S, Rỳger I, Benčurová A (2011) Algan/gan c-hemt structures for dynamic stress detection. Sensors Actuators Phys 172(1):98–102
Vitanov S, Palankovski V, Maroldt S, Quay R (2010) High-temperature modeling of algan/gan hemts. Solid-State Electron 54(10):1105–1112
Vitusevich S, Klein N, Belyaev A, Danylyuk S, Petrychuk M, Konakova R, Kurakin A, Rengevich A, Avksentyev AY, Danilchenko B et al. (2003) Effects of γ-irradiation on algan/gan-based hemts. Physica Status Solidi (A) 195(1):101–105
Wang X-D, Hu W-D, Chen X-S, Lu W (2012) The study of self-heating and hot-electron effects for algan/gan double-channel hemts. IEEE Trans Electron Dev 59(5):1393–1401
Wang A, Tadjer M, Calle F (2013) Simulation of thermal management in algan/gan hemts with integrated diamond heat spreaders. Semiconductor Sci Technol 28(5):055010
Wang P, Manno M, Bar-Cohen A (2012) Quantum-well si/sic self-cooling for thermal management of high heat flux gan hemt semiconductor devices. In: International conference on micro/nanoscale heat transfer, vol 54778. American Society of Mechanical Engineers, pp 615–624
Werquin M, Vellas N, Guhel Y, Ducatteau D, Boudart B, Pesant J, Bougrioua Z, Germain M, De Jaeger J-C, Gaquière C (2005) First results of algan/gan hemts on sapphire substrate using an argon-ion implant-isolation technology. Microwave Opt Technol Lett 46(4):311–315

Wijesundara MB, Azevedo R, Wijesundara MB, Azevedo RG (2011) System integration. Silicon Carbide Microsyst Harsh Environ, 189–230

Won Y, Cho J, Agonafer D, Asheghi M, Goodson KE (2015) Fundamental cooling limits for high power density gallium nitride electronics. IEEE Trans Components Packaging Manuf Technol 5(6):737–744

Wu Y-F, Thibeault B, Keller B, Keller S, Denbaars S, Mishra U (1998) 3-watt algan-gan hemts on sapphire substrates with thermal management by flip-chip bonding. In: 56th Annual device research conference digest (Cat. No. 98TH8373). IEEE, pp 118–119

Wu T-L, Kutub SB (2020) Machine learning-based statistical approach to analyze process dependencies on threshold voltage in recessed gate algan/gan mis-hemts. IEEE Trans Electron Dev 67(12):5448–5453

Wu Y-F, Keller B, Keller S, Kapolnek D, Kozodoy P, Denbaars S, Mishra U (1997) High power algan/gan hemts for microwave applications. Solid-State Electron 41(10):1569–1574

Wu Y-F, Kapolnek D, Ibbetson JP, Parikh P, Keller BP, Mishra UK (2001) Very-high power density algan/gan hemts. IEEE Trans Electron Dev 48(3):586–590

Wu Q, Xu Y, Chen Y, Wang Y, Fu W, Yan B, Xu R (2017) A scalable multiharmonic surface-potential model of algan/gan hemts. IEEE Trans Microwave Theory Techn 66(3):1192–1200

Yamakawa S, Goodnick S, Aboud S, Saraniti M (2004) Quantum corrected full-band cellular monte carlo simulation of algan/gan hemts. J Comput Electron 3:299–303

Yang J, Feng S, Shi D, Yang C (2015) Thermal time-constant spectrum extraction method in algan/gan hemts. J Semiconductors 36(8):084003

Zehetner J, Kraus S, Lucki M, Vanko G, Dzuba J, Lalinsky T (2016) Manufacturing of membranes by laser ablation in sic, sapphire, glass and ceramic for gan/ferroelectric thin film mems and pressure sensors. Microsyst Technol 22:1883–1892

Zhang Y, Smorchkova I, Elsass C, Keller S, Ibbetson JP, Denbaars S, Mishra UK, Singh J (2000) Charge control and mobility in algan/gan transistors: Experimental and theoretical studies. J Appl Phys 87(11):7981–7987

Zhang G-C, Feng S-W, Zhou Z, Li J-W, Guo C-S (2011) Evaluation of thermal resistance constitution for packaged algan/gan high electron mobility transistors by structure function method. Chinese Phys B 20(2):027202

Zhang Y-Y, An S, Zheng Y, Lai J, Seo J-H, Lee KH, Kim M (2022) Releasable algan/gan 2d electron gas heterostructure membranes for flexible wide-bandgap electronics. Adv Electron Mater 8(2):2100652

Zhao M, Liu X, Zheng Y, Peng M, Ouyang S, Li Y, Wei K (2013) Thermal analysis of algan/gan high-electron-mobility transistors by infrared microscopy. Optics Commun 291:104–109

Zhou Y, Ramaneti R, Anaya J, Korneychuk S, Derluyn J, Sun H, Pomeroy J, Verbeeck J, Haenen K, Kuball M (2017) Thermal characterization of polycrystalline diamond thin film heat spreaders grown on gan hemts. Appl Phys Lett 111(4)

Zhu J, Jing S, Ma X, Liu S, Wang P, Zhang Y, Zhu Q, Mi M, Hou B, Yang L et al (2020) Improvement of electron transport property and on-resistance in normally-off Al_2O_3/algan/gan mos-hemts using post-etch surface treatment. IEEE Trans Electron Dev 67(9):3541–3547

Zhu J, Zhou X, Jing L, Hua Q, Hu W, Wang ZL (2019) Piezotronic effect modulated flexible algan/gan high-electron-mobility transistors. ACS Nano 13(11):13 161–13 168

Charge Based Compact Modelling of Gate Leakage Current

Arghyadeep Sarkar

Abstract Gate leakage current is a critical parameter for GaN HEMTs as it restricts the device performance. Gate current is affected by the quality of gate contact formed between the metal and GaN. A Schottky contact is expected to be formed when a metal is deposited on top of the GaN layer. Additionally, gate current can be used to study forward gate breakdown which is then subsequently used to analyse time dependent gate breakdown mechanisms. Therefore, the gate leakage current is crucial to the GaN HEMTs device designs and there has been some research modeling study to illustrate the gate leakage current. A numbers of models have been reported to analyze the leakage currents in HEMTs. They include Thermionic Field Emission (TFE), Thermionic Emission (TE), Poole Frenkel Emission (PFE), and Fowler–Nordheim Tunneling (FNT). For smaller level of leakage current, models such as Generation-Recombination (G-R), Defect Assisted Tunneling (DAT), Trap Assisted Tunneling (TAT), and two-dimensional variable range hopping (2D-VRH) are used. In lot of cases, different mechanisms are proposed to model the behavior of different devices without giving vivid explanation for their choice of a particular mechanism. In some cases, leakage at different gate bias voltages is modeled with different mechanisms, even at high bias. This chapter deals with Charge-Based EPFL HEMT Model, Charge-Based Compact Model of Gate Leakage and Surface-Potential-Based Modeling approach for the Gate Current in pGaN Gate HEMTs.

Keywords Schottky contact · Field emission · Tunnelling · Thermionic emission · p-GaN HEMT · Compact models

---

A. Sarkar (✉)
Department of Electrical and Computer Engineering, McMaster University, Hamilton, Canada
e-mail: sarkar.arghya4@gmail.com

D. Nirmal and J. Ajayan (eds.), *Modeling of AlGaN/GaN High Electron Mobility Transistors*, Springer Tracts in Electrical and Electronics Engineering,
https://doi.org/10.1007/978-981-97-7506-4_7

1 Introduction

Gate leakage current is acritical parameter for GaN HEMTs as it restricts the device performance. Gate current is affected by the quality of gate contact formed between the metal and GaN. A Schottky contact is expected to be formed when a metal is deposited on top of the GaN layer. Despite this Schottky gate formation, a high amount of leakage current is reported at positive gate bias voltages (Tallarico et al. 2020; Ťapajna et al. 2015; Ruzzarin et al. 2018; Xu et al. 2018). Compared to Silicon MOSFETs, the leakage current in GaN HEMTs is significantly higher (by approximately four orders of magnitude) (Pal and Narayanan 2014). A severe problem as a result of this high leakage current is that it can lower the permitted gate voltage swing which in turn create false turn-on events in power converters with 2 GaN switches (Tang et al. 2013). In terms of device reliability, gate current has been used as a degradation parameter to develop lifetime acceleration models (Rossetto et al. 2016). Additionally, gate current can be used to study forward gate breakdown which is then subsequently used to analyse time dependent gate breakdown mechanisms (Meneghini et al. 2017). Therefore, the gate leakage current is crucial to the GaN HEMTs device designs and there has been some research modeling study to illustrate the gate leakage current (Ťapajna et al. 2015; Xu et al. 2018; Stockman et al. 2018; He et al. 2020). Quite a few numbers of models have been investigated to analyze the leakage experimental result. They include Thermionic Field Emission (TFE), Thermionic Emission (TE), Poole Frenkel Emission (PFE), and Fowler–Nordheim Tunneling (FNT). For smaller level of leakage current, models such as Generation-Recombination (G-R), Defect Assisted Tunneling (DAT), Trap Assisted Tunneling (TAT), and two-dimensional variable range hopping (2D-VRH) are used. In lot of cases, different mechanisms are proposed to model the behavior of different devices without giving vivid explanation for their choice of a particular mechanism. In some cases, leakage at different gate bias voltages is modeled with different mechanisms, even at high bias (Xu et al. 2018; Stockman et al. 2018; He et al. 2020).

Device modeling can be broadly classified to be either potential based or charge based. In this chapter we look at strategies to develop compact model for the gate current in viewpoint of circuit applications. Without availability of proper compact models, it can lead to erroneous circuit operation especially in low power digital circuits (Bertuccio et al. 1997). There have been quite few empirical models—Statz model (Statz et al. 1987), Curtice model (Curtice 2010) as well as industry adopted compact model models such as MIT virtual source model (Radhakrishna 2016) and ASM model (Khandelwal et al. 2013). However, most of these models focus primarily on the drain current. In this chapter, our intention is to look at compact models with an emphasis on the gate leakage current. In Fig. 1, a standard structure of a GaN based HEMT device is shown.

This chapter is organized as follows. In Sect. 2, we look at the current conduction mechanisms currently established for metal/semiconductor interface responsible for gate leakage process. In Sects. 3, 4 and 5, we look at the charge based EPFL HEMT model with a focus on gate charge computation, another charge based compact gate

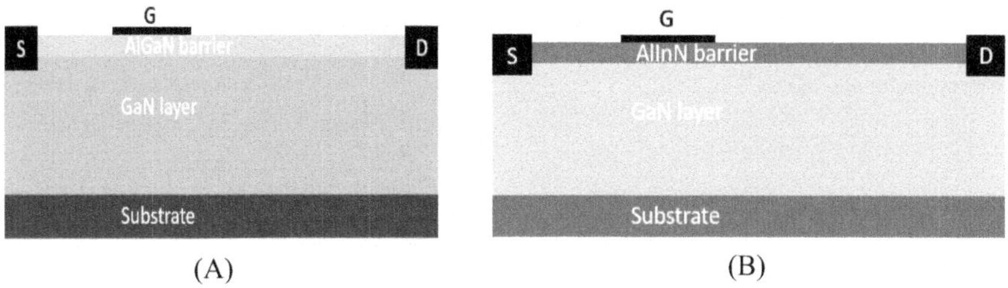

Fig. 1 Standard structure of a GaN based HEMT device. **a** AlGaN barrier **b** AlInN barrier

current model and a surface potential-based gate current model, respectively. In Sect. 6, we discuss the findings from model explained from the previous section. Finally, Sect. 7 is the conclusion.

2 Current Through the Schottky Contact

GaN HEMTs are inherently depletion mode devices with a negative threshold voltage. To have enhancement operation, a small p doped GaN cap layer is added on top of the AlGaN barrier layer which moves the fermi-level away from the conduction band edge. This gives rise to a positive threshold voltage. In Fig. 2, we show the structure of a pGaN/AlGaN/GaN HEMT. There are 3 main ways of carrier transport through the Schottky contact between the metal and semiconductor (GaN in our case). For example, in the case of metal contact with pGaN, this process is depicted in Fig. 3.

Thermionic emission implies that carriers gain sufficient energy to overcome the effective energy barrier to moving over to the other side of the junction. This energy barrier is known as the Schottky barrier height. Field emission refers to the tunneling of carriers through the forbidden gap of GaN. Thermionic Field Emission is the combination between the above two processes. Here, the holes in the case of metal/

Fig. 2 Structure of pGaN/AlGaN/GaN HEMT

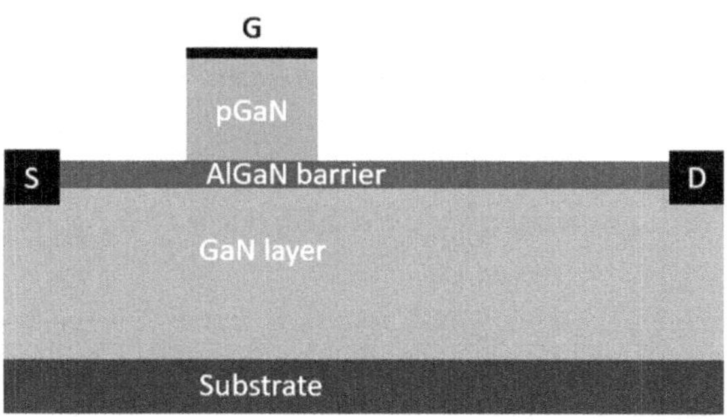

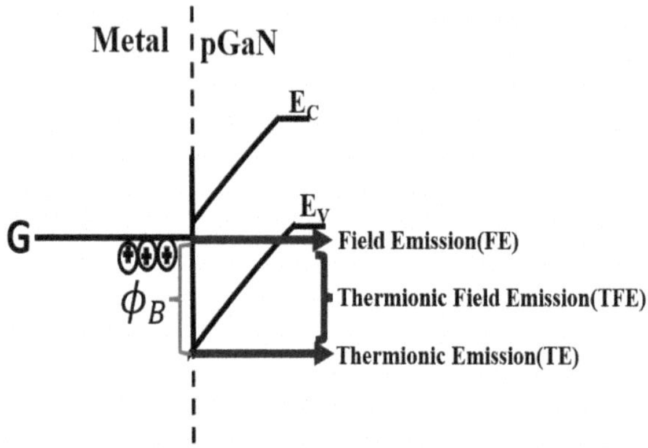

Fig. 3 Carrier transport process in Schottky diode between metal and pGaN

pGaN can tunnel above the valence band edge but do not have enough energy to overcome the Schottky barrier height. Each process is described below.

2.1 *Thermionic Emission (TE)*

As stated, earlier TE process requires carriers to have enough energy to overcome the energy barrier. This process is highly temperature dependent. The current density derived from the TE process is given by Kiziroglou et al. (2008)

$$J_{TE} = J_S\left(\frac{eV_A}{nKT} - 1\right). \tag{1}$$

where V_A is the applied voltage, K is the Boltzmann constant, n is the ideality factor and T is the absolute temperature. J_S is the reverse saturation current density given by

$$J_S = A^* T^2 e^{-q\phi_B/KT}. \tag{2}$$

Here, A^* represent Richardson's constant and ϕ_B is the Schottky barrier height.

2.2 *Thermionic Field Emission (TFE)*

In the TFE process, the tunneling occurs at an energy above the fermi-level (E_F). This process has a strong dependence both on temperature as well as the electric field. Since the Schottky diode is reverse biased during forward operation, only reverse TFE current density is being considered here, which is given by Kiziroglou et al.

(2008)

$$J_{TFE,r} = J_{s,TFE,r} e^{\frac{V}{\varepsilon'}}. \tag{3}$$

Here, V is applied forward bias voltage.

The reverse TFE saturation current is given by.

$$J_{\mathrm{s,TFE,\,r}} = \frac{A^* \cdot T^2 \cdot \sqrt{\pi E_{00}}}{V_{\mathrm{t}} \cdot \cosh(E_{00}/V_{\mathrm{t}})} \cdot \mathrm{e}^{-\frac{\phi_B}{E_0}} \cdot \sqrt{\phi_{Bn} - V \cdot \cosh^2\left(\frac{E_{00}}{V_{\mathrm{t}}}\right)}. \tag{4}$$

where,

$$E_0 = \frac{E_{00}}{\tanh\left(\frac{E_{00}}{V_t}\right)}, \tag{5}$$

$$E_{00} = \frac{h}{2}\sqrt{\frac{N_A}{m\varepsilon_s}}, \tag{6}$$

and

$$\varepsilon' = \frac{E_{00}}{E_{00}/V_{\mathrm{t}} - \tanh(E_{00}/V_{\mathrm{t}})}. \tag{7}$$

Here, E_0, ε', and E_{00} are energy parameters associated with the TFE model, A^* represent Richardson's constant, T is the absolute temperature, V_t is the thermal voltage, ε_s is dielectric constant of GaN, ϕ_B is Schottky barrier height, m is the effective mass of holes, h is the Planck's constant, N_V is density of state in the valence band and N_A is the acceptor doping.

2.3 *Field Emission (FE)*

FE process has a very strong dependence on the electric field and weak temperature dependence. At high electric field, the potential barrier becomes narrower. This allows tunneling of carriers through the barrier by quantum mechanical tunneling. The shape of the barrier is important in determining the amount of tunneling taking place. There are several categories of tunneling which are shown in Fig. 4.

Fowler–Nordhiem Tunneling

When potential is very narrow and has triangular shape at the presence of very high electric field, tunneling of carriers occurs via Fowler–Nordheim Tunneling (FNT)

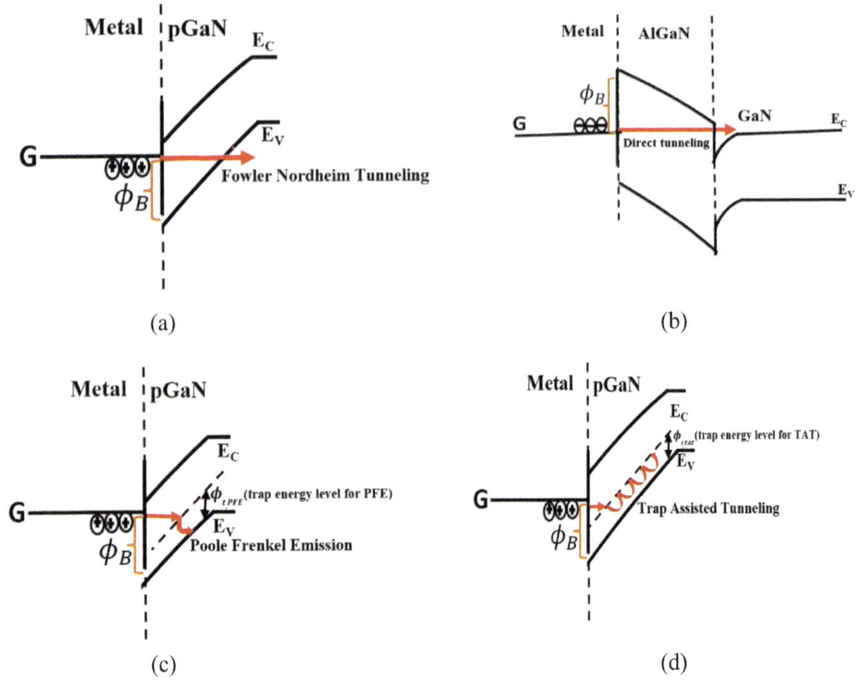

Fig. 4 Schematic of different tunneling process: **a** Fowler Nordhiem tunneling. **b** Direct tunneling. **c** Poole Frenkel emission. **d** Trap assisted tunneling

shown in Fig. 4a. The current density from FNT is given by Turuvekere et al. (2013) and Ranuárez et al. (2006)

$$J_{FNT} = \frac{A}{\phi_B} E_{j,Sch^2} exp\left(-\frac{B}{E_{j,Sch}}\phi_B^{3/2}\right) \tag{8}$$

where, *A* and *B* are *FNT* constants given below:

$$A = \frac{q^3}{8\pi h} \tag{9}$$

And

$$B = \frac{8\pi (2m^*)^{1/2}}{3qh}. \tag{10}$$

Here, $E_{j,Sch}$ is the electric field at the Schottky junction, ϕ_B is the Schottky barrier height, m^* is effective mass of holes, *h* is the Planck's constant.

Direct Tunneling

As already state, the shape of the barrier is important in determining the choice of tunneling mechanism. So, if the depletion region is extended, barrier becomes wider. The FNT is no longer valid in this case as the barrier shape is not triangular anymore. The carriers need to travel the entire width of the potential barrier which has a trapezoidal shape, hence direct tunneling takes place (Fig. 4b). The direct tunneling current density is given by Ranuárez et al. (2006):

$$J_{DT} = \frac{q^3}{16\pi^2\hbar\phi_b}\frac{1}{\left[1-\sqrt{1-\frac{qV_{j,Sch}}{\phi_b}}\right]^2}E_{j,Sch}^2 \times \exp\left\{-\frac{4}{3}\frac{\sqrt{2m^*}\phi_b^{3/2}}{\hbar q E_{j,Sch}}\left[1-\left(1-\frac{qV_{j,Sch}}{\phi_b}\right)^{3/2}\right]\right\} \quad (11)$$

where $E_{j,Sch}$ is the electric field at the Schottky junction, is the voltage drop across the Schottky junction, ϕ_B is Schottky barrier height, m^* is effective mass of holes and $\hbar$ is the reduced Planck's constant.

Poole Frenkel Emission

In Poole Frenkel Emission (PFE), the tunneling process occurs via trap at the presence of a large electric field (Fig. 4c). The tunneled carriers are initially trapped in localised trapped state characterised by a particular trap level ($\phi_{t,PFE}$). At high electric field, they have energy to break free from its localised state and move into valence band, for example if it is a hole trap. The PFE current density is given by Stockman et al. (2018)

$$J_{PFE} = CE_{j,Sch}\exp\left(-\frac{q\left(\phi_{t,PFE}-\beta\sqrt{E_{j,Sch}}\right)}{kT}\right) \quad (12)$$

where, C is a constant associated with PFE and $\beta = \sqrt{q\pi/\varepsilon}$ is the Schottky factor.

Trap Assisted Tunneling

Traps aid carrier conduction through hopping mechanism and random fluctuation within an atom (Fig. 4d). It primarily occurs at low electric field. The TAT current density is given by Dutta et al. (2017)

$$J_{TAT} = C_1 exp\left(-\frac{C_2}{E_{j,Sch}}\varphi_{t,TAT}^{3/2}\right) \quad (13)$$

where, C_1 is a temperature dependent parameter, $E_{j,Sch}$ is the electric field at the Schottky junction, $\varphi_{t,TAT}$ is the trap depth level measured from the top of valence band and C_2 is given by.

$$C_2 = \frac{8\pi\sqrt{2qm^*}}{3h}. \tag{14}$$

3 Charge-Based EPFL HEMT Model

This model was developed by a team at EPFL in Switzerland (Jazaeri and Sallese 2019). This model was adopted from regular silicon FETs with some additional physical parameters needed to extend this model for GaN HEMTs. Thus, all the features of the compact modeling in MOSFET are included such as convergence and accuracy in predictions. This model shows that the charge density has a linear dependence on the surface potential which can be used to derive channel charge and drain current expressions for all operating conditions. In this section of this chapter, we will only focus on the computation of the gate charge which can be calibrated from this model. Similar to any compact model in MOSFET, Poisson's equation must be solved which is given as follows.

$$\frac{d^2\Psi(x)}{d\mathrm{x}^2} = \frac{\rho}{\varepsilon_{1,2}} \tag{15}$$

where Ψ is the electrostatic potential, ρ is the local charge density ε_1 and ε_2 is the permittivity of AlGaN and GaN respectively. Here, x represents the direction going from the GaN surface going into the GaN bulk. In this model, it is assumed that the AlGaN layer is fully depleted and part of the GaN layer close to the surface is also fully depleted. Thus, x_1 and x_2 represent the thickness of depleted AlGaN and GaN layers. The electric fields in the depleted AlGaN and GaN layer are given by:

$$E_n(x) = \frac{qN_D}{2\varepsilon_1}x - a \tag{16}$$

$$E_p(x) = -\frac{qN_A}{2\varepsilon_1}x - a'. \tag{17}$$

Integrating the above electric field will give rise to an electrostatic potential stated below.

$$\Psi_n(x) = -\frac{qN_D}{2\varepsilon_1}x^2 + ax + b \tag{18}$$

$$\Psi_p(x) = \frac{qN_A}{2\varepsilon_2}x^2 + a'x + b'. \tag{19}$$

Here, q is the electronic charge, N_D and N_A represents the doping in the AlGaN and GaN layers respectively. Coefficients a, a', b and b' are evaluated from the boundary conditions.

Since we assume upto x_2 thickness in the GaN layer is fully depleted, the electric field disappears inside the GaN bulk after this point. So, we have $E_p(x_2) = 0$. Thus, we get $a' = -\frac{qN_A}{2\varepsilon_1}$. Therefore, the potential drop in GaN bulk is given by,

$$\Psi_2 = \Psi_p(0) - \Psi_p(x_2) = \frac{qN_A}{2\varepsilon_2}x_2^2 = \Psi_s. \tag{20}$$

where, Ψ_s is the surface potential of the GaN layer.

Computation of the gate charge presumes the charge in the gate electrode is being compensated by the total charge in the semiconductor. So, $E_G(x_1) = -\frac{Q_G}{2\varepsilon_1}$ where $Q_G = qn_G$. Here, n_G is charge per density in the gate electrode. Applying this condition when applied in (16), gives $a = \frac{qN_D}{\varepsilon_1}x_1 - \frac{qn_G}{\varepsilon_1}$. So, Eq. (18) can be written as

$$\Psi_n(x) = -\frac{qN_D}{2\varepsilon_1}x^2 + \frac{q}{\varepsilon_1}(N_D x_1 - n_G)x + b. \tag{21}$$

The maximum value of $\Psi_n(x)$ can be found at ($x = x_{max}$) such that $\frac{d\Psi_n(x)}{dx} = 0$. This occurs at $x_{max} = x_1 - \frac{n_G}{N_D}$. . So, the potential at the AlGaN barrier (Ψ_1) can be obtained by subtracting the potential at the gate electrode ($\Psi_n(x_1)$) from the surface potential in the GaN layer ($\Psi_n(0)$). So, this results in.

$$\Psi_1 = \Psi_n(x_1) - \Psi_n(0) = \frac{qN_D}{2\varepsilon_1}x_1^2 - \frac{qn_G}{\varepsilon_1}x_1. \tag{22}$$

Applying KVL in the gate stack, we can write,

$$V_G = \phi_B + \Psi_1 + \Psi_s - \frac{\Delta E_C}{q}. \tag{23}$$

Now from Eq. (9), the gate charge density can be written as

$$n_G = -\frac{\varepsilon_1}{qx_1}\left(\Psi_1 - \frac{qN_D}{2\varepsilon_1}x_1^2\right). \tag{24}$$

Using Eq. (10) in Eq. (11) results in

$$n_G = -\frac{\varepsilon_1}{qx_1}\left(V_G - \phi_B - \Psi_s + \frac{\Delta E_C}{q} - \frac{qN_D}{2\varepsilon_1}x_1^2\right). \tag{25}$$

Therefore, it can be deduced that the gate charge density has linear dependency on the surface potential (Ψ_s).

4 Charge-Based Compact Model of Gate Leakage

This model is based on the computation of the gate current from electric charge (Debnath et al. 2020). As stated earlier, many conventional Schottky conduction mechanisms may be working in tandem during the whole operating bias conditions. This study is used for both standard AlInN/GaN HEMTs and AlGaN/GaN HEMTs (Fig. 1). Therefore, gate current can be calculated as the summation of individual currents from each mechanism in a particular device. So, it can be written as,

$$I_G = I_{FNT} + I_{PFE} + I_{TE} + I_{DAT}. \tag{26}$$

where, I_{FNT}, I_{PFE}, I_{TE} and I_{DAT} are the current components from Fowler-Nordhiem Tunneling, Poole Frenkel Emission, Thermionic Emission, and Defect Assisted Tunneling mechanisms respectively. The gate current can be derived from its current density (J_G) which is given by:

$$I_G = W\int_0^L J_G dx = W\int_{Q_{ns}}^{Q_{nd}} J_G(Q_n)\frac{dx}{dQ_n}dQ_n. \tag{27}$$

where, L and W are the length and width of the device, respectively. Q_n is the charge density calculated at a distance x from the source. Q_{ns} and Q_{nd} are the charge densities at the source and the drain terminal, respectively. To compute the current from the integral above, Q_n need to be calculated first. To solve for $\frac{\text{dx}}{dQ_n}$, the drain current linearization method is used (Karumuri et al. 2016). So, the drain current (I_{DS}) is given as,

$$I_{\text{DS}} = W\mu_n\left(\frac{-Q_n}{\alpha_m} + V_{\text{t}}\right)\frac{dQ_n}{\text{dx}}. \tag{28}$$

Here, μ_n is the 2DEG mobility, α_m is the charge linearization factor and V_{t} is the thermal voltage. The above equation can be simplified to be written as:

$$I_{DS} = W\mu_n P. \tag{29}$$

where,

$$P = \frac{Q_{ns}^2 - Q_{nd}^2}{2\alpha_m} - V_{\text{th}}(Q_{ns} - Q_{nd}). \tag{30}$$

Using (28) and (29) we can write $\frac{dx}{dQ_n}$ as

$$\frac{dx}{dQ_n} = \frac{L\left(V_{th} - \frac{Q_n}{\alpha_m}\right)}{P}. \tag{31}$$

The above equation needs to be substituted in (27) to obtain J_G. Now, this will be used to derive individual current components of the leakage current.

A. *Fowler Nordhiem Tunneling*

As stated earlier, FNT is a strong field-dependent phenomenon and has a weak temperature dependence. Thus, it is expected to play a strong role during the reverse bias (RB) conditions. The FNT current density (Sarkar and Haddara 2022) can be written as,

$$J_{FNT} = AE^2 \exp\left(-\frac{B}{E}\right). \tag{32}$$

where, A and B are the Fowler constants. The electric field in the barrier layer is given as

$$E = \frac{q(\sigma - n_S)}{\varepsilon} = \frac{q\sigma + Q_n}{\varepsilon}. \tag{33}$$

Here, σ is polarization charge, n_S is the 2DEG density and ε is the dielectric constant. So, substituting (33) in (32), J_{FNT} becomes

$$J_{FNT} = AE^2 \exp\left\{\frac{-B}{\frac{q\sigma}{\varepsilon}\left(1 + \frac{Q_n}{q\sigma}\right)}\right\}. \tag{34}$$

Further, J_{FNT} can be rewritten as

$$J_{FNT}(Q_n) = AE^2 exp(-b_1)(b_2 Q_n). \tag{35}$$

where,

$$b_1 = \frac{B\varepsilon}{q\sigma}, b_2 = \frac{b_1}{q\sigma}. \tag{36}$$

Now, I_{FNT} current can be obtained by substituting (31) and (35) into (27).

$$I_{FNT} = k_{fn} E^2 \frac{M_1 \exp(b_2 Q_{nd}) - M_2 \exp(b_2 Q_{ns})}{b_2 P}. \tag{37}$$

where,

$$k_{fn} = WLA\exp(-b_1) \tag{38}$$

$$M_1 = V_t - \frac{b_2 Q_{nd} - 1}{b_2 \alpha_m} \tag{39}$$

$$M_2 = V_t - \frac{b_2 Q_{ns} - 1}{b_2 \alpha_m}. \tag{40}$$

When there is no bias at the drain terminal, it can be deduced that $Q_{nd} = Q_{ns} = Q_A$. In this case $P = 0$, so (24) becomes indeterminate. To resolve this problem, L' Hospital rule is applied (Debnath et al. 2020; Ishiwaki et al. 2017) to (37) to evaluate the FNT current. Therefore, the FNT current given by

$$I^0_{FNT} = k_{\mathrm{fn}} E^2 exp(-b_2 Q_A). \tag{41}$$

Further details of some these parameters can be found in Karumuri et al. (2016). Now, since the FNT mechanism is already assumed for reverse bias regions. The parameters k_{fn} and b_2 can be extracted from the logarithmic curve of I^0_{FNT} and Q_A written below.

$$ln\left(\frac{I^0_{FNT}}{E^2}\right) = ln\left(k_{fn}\right) + b_2 Q_a. \tag{42}$$

The values k_{fn} and b_2 are the intercept and the slope of the above equation respectively.

B. *Poole Frenkel Emission*

Poole Frenkel Emission (PFE) has both field and temperature dependence. The PFE current density is given by Stockman et al. (2018)

$$J_{PFE} = CE\exp\left(\alpha + \beta\sqrt{E}\right). \tag{43}$$

where,

$$\alpha = -\frac{\phi_t}{V_t}, \beta = \frac{1}{V_t}\sqrt{\frac{q}{\pi\varepsilon}} \tag{44}$$

C is a constant associated with PFE. ϕ_t is the trap energy level. Now, writing (43) in terms of charge density Q_n.

$$J_{PFE}(Q_n) = CEexp(\alpha + d_1)(d_2 Q_n). \tag{45}$$

where,

$$d_1 = \beta\sqrt{\frac{q\sigma}{\varepsilon}}, d_2 = \frac{d_1}{2q\sigma} \tag{46}$$

Substituting (31) and (44) into (27), the PFE current expression is written as

$$I_{PFE1} = k_{pf}E\frac{N_1\exp(d_2Q_{nd}) - N_2\exp(d_2Q_{ns})}{d_2P}. \tag{47}$$

where,

$$\begin{aligned} k_{pf} &= WLC\exp(\alpha + d_1) \\ N_1 &= V_t - \frac{d_2Q_{nd} - 1}{d_2\alpha_m} \\ N_2 &= V_t - \frac{d_2Q_{ns} - 1}{d_2\alpha_m}. \end{aligned} \tag{48}$$

Similar to FNT case earlier, L' Hospital rule is applied to (47). So, PFE current is expressed as

$$I^0_{PFE} = k_{fp}Eexp(d_2Q_A). \tag{49}$$

The parameters k_{pf} and d_2 can be extracted from the logarithmic curve of $\frac{I^0_{PFE}}{E}$ and Q_A written below.

$$ln\left(\frac{I^0_{PFE}}{E}\right) = ln\left(k_{pf}\right) + d_2Q_a. \tag{50}$$

Further details of this analysis can be found in Karumuri et al. (2014). The values k_{pf} and d_2 are the intercept and the slope of the above equation respectively.

C. *Thermionic Emission*

Thermionic Emission is a thermally activated process. It is used here in the forward bias (FB) conditions. The TE current is written as

$$J_{TE} = J_{TE0}\exp\left\{\frac{V}{\eta V_t} - 1\right\}. \tag{51}$$

Here, J_{TE0} is the TE saturation current and η is the ideality factor. The charge can be estimated by the expression below in FB case by the expression below.

$$Q_n = -C_g(V - V_{th}). \tag{52}$$

Here, C_g is the gate capacitance and V_{th} is threshold voltage. Like previous cases in FNT and PFE, I_{TE} can be written as,

$$I_{TE} = I_{TE0}\left[\frac{A_1 exp(-k_{te}Q_{nd}) - A_2 exp(k_{te}Q_{ns})}{P}\right]. \tag{53}$$

where,

$$\begin{aligned} A_1 &= \tfrac{k_1}{\alpha_m k_{te}^2}\left[1 - \alpha_m k_{te} V_t + k_{te} Q_{nd}\right] \\ A_2 &= \tfrac{k_1}{\alpha_m k_{te}^2}\left[1 - \alpha_m k_{te} V_t + k_{te} Q_{ns}\right]. \end{aligned} \tag{54}$$

Again, applying the L' Hospital's rule in this case like previous two cases, the TE current is stated as

$$I_{TE}^0 = I_{TE0}\left[exp\{-k_{te}(Q_a - C_g V_{OFF})\} - 1\right]. \tag{55}$$

The parameters k_{te} and I_{TE0} can be extracted from the logarithmic curve of I_{TE}^0 and Q_A written below.

$$ln\left(I_{TE}^0\right) = ln(I_{TE0}) - k_{te}\left(Q_a - C_g V_{OFF}\right). \tag{56}$$

Further details of this analysis can be found in Karumuri et al. (2014). The values I_{TE0} and k_{te} are the intercept and the slope of the above equation respectively.

D. *Defect Assisted Tunneling*

Defect Assisted Tunneling (DAT) occurs when charge transfer occurs between defects through hopping. This process is similar to the TAT process described earlier. The DAT current is expressed as

$$J_{DAT} = J_{DAT0}\left\{\exp\left(\frac{V - V_0}{\eta_{DAT} V_t}\right) - 1\right\}. \tag{57}$$

Here, J_{DAT0} is the reverse saturation DAT current and η_{DAT} is the ideality factor. V_0 is an adjustable parameter for zero bias conditions such that $I_G = 0$ when $V_G = 0$. Now, DAT current density in terms of Q_n,

$$J_{DAT}(Q_n) = J_{DAT0}\{k_2 exp(-k_{dat}Q_n) - 1\} \tag{58}$$

$$I_{DAT} = I_{DAT0}\left[\frac{B_1 exp(-k_{dat}Q_{nd}) - B_2 exp(-k_{dat}Q_{ns})}{P}\right]. \tag{59}$$

Like previous cases in FNT, PFE, and TE, I_{DAT} can be written as,Where,

$$\begin{aligned} B_1 &= \frac{k_2}{\alpha_m k_{dat}^2}\left[1 - \alpha_m k_{dat} V_t + k_{dat} Q_{nd}\right] \\ B_2 &= \frac{k_2}{\alpha_m k_{dat}^2}\left[1 - \alpha_m k_{dat} V_t + k_{dat} Q_{ns}\right] \end{aligned} \tag{60}$$

Again, applying the L' Hospital's rule in this case like previous three cases, the DAT current is given as follows.

$$I_{DAT}^0 = I_{DAT0}\{k_2 \exp(-k_{dat} Q_A) - 1\}. \tag{61}$$

where,

$$I_{DAT0} = \frac{k_{\text{pf}} E_0 \exp(d_2 Q_{a0})}{k_2 \exp(-k_{\text{dat}} Q_{a0}) - 1}. \tag{62}$$

Q_{a0} is channel charge 2DEG density and E_0 is the electric field in the barrier layer at equilibrium condition ($V_G = V_{DS} = 0, I_G \sim I_{PFE} + I_{DAT} = 0$).

The parameters k_{dat} and $I_{DAT0}k_2$ can be extracted from the logarithmic curve of I_{DAT}^0 and Q_A written below.

$$\ln\left(I_{DAT}^0\right) = \ln(I_{DAT0}k_2) - k_{dat} Q_a. \tag{63}$$

Further details of this analysis can be found in Karumuri et al. (2014). The values $I_{DAT0}k_2$ and k_{dat} are the intercept and the slope of the above equation respectively. Additionally, using (62), I_{DAT0} and k_2 can be obtained.

5 Surface-Potential-Based Modeling Approach for the Gate Current in pGaN Gate HEMTs

Unlike the previous two models, this model is not based on charge computation (Wang et al. 2020). This model is simpler compared to the other 2 models as it involves fewer complex calculations. This model includes explicitly solves Poisson's equation in the GaN channel by considering all potential charges inside the device structure (Wang et al. 2020, 2021). Similar to derivation in long channel MOSFETs, gradual channel approximation is considered (i.e., the electric field perpendicular to the channel is negligible compared to the channel direction). This model has been developed for pGaN based HEMT device (Fig. 2). So, Poisson's equation can be written as,

$$\frac{d^2\phi}{dx^2} = \frac{q}{\varepsilon_{\text{GaN}}}\left(n_{\text{GaN}} + N_{\text{Mg}} - p_{\text{GaN}}\right). \tag{64}$$

Here, ϕ is the electrostatic potential, q is the electronic charge and ε_{GaN} is the dielectric constant of GaN. The charge density consists of N_{Mg} (the unintended Mg doped channel arising from out-diffusion), n_{GaN} and p_{GaN} are the electrons and holes charge density in the GaN channel. The electron and hole densities are given by,

$$n_{\mathrm{GaN}} = n_{\mathrm{i}} \cdot \exp((-\varphi_{\mathrm{F}} + \phi - V_{\mathrm{C}})/V_{\mathrm{T}}) \tag{65}$$

$$p_{\mathrm{GaN}} = n_{\mathrm{i}} \cdot \exp\left(\frac{(\varphi_{\mathrm{F}} - \phi)}{V_{\mathrm{T}}}\right). \tag{66}$$

where, n_{i} is the intrinsic carrier density, φ_{F} is the quasi-fermi potential [$\varphi_F = V_T\left(\ln(N_0/n_i) + 2^{-1.5}(N_0/n_i)\right)$] and V_{C} is the voltage between the channel and the substrate. N_0 is approximately equal to the total hole and electron densities.

Some boundary conditions need to be considered to solve the above equations. These two conditions are given as follows.

$$\phi(x) = 0 \text{ for } x \to \infty \tag{67}$$

$$\frac{d\phi}{dx} = 0 \text{ for } x \to \infty \tag{68}$$

So, $\phi(x = 0) = \phi_{\mathrm{GaN}}.$. Now, surface potential equation (SPE) (Wang et al. 2021) can be formed which is given below.

$$V_{\mathrm{G}} - V_0 - \phi_{\mathrm{GaN}} = \gamma \sqrt{\begin{matrix} \frac{\phi_{\mathrm{GaN}}}{V_{\mathrm{T}}} N_{\mathrm{Mg}} + n_{\mathrm{i}} \cdot \exp\left(\frac{-\varphi_{\mathrm{F}} - V_{\mathrm{C}}}{V_{\mathrm{T}}}\right)\left[\exp\left(\frac{\phi_{\mathrm{GaN}}}{V_{\mathrm{T}}}\right) - 1\right] \\ + n_{\mathrm{i}} \cdot \exp\left(\frac{\varphi_{\mathrm{F}}}{V_{\mathrm{T}}}\right)\left[\exp\left(-\frac{\phi_{\mathrm{GaN}}}{V_{\mathrm{T}}}\right) - 1\right] \end{matrix}}. \tag{69}$$

Here, γ is the body factor, ϕ_{GaN} is GaN channel surface potential and V_0 is fitting parameter such that when $\mathrm{V_{GS}} = \mathrm{V_0}$.

The above equation works for all biasing conditions. This equation needs to be solved by numerical methods to result in an analytical solution. Full details of the methodology used can be found in Wang (2013). This surface potential model can be used in the same way to calculate potential in the pGaN layer interface. Nonetheless, this model avoids some complex calculations from the charge-based model.

For the device used in this study, it was assumed that the TE mechanism is sufficient for forward leakage characteristics and hopping transport (HT) mechanism from the surface trap which is strong influenced by temperature to be the dominant mechanism for reverse bias conditions. The TE and HT current equations derived using this model are given below.

$$J_{\mathrm{TE}} = \mathrm{Area} \cdot J_{\mathrm{TE0}} \cdot T^2 \cdot \left[\exp\left(\frac{\varphi_{\mathrm{pGaN}} - \varphi_{\mathrm{GaN}}}{n_{\mathrm{TE}} \cdot V_T}\right) - 1\right] \tag{70}$$

$$J_{\mathrm{HT}} = \mathrm{Peri} \cdot J_{\mathrm{HT0}} \cdot \left[\exp\left(\frac{\varphi_{\mathrm{pGaN}}}{n_{\mathrm{HT}} \cdot V_T}\right) - 1\right] . \exp\left[-T^{\alpha}\right]. \tag{71}$$

Here, J_{TE0} is the TE current saturation density and n_{TE} is the ideality factor. The values that can be extracted from experimental measurements were 10^{-15} A and 0.12 respectively. Also, Peri is the gate perimeter, J_{HT0} is the HT reverse current saturation density, n_{HT} is the ideality factor. The values that can be extracted from experimental measurements for J_{HT0} and n_{HT} were 2×10^{-12} A and 14 respectively. Finally, α is the temperature coefficient ($\alpha = 0.25$).

6 Discussions

Now we are going to look at some of the results from the models that have been discussed earlier. Using Eq. (68) in the pGaN/AlGaN/GaN structure, potential at GaN and pGaN can be calculated which is represented in Fig. 5.

As indicated earlier in the SP modeling section, TE and HT mechanisms have been considered dominant mechanisms for forward and reverse bias conditions, respectively. To obtain the model parameters for Eqs. (70) and (71), logarithmic current curve versus temperature at different biases have been used to extract these parameters as shown in Fig. 6.

Finally, the gate current computed from the surface potential model has been compared with experimental measurements. The model shows good agreement as demonstrated in Fig. 7.

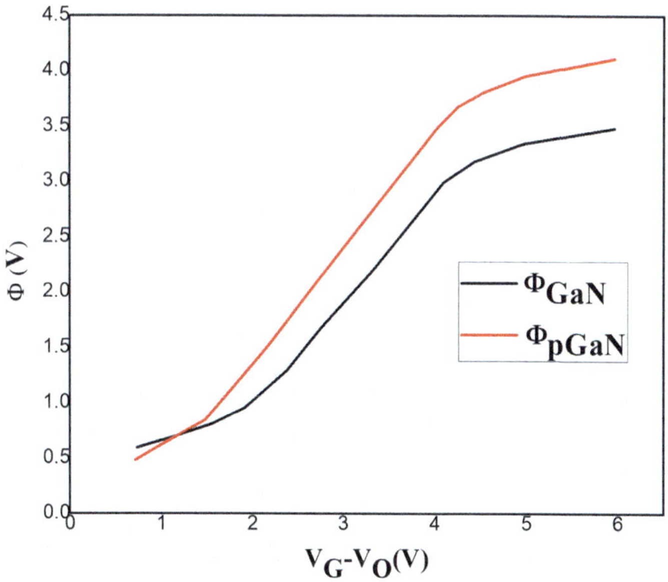

Fig. 5 Potential variation at GaN and pGaNGaN and pGaN with respect to the gate bias. (Redrawn from Wang et al. 2020)

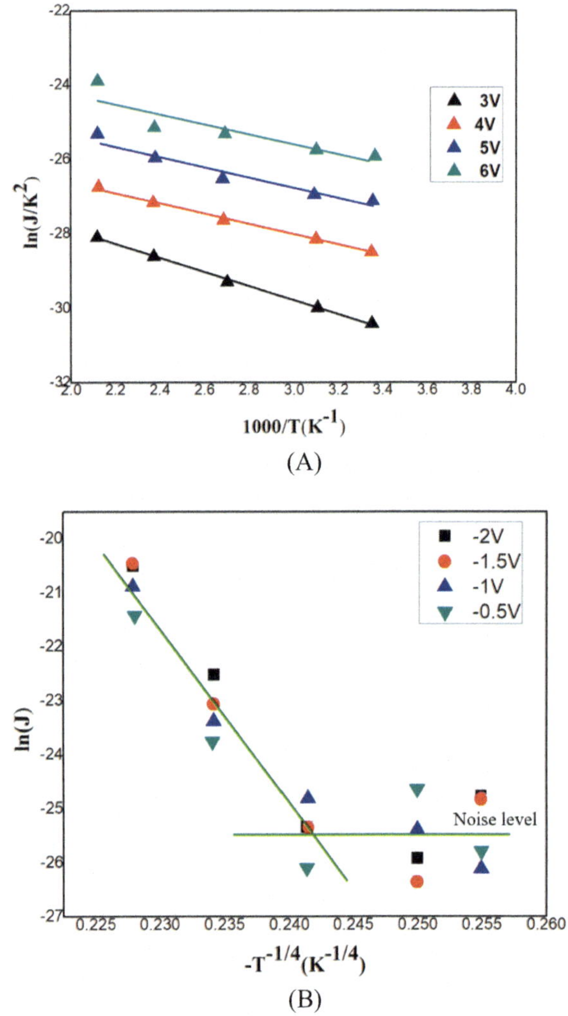

Fig. 6 Logarithmic gate current density versus temperature plots. **a** $V_G > 0$ V. **b** $V_G < 0$ V. (Redrawn from Wang et al. 2020)

The gate current consists of many components as shown charge-based model in Sect. 4 of this chapter. This model has been compared with experimental results in Fig. 8. In Fig. 8a, the model with different current components is compared to gate leakage measurements at 423 K, while Fig. 8b looks at the model with all components being summed up compared to measurement data at different temperatures.

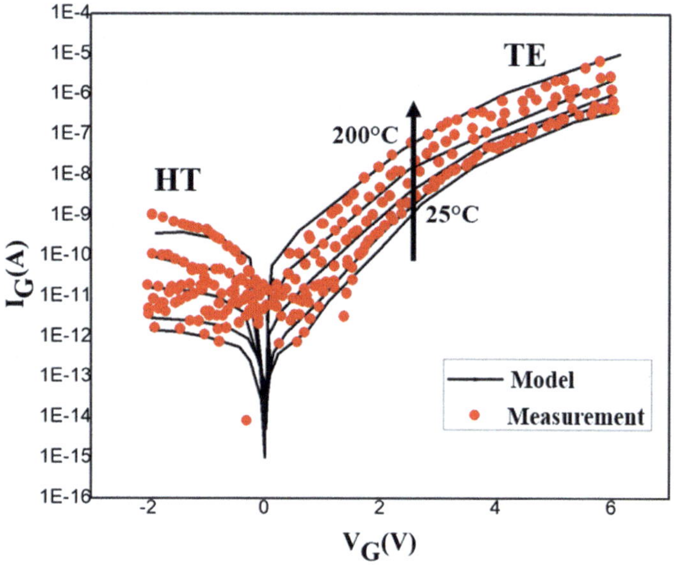

Fig. 7 Comparison of surface potential-based gate leakage current model for the pGaN/AlGaN/GaN HEMTs with experimental measurements. (Redrawn from Wang et al. 2020)

In Fig. 9, the variation of gate current has been evaluated in terms of AlGaN barrier thickness as well mole fraction. It can be seen that increasing the thickness as well as the mole fraction causes an increase in the gate current which could be due to an increase in stress created inside the barrier layer from changing those two parameters.

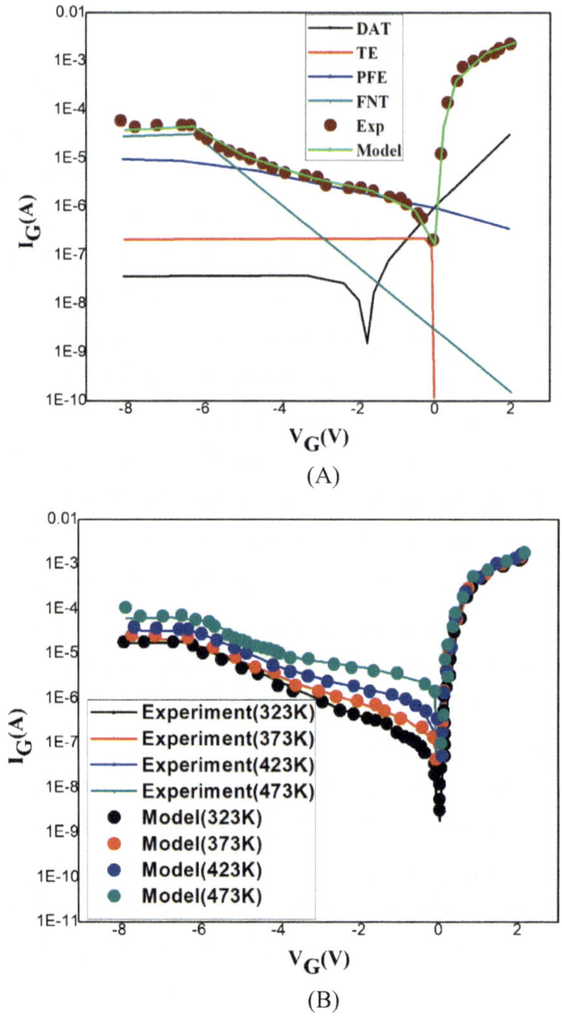

Fig. 8 Comparison of the charge-based gate current model with experimental data. **a** Gate current model with each current component being shown separately at 423 K. **b** Gate current model as whole shown at different temperatures. (Redrawn from Debnath et al. 2020)

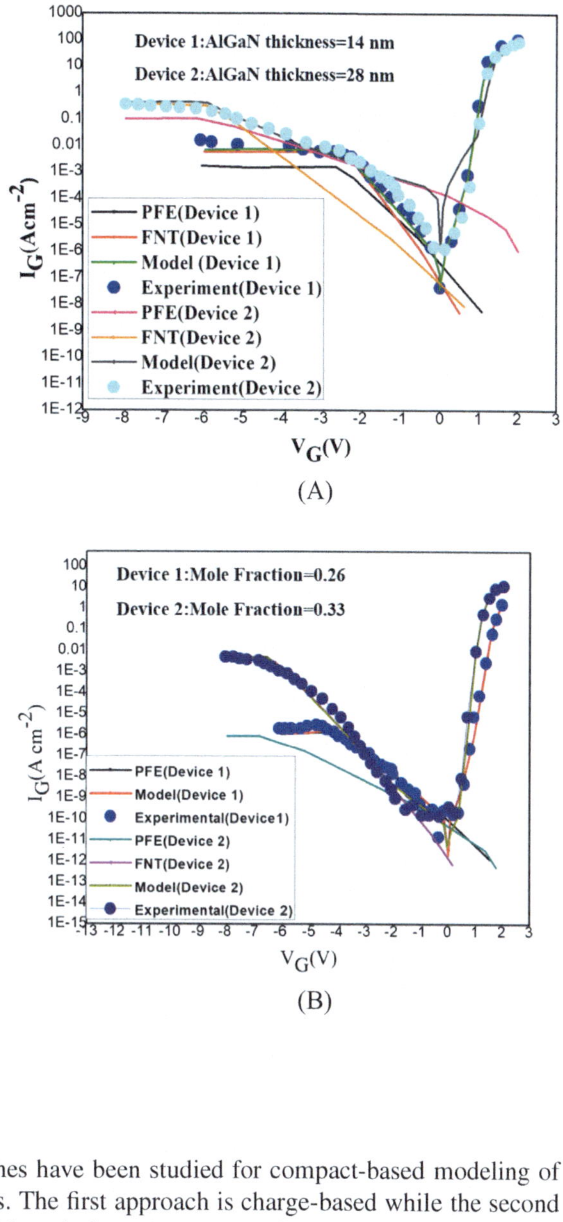

Fig. 9 Effect the gate leakage current with different device parameters. **a** Barrier layer thickness. **b** Mole fraction. (Redrawn from Debnath et al. 2020)

7 Conclusion

In this chapter, two approaches have been studied for compact-based modeling of gate leakage in GaN HEMTs. The first approach is charge-based while the second approach is surface potential based. Gate current modeling is mostly based on the choice of different assumed mechanisms for each component contributing to the total gate current in a particular device. As seen from charge based, the gate current had 4 different components while in the SP based model, the gate current had 2 different components. Both these models are excellent in each way and can be developed further to be put beside other compact models available for GaN HEMTs.

References

Bertuccio G, Fasoli L, Berroth M (1997) A SPICE model for the gate current of HEMTs

Curtice WR (2010) Nonlinear modeling of compound semiconductor HEMTs state of the art. In: 2010 IEEE MTT-S international microwave symposium. IEEE

Debnath A, DasGupta N, DasGupta A (2020) Charge-based compact model of gate leakage current for AlInN/GaN and AlGaN/GaN HEMTs. IEEE Trans Electron Devices 67(3):834–840

Dutta G, DasGupta N, DasGupta A (2017) Gate leakage mechanisms in AlInN/GaN and AlGaN/GaN MIS-HEMTs and its modeling. IEEE Trans Electron Devices 64(9):3609–3615

He J et al (2020) Characterization and analysis of low-temperature time-to-failure behavior in forward-biased Schottky-type p-GaN gate HEMTs. Appl Phys Lett 116(22):223502

Ishiwaki S et al (2017) Analysis of false turn-on phenomenon of GaN HEMT with parasitic inductances for propose novel design method focusing on peak gate voltage. In: 2017 IEEE energy conversion congress and exposition (ECCE). IEEE

Jazaeri F, Sallese J-M (2019) Charge-based EPFL HEMT model. IEEE Trans Electron Devices 66(3):1218–1229

Karumuri N et al (2014) A continuous analytical model for 2-DEG charge density in AlGaN/GaN HEMTs valid for all bias voltages. IEEE Trans Electron Devices 61(7):2343–2349

Karumuri N et al (2016) A compact model of drain current for GaN HEMTs based on 2-DEG charge linearization. IEEE Trans Electron Devices 63(11):4226–4232

Khandelwal S et al (2013) Robust surface-potential-based compact model for GaN HEMT IC design. IEEE Trans Electron Devices 60(10):3216–3222

Kiziroglou ME et al (2008) Thermionic field emission at electrodeposited Ni–Si Schottky barriers. Solid-State Electron 52(7):1032–1038

Meneghini M et al (2017) Reliability and failure analysis in power GaN-HEMTs: an overview. In: 2017 IEEE international reliability physics symposium (IRPS). IEEE

Pal A, Narayanan G (2014) Comparative study of enhancement-mode gallium nitride FETs and silicon MOSFETs for power electronic applications. In: 2014 IEEE 6th India international conference on power electronics (IICPE). IEEE

Radhakrishna U (2016) Modeling gallium-nitride based high electron mobility transistors: linking device physics to high voltage and high frequency circuit design. Dissertation, Massachusetts Institute of Technology

Ranuárez JC, Jamal Deen M, Chen C-H (2006) A review of gate tunneling current in MOS devices. Microelectron Reliab 46(12):1939–1956

Rossetto I et al (2016) Time-dependent failure of GaN-on-Si power HEMTs with p-GaN gate. IEEE Trans Electron Devices 63(6):2334–2339

Ruzzarin M et al (2018) Degradation mechanisms of GaN HEMTs with p-type gate under forward gate bias overstress. IEEE Trans Electron Devices 65(7):2778–2783

Sarkar A, Haddara YM (2022) Modeling of forward gate leakage current for normally off pGaN/AlGaN/GaN HEMTs. Solid-State Electron 196:108420

Statz H et al (1987) GaAs FET device and circuit simulation in SPICE. IEEE Trans Electron Devices 34(2):160–169

Stockman A et al (2018) Gate conduction mechanisms and lifetime modeling of p-gate AlGaN/GaN high-electron-mobility transistors. IEEE Trans Electron Devices 65(12):5365–5372

Tallarico AN et al (2020) Role of the AlGaN barrier on the long-term gate reliability of power HEMTs with p-GaN gate. Microelectron Reliab 114:113872

Tang Z et al (2013) 600-V normally off SiNx/AlGaN/GaN MIS-HEMT with large gate swing and low current collapse. IEEE Electron Device Lett 34(11):1373–1375

Ťapajna M et al (2015) Investigation of gate-diode degradation in normally-off p-GaN/AlGaN/GaN high-electron-mobility transistors. Appl Phys Lett 107(19):193506

Turuvekere S et al (2013) Gate leakage mechanisms in AlGaN/GaN and AlInN/GaN HEMTs: comparison and modeling. IEEE Trans Electron Devices 60(10):3157–3165

Wang J (2013) Study on the surface potential-based model of GaN HEMT. Dissertation, Department of Electrical Engineering, Zhejiang University

Wang J et al (2020) Surface-potential-based compact model for the gate current of p-GaN gate HEMTs. IEEE Trans Electron Devices 67(9):3564–3567

Wang J et al (2021) Surface-potential-based compact modeling of p-GaN gate HEMTs. Micromachines 12(2):199

Xu N et al (2018) Gate leakage mechanisms in normally off p-GaN/AlGaN/GaN high electron mobility transistors. Appl Phys Lett 113(15):152104

Modelling of Kink Effects and Current Collapse in GaN HEMTS

Sneha Kabra and Neha Garg

Abstract GaN HEMTs are attractive devices for high-power and high-frequency applications due to their superior electron mobility and wide bandgap characteristics. However, they are also susceptible to other phenomena, such as current collapse and kink effects, which can reduce their performance and dependability. Precise material models are crucial to accurately represent the fundamental physics of GaN HEMTs. These model stake into account impact ionization, trap capture and emission kinetics, and other related physical procedures. Additionally, it is necessary to properly incorporate surface and interface effects into the models because they are crucial to both kink effects and current collapse. Understanding the behavior of the device and developing optimized circuits depend on modeling the kink effect in GaN HEMTs. Various methods have been devised to capture and reproduce the kink effect, including empirical modeling and physical models. To study the kink effect detected in S_{22} for AlGaN/GaN HEMTs using a physics-based model AlGaN/GaN HEMT must first undergo thorough RF characterization at a range of bias settings and RF power levels. The S-parameters (S_{11} and S_{22}) must be measured to examine the device's transmission and reflection characteristics. At higher drain bias levels or RF power levels, the kink effect is often seen as an abrupt decrease in the S_{22} parameter (reflection coefficient).Then, a physics-based model is created to comprehend the underlying physical principles driving the kink effect. The kink phenomena in AlGaN/GaN high-electron-mobility transistors require the development of a large-signal model that encompasses all aspects of device behavior, including the kink phenomenon under high-power and high-voltage situations.

Keywords Kink effects · Impact Ionization · Passivation · RF dispersion · Surface states · Traps

S. Kabra (✉) · N. Garg
Department of Instrumentation, Shaheed Rajguru College of Applied Sciences for Women, University of Delhi), Vasundhara Enclave Delhi, New Delhi 110096, India
e-mail: snehakabra1@gmail.com

N. Garg
e-mail: neha.garg@rajguru.du.ac.in

D. Nirmal and J. Ajayan (eds.), *Modeling of AlGaN/GaN High Electron Mobility Transistors*, Springer Tracts in Electrical and Electronics Engineering,
https://doi.org/10.1007/978-981-97-7506-4_8

1 Origin of Kink Effect in Current–Voltage Characteristics

High Electron Mobility Transistors (HEMTs) based on Gallium Nitride (GaN) have exhibited impressive capabilities like high power, high frequency, and high-temperature applications, due to the exceptional breakdown voltage (Pengelly et al. 2012; Raut et al. 2021), abundant concentration of two-dimensional electron gas (2-DEG) (Bhat et al. 2023), and high saturation electron velocity (Chen et al. 2021). Despite these notable advantages, the practical implementation of GaN-based HEMTs has been constrained by critical concerns such as current dispersion and device reliability (Roccaforte et al. 2019).

The drain current collapse is a persistent issue observed in AlGaN/GaN HEMTs, which leads to device degradation (Rawal et al. 2019). The origin of this phenomenon is associated with various factors, including the presence of traps in the buffer layer and barrier layer (Khade et al. 2023), as well as the formation of a virtual gate (Wang et al. 2021). Different terminologies such as knee walkout, RF dispersion, and kink in the output current–voltage (I–V) characteristics have been used to describe drain current collapse phenomenon.

The Kink effect has been identified in diverse MOSFET variations and is predominantly associated with the slow trapping of charge carriers. It manifests as a notable decline in the output current, particularly within the linear region of the output characteristic, succeeded by a sudden upsurge in the drain current once a specific drain-source voltage (V_{DS}) threshold is exceeded. This abrupt surge in the output current generates a characteristic kink in the I-V output characteristic, hence the name "kink effect".

The kink effect observed in GaN HEMTs arises due to:

1. Impact Ionization: Under high electric field conditions, GaN HEMTs experience carrier (electrons or holes) acceleration, enabling them to attain sufficient energy to initiate impact ionization. Consequently, this phenomenon generates additional electron–hole pairs, resulting in an augmented drain current.
2. Traps: GaN HEMTs may contain defects, impurities, or traps within their structure. These traps have the capability to capture and release charge carriers. As impact ionization generates carriers, some of them can be captured by these traps, thereby reducing the number of available carriers for conduction.

The interaction between charge trapping and impact ionization results in the kink effect. Impact ionization produces more carriers thereby increasing the drain current. This boosts the electric field everywhere around the device, speeding carriers even more and accelerating impact ionization. The existence of traps, however, captures carriers that cause the drain current to fall past the kink threshold.

The kink effect, observed by various researchers, has been associated with different mechanisms underlying its origin. Ando et al. (n.d.) observed a kink in the output I-V characteristic of passivated AlGaN/GaN HEMTs at higher drain currents but did not provide an explanation for its occurrence. Xu et al. (n.d.) described a kink in the dc output I-V characteristics of GaN HEMTs, which could be reduced by using Al_2O_3 passivation instead of Si_3N_4. Lin et al. (2005) proposed a time-constant-related mechanism that contributes to an enhanced kink at low temperatures. Killat et al. (2012) put forward a complex multistage trapping mechanism as the cause of the kink in AlGaN/GaN HEMTs. Binari et al. (2001) and Fu et al. (2011) attributed the kink to traps in the buffer layer. Wang et al. (2011) suggested that the kink may be induced by hot electron trapping and field-dependent detrapping of electrons from donor-like buffer traps. Meneghesso et al. (2009) proposed a mechanism involving negative charge buildup in the buffer layer, which is responsible for the kink in GaN HEMTs. Sun et al. (2007) suggested that impact ionization at higher drain biases contributes to the kink.

Vetury et al. (2001) proposed the concept of virtual gate formation as an explanation for the collapse in drain current. This formation of the virtual gate, comprised of the neutralized surface states and the negative polarization charge which affect the 2-DEG concentration in the channel. The concentration is reduced by an amount equivalent to the density of neutralized surface states through internal charge transfer. This process alters the charge distribution and overall behavior of the HEMT. When a drain-to-gate voltage (V_{DG}) is applied to a GaN HEMT, the lateral electric field disrupts the equilibrium between the surface states and the 2-DEG channel. This disturbance affects the charge transfer dynamics, leading to a collapse in the drain current.

GaN HEMT performance must be optimized, especially in high-power and high-frequency applications. This requires an understanding of the kink effect. In order to reduce or mitigate the kink effect and improve the general performance of GaN HEMTs, researchers are actively looking into techniques including trap engineering, material advancements and device design improvements.

2 Impact Ionization

Impact ionization occurs when electrons, moving through a region with a strong electric field, experience scattering events that result in the transfer of excess energy to valence electrons. This energy transfer elevates the valence electrons into the conduction band, leading to the formation of new electron-hole pairs and a subsequent increase in the density of free carriers. It is worth noting that the newly generated holes can also contribute to impact ionization processes.

In the process of drain impact ionization, the holes generated can either accumulate at the gate or the source. This accumulation leads to the buildup of positive charge beneath the gate, resulting in a shift in the threshold voltage and the appearance of a "kink" effect in the current-voltage characteristics, along with an increase in gate current.

Figure 1 shows the energy band diagram along the channel which depicts the phenomena of impact ionization, when electrons within the channel reach position 1, they enter the high-field drift region situated between the gate and drain. At this point, these electrons acquire energy equivalent to the bandgap of the GaN channel

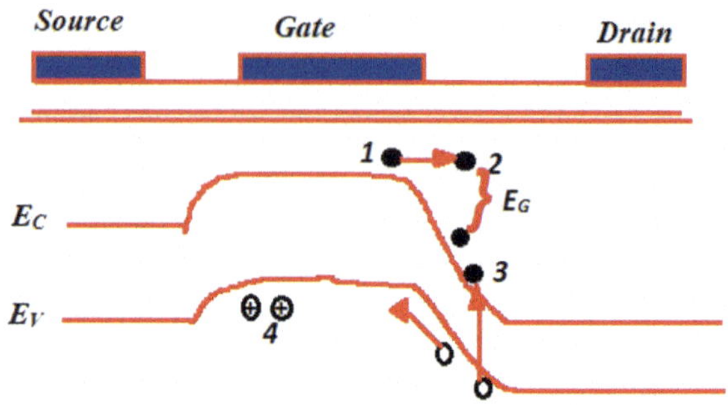

Fig. 1 Energy Band diagram along the channel

at position 2. As a result of impact ionization, these energized electrons undergo a process where two conduction band electrons and a hole in the valence band are generated. It is important to note that the collection of these electrons by the drain alone does not account for the significant increase in the drain conductance. The holes positioned at 3, are attracted to both the source and the gate. In certain cases, these holes may get trapped within the barrier layer. The accumulation of positive charge contributed by these holes, which occurs at position 4, reduces the potential barrier that exists between the source and the channel. Consequently, this leads to an amplification in the drain conductance.

The impact ionization process in these devices results in the generation of holes, which in turn leads to an increase in both the output conductance and gate leakage. This effect is observed when the devices are biased close to the threshold associated with the bandgap of the GaN channel. At lower temperatures, these effects are enhanced due to an elevated impact ionization rate of the hot electrons.

The impact ionization coefficient in gallium nitride has been estimated through Monte Carlo calculations (Jacoboni 2010) and verified experimentally in AlGaN/GaN heterostructure field-effect transistors (Cao et al. 2018). These experimental studies employed gate/drain current measurements and utilized the Hui method (Kunihiro et al. 1999) to analyze the impact ionization phenomenon. Additionally, numerical simulations have been conducted by Manel et al. to predict impact ionization in lateral AlGaN/GaN high-electron-mobility transistors (Charfeddine et al. 2013).

3 RF Dispersion

Dispersion refers to the phenomenon where the dynamic behavior of a device deviates from its static characteristics. Dispersion in GaN HEMTs structures arises due to the leakage of electrons from the gate onto the semiconductor surface located on the drain side of the device (Roff et al. 2009). This phenomenon is particularly pronounced in GaN devices that operate at high drain bias voltages. The electron leakage gives rise to what is known as the "virtual gate" effect, wherein the accumulated electrons along the surface create an electrostatic charge that partially depletes the conducting channel (Zhu et al. 2022). Therefore, the active channel carries reduced current, resulting in a decrease in RF output power.

Dispersion in AlGaN/GaN HEMTs arises from two main sources: trapped charge and self-heating (Liu and Reese 2013). While electron trapping can occur throughout the device, the surface is identified as a significant contributor to dispersion in these HEMTs. For frequencies below the Ka-band range, the dispersion can be managed through a combination of surface passivation, typically using Silicon Nitride (SiN), and the implementation of a field plate structure.

However, as the operating frequency surpasses 30GHz, the use of field plate structures becomes limited due to the additional capacitance they introduce. Moreover, when the gate length is reduced to enhance the operating frequency, the peak electric field in the channel increases, resulting in higher dispersion and reduced reliability.

Achieving excellent power performance and reliability in AlGaN/GaN HEMTs relies on comprehending and controlling dispersion. This involves addressing the effects of trapped charges and self-heating and finding suitable methods to mitigate dispersion.

There have been five distinct methods developed and demonstrated to address RF dispersion. These methods are as follows:

1. Surface treatment: This approach aims to reduce the surface state density in order to minimize the likelihood of surface states trapping electrons. By diminishing surface states, the trapping effect is mitigated.
2. Passivation: The goal of passivation is to bury surface traps by depositing dielectrics onto the surface of the device. This protective layer acts as a barrier, preventing gate leakage electrons from becoming trapped in these surface traps.
3. Field plate structure: The utilization of a field plate structure aims to alleviate the crowding of the gate edge electric field. By doing so, the emission rate of trapped electrons is enhanced, improving the overall performance of the device.
4. Addition of a GaN cap: This method involves adding a GaN cap to the device, specifically to address gate lag caused by leakage current. The GaN cap helps regulate the flow of current and mitigates the detrimental effects on device operation.
5. Si surface doping: In this approach, the intention is to compensate for the trapping energy level by introducing silicon surface doping. By doing so, the 2DEG is shielded from the traps, reducing their impact.

Different techniques to measure RF dispersion in AlGaN/GaN HEMTs are as follows:

1. S-parameter measurements: S-parameters are widely used for describing RF devices. In this measurement, the scattering parameters (S-parameters) of the HEMT are measured over a range of frequencies using a vector network analyzer (VNA). S-parameters provide information about the device input–output characteristics and can show the frequency-dependent behavior of the device.

2. Small-signal measurements: In Small-signal analysis low-power, small-amplitude signals are applied to the HEMT and small-signal gain and phase responses are measured. The device dispersion characteristics can be studied by varying the frequency of the input signal.
3. Large-signal pulsed measurements: In this method, high-power, pulsed signals are applied to the HEMT, and the output response is analyzed. The device behavior under large-signal conditions, such as output power compression and distortion, can indicate RF dispersion effects.
4. Time-domain measurements: Dispersion-related effects such as trapping and relaxation processes, can be found by applying fast-rise time pulses and analyzing the device output waveform.
5. Frequency-dependent noise measurements: By studying the noise characteristics across a frequency range, the presence of RF dispersion effects can be identified.

4 Buffer Layer Traps

Trap states are deep-level states that exist within the bandgap of a material and have a significant impact on the frequency response and may introduce delays. These trap states also contribute to power loss. They can manifest in various regions such as the AlGaN surface, the AlGaN barrier layer, the AlGaN/GaN heterointerface, or the GaN buffer layer

In AlGaN/GaN high-electron-mobility transistors (HEMTs), the presence of trap states leads to voltage delays in device operation due to the trapping and de-trapping processes. This can result in a degradation of the device's power handling capability (Nautiyal et al. 2022).

Bulk traps refer to electron traps present in various parts of the GaN buffer, AlGaN barrier, or other regions within the heterostructure of the device, as depicted in Fig. 2. Joh and Alamo et al. (2011) demonstrated that electron injection from the gate can induce bulk trapping effects. Zheng et al. (2016) illustrated that varying the gate-to-source voltage (V_{GS}) has significant effects on two types of traps: TP1 in the GaN channel layer and TP2 in the AlGaN barrier.

Some common sources of buffer traps in AlGaN/GaN HEMTs:

1. Dislocations: Dislocations are imperfections in the crystal lattice that can develop during the formation of the AlGaN/GaN heterostructure. AlGaN/GaN HEMTs are particularly prone to screw and edge dislocations, which can result in trap-induced dispersion and poor device performance.
2. Surface States: Surface states can be produced at the interface between the semiconductor and the dielectric layer or the surrounding environment. Contaminant adsorption, insufficient passivation, or other interface-related processes can all result in surface states. These surface states have the potential to operate as trap sites, causing carrier trapping and release that can change the properties of the device.

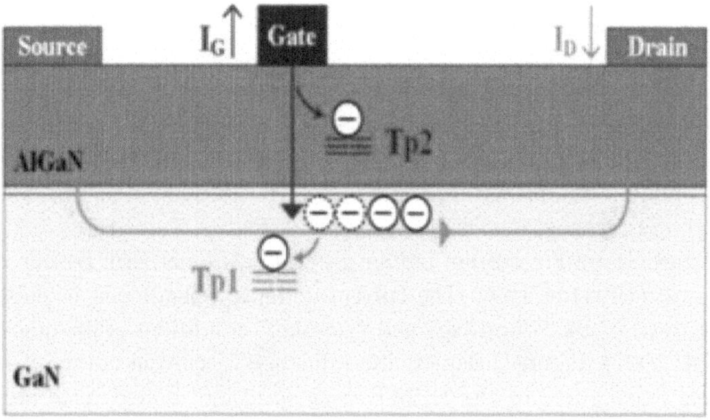

Fig. 2 Bulk Traps in AlGaN/GaN HEMTs (Zheng et al. 2016)

3. Buffer Layer Defect: Defects like point defects, vacancies, or impurities may be present in the buffer layer, which acts as the framework for the AlGaN/GaN heterostructure. These flaws have the potential to produce trap states that impede carrier movement through the device.
4. Interface roughness: Localised trap states can develop as a result of non-idealities in the interface between the buffer layer and the AlGaN/GaN layers, such as interface roughness or fluctuations. These traps have the potential to affect carrier transport and alter device dispersion effects.
5. Alloy fluctuations: The heterostructure's AlGaN composition might not be completely uniform, which could cause regional changes in the alloy's composition. These changes in composition can bring about trap states that affect device performance.
6. Piezoelectric polarization: The lattice mismatch between the AlGaN and GaN layers can cause strain that can lead to piezoelectric polarization effects. These polarization effects can alter the charge distribution and cause potential changes, which can result in trap states.

In order to reduce the impact of buffer traps and improve device performance, techniques such as enhanced material growth procedures, surface passivation, and engineering of heterostructures have been developed.

5 Current Collapse

Gallium Nitride High Electron Mobility Transistors (GaN HEMTs) experience an unwanted phenomenon referred to as current collapse. This phenomenon leads to a reduction in the operational current level compared to the device's inherent current capacity. The occurrence of current collapse is linked to electron entrapment, which can manifest either on the surface or within the bulk of the epitaxial layers (Huang et al. 2018). It is assumed that the trap levels within a particular heterostructure originate during the crystal growth phase and subsequent device processing (Jones et al. 2015).

The consequence of current collapse is an increase in the device's on-resistance, which poses a significant challenge for GaN devices intended to be used in power-switching applications. Furthermore, it also leads to a reduction in transconductance for RF (Radio Frequency) devices.

The first article on the current collapse in AlGaN/GaN HFETs was published by Khan et al. (1994) in 1994. The current collapse phenomena in these devices are characterized by the "drain-lag" and "gate-lag" conditions at the quiescent bias (Binari et al. 2001). Figure 3 depicts the influence of current collapse on the ID-VD characteristics curve of the HEMT, with solid and dotted lines representing the curve before and after the occurrence of current collapse, respectively. As an outcome of current collapse, the knee voltage (V_{knee}) increases while the maximum current (I_{Dmax}) decreases. This reduction in I_{Dmax} leads to a decrease in the maximum achievable output power (P_{max}).

Surface passivation is a technique employed to reduce the density of electron traps present on the surface of AlGaN. Several dielectric materials have been examined for this objective, encompassing silicon nitride (SiN_X), silicon dioxide (SiO_2), SiON, zirconium dioxide (ZrO_2), aluminum oxide (A12O_3), hafnium dioxide (HfO_2) (Cheng et al. 2020; Liu et al. 2007), and various others. While surface passivation with dielectric films has shown improvement in reducing current collapse levels, it does not remove this problem completely.

Field plates are commonly employed to suppress current collapse in GaN devices. A field plate consists of a metal plate strategically positioned above the gate-drain

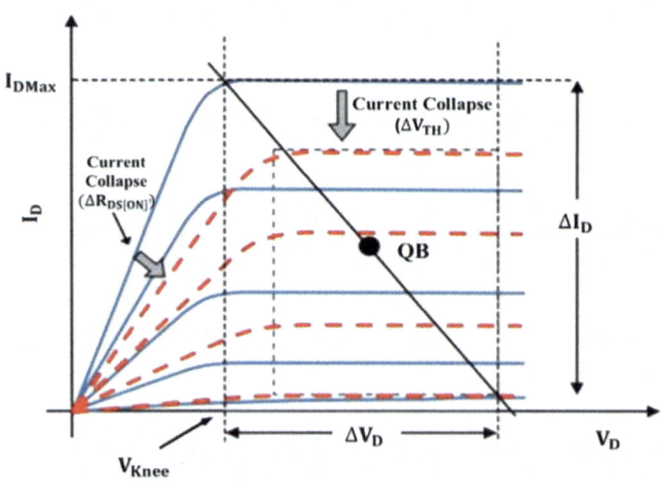

Fig. 3 Impact of the current collapse on the I_D-V_D characteristics curve of the HEMT (Hasegawa et al. 2003)

region of the device, where the electric field is most intense. This metal plate is elevated by dielectric passivation films, which serve as insulation between the field plate and the device. The field plate effectively redistributes the electric field, mitigating the effects of current collapse and enhancing device performance. By utilizing this technique, the effect of the current collapse on GaN devices can be significantly reduced (Hasan et al. 2013).

Nevertheless, the incorporation of field plates in AlGaN/GaN HEMTs introduces a rise in drain-to-source capacitance (C_{ds}) and gate-to-drain capacitances (C_{gd}), leading to a reduced current gain frequency (f_T). Consequently, a trade-off arises between achieving high breakdown voltages and operating at high frequencies. Furthermore, the implementation of an extra passivation layer for field plate fabrication influences the strain in the barrier layer, thereby diminishing power performance.

Field plates can be arranged in different configurations, such as source-connected field plates, gate-connected field plates, and unconnected field plates. In cases where multiple field plates are employed, they are typically separated by a dielectric film, such as SiN_X or SiO_2, with a chosen thickness. Saito et al. have reported that the implementation of field plates can enhance the breakdown voltage and mitigate the increase in dynamic on-resistance (RON) and current collapse caused by electric field concentration at the gate edge on the drain side (Saito et al. 2007). Improved breakdown voltage allows for higher drain bias, resulting in enhanced output power performance and improved reliability.

Moreover, field plates have demonstrated their effectiveness in reducing electron trapping by limiting the tunneling injection of electrons into surface traps located in the gate-drain region (Brannick et al. 2009). However, when employing gate field plates, since they effectively extend the gate using a dielectric layer underneath, they result in an increased effective gate length, resulting in a degradation of saturated drain current and transconductance (Shi et al. 2022). Additionally, this leads to an elevation in the drain-to-gate capacitance, which acts as a drain-to-gate feedback capacitor, influencing signals at the input and output of the device and negatively impacting RF performance, such as a decrease in cut-off frequency.

6 Modeling of Kink-Effect in RF Behaviour of GaN HEMTs

Gallium Nitride High Electron Mobility Transistors (GaN HEMTs) exhibit the kink effect, a phenomenon that alters their RFbehavior. During operation in the saturation region, the device exhibits a notable increase in the drain current accompanied by a decrease in the output resistance. The kink effect can be modeled using a variety of methods, including physics-based and empirical models.

Curve-fitting methods are used in empirical models to represent the behaviour of the kink effect based on experimental data. These models can be used in circuit simulations or device design since they frequently use additional parameters or equations

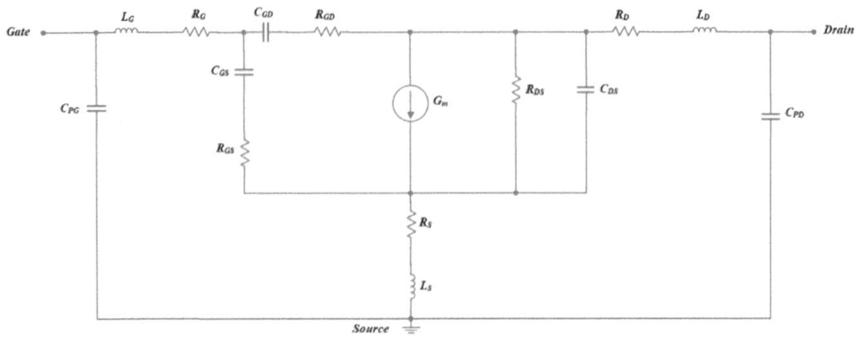

Fig. 4 Small Signal Model to study the kink effect observed in S_{22} for AlGaN/GaN HEMTs

to characterize the kink effect. On the other hand, physics-based models consider the device material characteristics and the trapping and untrapping of charge carriers to explain the fundamental mechanisms creating the kink effect.

Study of kink effect detected in S_{22} for AlGaN/GaN HEMTs using a physics-based model:

The kink effect arises from the dual nature of the output impedance in GaN HEMTs. It behaves as a series-RC network at low frequencies, while at high frequencies, it transforms into a parallel-RC network. This transition gives rise to a distinctive kink at the frequency of intersection between the contours representing the low and high-frequency approximations. The occurrence of the kink effect is solely ascribed to the intrinsic characteristics of the device itself, while the impact of extrinsic parasitic elements is confined to modifying the shape of the kink.

Figure 4 illustrates small signal equivalent (SSE) circuit of AlGaN/GaN HEMTs. The SSE incorporates various bias-independent extrinsic parasitic elements resulting from the device packaging, pads, and contacts. These elements are denoted as C_{PG}, C_{PD}, R_G, R_D, R_S, L_G, L_D, and L_S.

In the small signal equivalent circuit of the intrinsic device, the bias-dependent intrinsic capacitances are represented by C_{GS}, C_{GD}, and $C_{DS,}$ and resistances are denoted as R_{GS}, R_{GD}, and R_{DS}. The device transconductance is denoted by G_m, while R_S and R_D signify the access region resistances within the intrinsic device model. The intrinsic components of the device's small-signal model play a crucial role in determining the value of Z_{out}, which represents the output impedance. As S_{22} strongly depends on Z_{out}, any variations in the intrinsic device characteristics, such as transconductance (G_m), capacitances (C_{GS}, C_{GD}, and C_{DS}), and others, can significantly impact S_{22}.

The kink effect becomes more pronounced with increasing values of G_m and C_{GD}. On the other hand, an increase in C_{GS} shifts the kink towards the open circuit load on the Smith chart. However, the presence of the kink effect diminishes as C_{DS} is increased.

The extrinsic elements were derived by analyzing the S-parameters under "cold" pinch-off condition, specifically when V_{DS} was set to 0 V and V_{GS} was set to -4 V. Upon acquiring the extrinsic elements, the intrinsic elements were subsequently calculated using the intrinsic admittance (Y) parameters at every bias point (Crupi et al. 2018).

7 Large Signal Model for Kink Effect on AlGaN/GaN HEMTs

Load-pull measurements and nonlinear simulation serve as the foundation for power amplifier design and optimization. However, nonlinear simulation holds an edge over load-pull measurements due to its convenience and cost-effectiveness. Therefore, it is the preferred method in most cases. To employ this approach effectively, it is essential to have a precise large-signal model specifically developed for an AlGaN/GaN HEMT device. This model serves as a prerequisite for conducting successful amplifier design activities.

Small signal model involves taking a circuit and linearizing its components on the basis of bias or operating point. In this model, it is assumed that the signal is sufficiently small, resulting in negligible changes to the operating point, gain, capacitance, and other circuit parameters.

Conversely, the large signal model assumes the impact of the actual large signal on the operating point. It considers non-linear characteristics of the elements in the circuit and also considers the limitation imposed by the power supply to prevent variations in the operating point. Unlike the small signal model, the large signal model acknowledges the simultaneous variations in both the gain and supply values.

To analyze the kink effect in AlGaN/GaN HEMTs, a large signal model can be used. The large signal model describes the device behavior using a set of equations that relate the terminal voltages and currents. Figure 5 represents a simplified large signal model for analyzing the kink effect in AlGaN/GaN HEMTs:

The behavior of the High Electron Mobility Transistor (HEMT) under large-signal conditions is represented by a current source and three nonlinear capacitances. These capacitances change in response to the instantaneous voltages applied across them.

The nonlinear current source, I_D, is employed to simulate drain current (DC) characteristics of a GaN HEMT using a predefined set of DC current–voltage (IV) curves. This current source utilizes a fitting function to accurately represent the behavior of the drain current (Aflaki et al. 2009).

$$I_{DS} = \frac{I_{DS0}}{1 + (\delta V_{DS} + \lambda V_{GS})I_{DS0}} V_G > V_P$$

$$I_{DS} = \frac{V_{DS}}{R_{OFF}} V_G \leq V_P$$

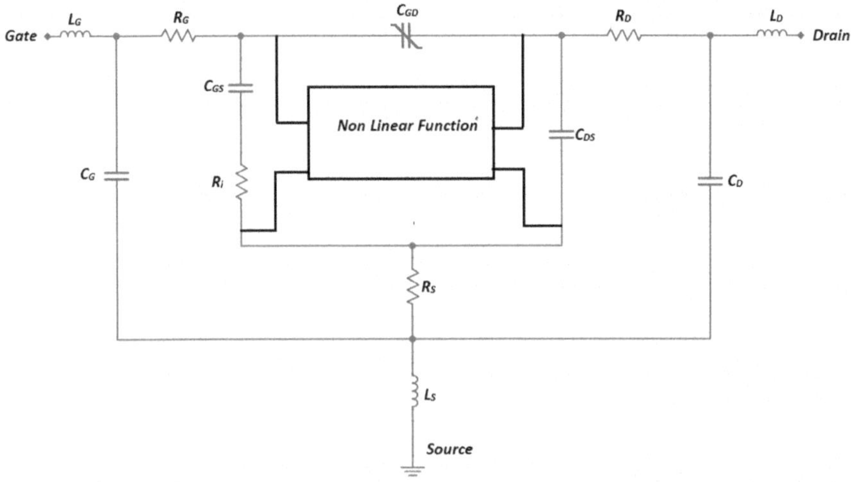

Fig. 5 Large signal model for analyzing the kink effect in AlGaN/GaN HEMTs

$$I_{DS0} = \beta(V_{GS} - V_P)^Q \tanh(\alpha V_{DS})$$

The control voltages V_{DS} and V_{GS} are used to regulate the operation of the device, while V_P represents the pinch-off voltage of the HEMT. The parameter α determines the voltage V_{DS} at which the drain current starts to saturate, and β acts as a scaling parameter. Additionally, the parameters δ and λ describe the reduction in drain current (I_{DS}) for high values of V_{DS} and V_{GS}, while the parameter Q captures the non-square law behavior commonly observed in HEMTs. When the device is in a pinched-off state, the parameter R_{OFF} represents the resistance of the entire region from the drain to the source terminal.

Through the analysis of extracted parasitic capacitance behavior in a GaN device, it has been determined that the gate-to-drain capacitance, C_{GD}, exhibits significant variations and plays a crucial role in modeling GaN devices. Due to its position between the input and output terminals of the device, C_{GD} influences both the input and output impedance. Therefore, exact modeling of its performance is essential for achieving accurate results. Recognizing the nonlinear variation of C_{GD} with respect to V_{GD}, it is more accurate to consider it as a nonlinear component instead of assuming it to be constant.

$$C_{GD} = \frac{C_{GD0}}{(1 - \frac{V_{GD}}{V_B})^n}$$

Here small-signal capacitance is represented by C_{GD0}. Additionally, there are two parameters, V_B and n, that can be determined through a curve fitting method using extracted data. These parameters are essential for accurately modeling the behavior of the capacitance and aligning it with the observed characteristics.

References

Aflaki P, Negra R, Ghannouchi FM (2009) Dedicated large-signal GaN HEMT model for switching-mode circuit analysis and design. IEEE Microw Wireless Compon Lett 19(11):740–742. https://doi.org/10.1109/lmwc.2009.2032023

Ando Y, Okamoto Y, Miyamoto H, Hayama N, Nakayama T, Kasahara K, Kuzuhara M (n.d.) A 110-W AlGaN/GaN heterojunction FET on thinned sapphire substrate. https://doi.org/10.1109/iedm.2001.979516

Bhat AM, Poonia R, Varghese A, Shafi N, Periasamy C (2023) AlGaN/GaN high electron mobility transistor for various sensing applications: a review. Micro and Nanostructures 176:207528. https://doi.org/10.1016/j.micrna.2023.207528

Binari SC, Ikossi K, Roussos JA, Kruppa W, Doewon Park, Dietrich HB, … Henry RL (2001) Trapping effects and microwave power performance in AlGaN/GaN HEMTs. IEEE Trans Electron Dev 48(3):465–471https://doi.org/10.1109/16.906437

Brannick A, Zakhleniuk NA, Ridley BK, Shealy JR, Schaff WJ, Eastman LF (2009) Influence of field plate on the transient operation of the AlGaN/GaN HEMT. IEEE Electron Dev Lett 30(5):436–438. https://doi.org/10.1109/led.2009.2016680

Cao L, Wang J, Harden G, Ye H, Stillwell R, Hoffman AJ, Fay P (2018) Experimental characterization of impact ionization coefficients for electrons and holes in GaN grown on bulk GaN substrates. Appl Phys Lett 112(26):262103. https://doi.org/10.1063/1.5031785

Charfeddine M, Echouchene F, Zaidi MA, Maaref H (2013) Numerical investigation of kink effect correlated with defects in AlGaN/GaN high electron mobility transistors. J Comp Theo Nano 10(6):1347–1353. https://doi.org/10.1166/jctn.2013.2854

Chen Z, Yue S, Peng C, Zhang Z, Liu C, Wang L, … Zhong X (2021) Hydrogen-related recovery effect of AlGaN/GaN high-electron-mobility transistors irradiated by high-fluence protons. IEEE Trans Nucl Sci 68(2):118–123.https://doi.org/10.1109/tns.2021.3051972

Cheng K-Y, Wu S-C, Yu C-J, Wang T-W, Liao J-H, Wu M-C (2020) Comparative study on performance of AlGaN/GaN MS-HEMTs with SiNx, SiOx, and SiNO surface passivation. Solid-State Electron 170:107824. https://doi.org/10.1016/j.sse.2020.107824

Crupi G, Raffo A, Vadala V, Avolio G, Schreurs DMM-P, Vannini G, Caddemi A (2018) Technology-independent analysis of the double current-gain peak in millimeter-wave FETs. IEEE Microw Wireless Compon Lett 28(4):326–328. https://doi.org/10.1109/lmwc.2018.2808418

Fu L, Lu H, Chen D, Zhang R, Zheng Y, Chen T, … Liu X (2011) Field-dependent carrier trapping induced kink effect in AlGaN/GaN high electron mobility transistors. Appl Phys Lett 98(17):173508. https://doi.org/10.1063/1.3584861

Hasan MdT, Asano T, Tokuda H, Kuzuhara M (2013) Current collapse suppression by gate field-plate in AlGaN/GaN HEMTs. IEEE Electron Dev Lett 34(11):1379–1381. https://doi.org/10.1109/led.2013.2280712

Hasegawa H, Inagaki T, Ootomo S, Hashizume T (2003) Mechanisms of current collapse and gate leakage currents in AlGaN/GaN heterostructure field effect transistors. J Vac Sci Technol B 21(4):1844. https://doi.org/10.1116/1.1589520

Huang H, Sun Z, Cao Y, Li F, Zhang F, Wen Z, … Hu L (2018) Investigation of surface traps-induced current collapse phenomenon in AlGaN/GaN high electron mobility transistors with schottky gate structures. J Phys D Appl Phys 51(34):345102. https://doi.org/10.1088/1361-6463/aad455

Jacoboni C (2010) Monte carlo simulation of bulk electron transport, 237–263. https://doi.org/10.1007/978-3-642-10586-9_14

Joh J, del Alamo JA (2011) A current-transient methodology for trap analysis for GaN high electron mobility transistors. IEEE Trans Electron Dev 58(1):132–140. https://doi.org/10.1109/ted.2010.2087339

Jones KA, Chow TP, Wraback M, Shatalov M, Sitar Z, Shahedipour F, … Tompa GS (2015) AlGaN devices and growth of device structures. J Mater Sci 50(9):3267–3307.https://doi.org/10.1007/s10853-015-8878-3

Khade RP, Sarkar S, Shanbhag A, DasGupta A, DasGupta N (2023) Substrate bias stress induced kink effect in GaN-on-silicon high-electron-mobility transistor. IEEE J Electron Dev Soc 11:294–302. https://doi.org/10.1109/jeds.2023.3275277

Khan MA, Kuznia JN, Shur MS, Chen QC (1994) Current/voltage characteristic collapse in AlGaN/GaN Heterostructure insulated gate field effect transistors at high drain bias 30(25):2175–2176. https://doi.org/10.1049/el:19941461

Killat N, Uren MJ, Wallis DJ, Martin T, Kuball M (2012) Origin of kink effect in AlGaN/GaN high electron mobility transistors: Yellow luminescence and Fe doping. Appl Phys Lett 101(15):153505. https://doi.org/10.1063/1.4757993

Kunihiro K, Kasahara K, Takahashi Y, Ohno Y (1999) Experimental evaluation of impact ionization coefficients in GaN. IEEE Electron Dev Lett 20(12):608–610. https://doi.org/10.1109/55.806100

Lin C-H, Wang W-K, Lin P-C, Lin C-K, Chang Y-J, Chan Y-J (2005) Transient pulsed analysis on GaN HEMTs at cryogenic temperatures. IEEE Electron Dev Lett 26(10):710–712. https://doi.org/10.1109/led.2005.856709

Liu C, Chor EF, Tan LS (2007) Enhanced device performance of AlGaN/GaN HEMTs using HfO2 high-k dielectric for surface passivation and gate oxide. Semicond Sci Technol 22(5):522–527. https://doi.org/10.1088/0268-1242/22/5/011

Liu Y, Reese E (2013b) AlGaN/GaN HEMT large signal nonlinear compact model accounting for thermal effects and trapping dispersion.https://doi.org/10.1109/csics.2013.6659214

Meneghesso G, Zanon F, Uren MJ, Zanoni E (2009) Anomalous kink effect in GaN high electron mobility transistors. IEEE Electron Dev Lett 30(2):100–102. https://doi.org/10.1109/led.2008.2010067

Nautiyal P, Pande P, Kundu VS, Moghadam HA (2022) Trap-assisted degradation mechanisms in E-mode p-GaN power HEMT: a review. Microelectron Reliab 139:114800. https://doi.org/10.1016/j.microrel.2022.114800

Pengelly RS, Wood SM, Milligan JW, Sheppard ST, Pribble WL (2012) A Review of GaN on SiC high electron-mobility power transistors and MMICs. IEEE Trans Microwave Theory Techn 60(6):1764–1783. https://doi.org/10.1109/tmtt.2012.2187535

Raut S, Sehra K, Mishra M, Rawal DS, Gupta M, Saxena M (2021) Proton irradiation effects on buffer-free gallium nitride on silicon carbide high electron mobility transistor-based radio frequency power amplifier. Semicond Sci Technol 36(4):045019. https://doi.org/10.1088/1361-6641/abe2df

Rawal DS, Amit, Sharma S, Kapoor S, Laishram R, Vinayak S (2019) Current collapse scaling in GaN/AlGaN/SiC high electron mobility transistors. Solid State Electron Lett 1(1):30–37. https://doi.org/10.1016/j.ssel.2019.04.002

Roccaforte F, Greco G, Fiorenza P, Iucolano F (2019) An overview of normally-off GaN-based high electron mobility transistors. Materials 12(10):1599. https://doi.org/10.3390/ma12101599

Roff C, Benedikt J, Tasker PJ, Wallis DJ, Hilton KP, MacleanJO, … Martin T (2009) Analysis of DC–RF dispersion in AlGaN/GaN HFETs using RF waveform engineering. IEEE Trans Electron Dev 56(1):13–19.https://doi.org/10.1109/ted.2008.2008674

Saito W, Nitta T, Kakiuchi Y, Saito Y, Tsuda K, Omura I, Yamaguchi M (2007) Suppression of dynamic on-resistance increase and gate charge measurements in high-voltage GaN-HEMTs with optimized field-plate structure. IEEE Trans Electron Devices 54(8):1825–1830. https://doi.org/10.1109/ted.2007.901150

Shi N, Wang K, Zhou B, Weng J, Cheng Z (2022) Optimization AlGaN/GaN HEMT with field plate structures. Micromachines 13(5):702. https://doi.org/10.3390/mi13050702

Sun HF, Bolognesi CR (2007) Anomalous behavior of AlGaN/GaN heterostructure field-effect transistors at cryogenic temperatures: From current collapse to current enhancement with cooling. Appl Phys Lett 90(12):123505. https://doi.org/10.1063/1.2715032

Vetury R, Zhang NQ, Keller S, Mishra UK (2001) The impact of surface states on the DC and RF characteristics of AlGaN/GaN HFETs. IEEE Trans Electron Dev 48(3):560–566. https://doi.org/10.1109/16.906451

Wang M, Chen KJ (2011) Kink effect in AlGaN/GaN HEMTs induced by drain and gate pumping. IEEE Electron Dev Lett 32(4):482–484. https://doi.org/10.1109/led.2011.2105460

Wang, Yan, Wu Q, Mao S, Xu R, Yan B, Xu Y (2021) Bendable microwave AlGaN/GaN high-electron-mobility transistor with output power density of 2.65 W/mm. IEEE Electron Dev Lett 42(5):677–680. https://doi.org/10.1109/led.2021.3068738

Xu D, Chu K, Diaz J, Zhu W, Roy R, Pleasant LMt, … Ye PD 0.2-μm AlGaN/GaN high electron-mobility transistors with atomic layer deposition Al_2O_3 passivation. IEEE Electron Dev Lett 34(6):744–746. https://doi.org/10.1109/led.2013.2255257

Zheng X, Feng S, Zhang Y, Yang J (2016) Identifying the spatial position and properties of traps in GaN HEMTs using current transient spectroscopy. Microelectron Reliab 63:46–51. https://doi.org/10.1016/j.microrel.2016.05.001

Zhu P, Ni X, Fan Q, Gu X (2022) Surface dispersion suppression in high-frequency GaN devices. Crystals 12(10):1461. https://doi.org/10.3390/cryst12101461

Numerical Modelling of GaN HEMTS

V. Sandeep and J. Charles Pravin

Abstract Numerical modeling has emerged as an indispensable tool in the design and optimization of AlGaN/GaN High-Electron-Mobility Transistors (HEMTs). These devices, based on the AlGaN/GaN heterostructure, exhibit remarkable properties such as high electron mobility, wide bandgap, and high breakdown voltage, making them highly promising for high-power and high-frequency applications. However, the complex interplay between material properties, device geometry, and operating conditions poses significant challenges in understanding and predicting the device behavior solely through experimental techniques. To address these challenges, numerical modeling techniques have become vital for gaining deeper insights into the underlying physics, optimizing device performance, and guiding the design process. This chapter provides an overview of the numerical modeling approaches employed in the study of AlGaN/GaN HEMTs, with a specific focus on Density Functional Theory (DFT), Finite Element Method (FEM), Monte Carlo simulation, and commercially available software packages such as COMSOL and Sentaurus TCAD. The chapter begins by introducing the AlGaN/GaN HEMT structure, highlighting its layered composition, including the AlGaN barrier, GaN channel, and various doping profiles. The unique characteristics of AlGaN/GaN HEMTs, such as the 2-Dimensional Electron Gas (2DEG) formed at the heterointerface and the polarization-induced charges, are discussed. These features contribute to the exceptional electronic properties of the devices but also introduce complexities that necessitate numerical modeling for comprehensive understanding.

Keywords Numerical modelling · Monte Carlo simulation · Density functional theory (DFT) · Finite element method (FEM) · 3D Modelling

V. Sandeep
Department of ECE, Mangalore Institute of Technology & Engineering, Mangalore, India

J. C. Pravin (✉)
Department of ECE, Kalasalingam Academy of Research and Education, Krishnankoil, Tamil Nadu, India
e-mail: jcharlespravin@gmail.com

D. Nirmal and J. Ajayan (eds.), *Modeling of AlGaN/GaN High Electron Mobility Transistors*, Springer Tracts in Electrical and Electronics Engineering,
https://doi.org/10.1007/978-981-97-7506-4_9

1 Introduction

Numerical modeling has emerged as an indispensable tool in the design and optimization of AlGaN/GaN high-electron-mobility transistors (HEMTs). These devices, based on the AlGaN/GaN heterostructure, exhibit remarkable properties such as high electron mobility, wide bandgap, and high breakdown voltage, making them highly promising for high-power and high-frequency applications. However, the complex interplay between material properties, device geometry, and operating conditions poses significant challenges in understanding and predicting the device behavior solely through experimental techniques. To address these challenges, numerical modeling techniques have become vital for gaining deeper insights into the underlying physics, optimizing device performance, and guiding the design process. This chapter provides an overview of the numerical modeling approaches employed in the study of AlGaN/GaN HEMTs, with a specific focus on density functional theory (DFT), finite element method (FEM), Monte Carlo simulation, and commercially available software packages such as COMSOL and Sentaurus TCAD. The chapter begins by introducing the AlGaN/GaN HEMT structure, highlighting its layered composition, including the AlGaN barrier, GaN channel, and various doping profiles. The unique characteristics of AlGaN/GaN HEMTs, such as the 2-dimensional electron gas (2DEG) formed at the heterointerface and the polarization-induced charges, are discussed. These features contribute to the exceptional electronic properties of the devices but also introduce complexities that necessitate numerical modeling for comprehensive understanding.

Density functional theory (DFT) is introduced as a powerful quantum mechanical approach that allows for accurate calculation of electronic properties. By solving the Schrödinger equation, DFT enables the determination of energy band structures, charge carrier densities, and transport characteristics of AlGaN/GaN HEMTs. The chapter highlights the application of DFT in studying material properties, such as band alignment, polarization effects, and strain distribution within the heterostructure, which directly impact device performance.

The Finite Element Method (FEM) is presented as a numerical technique for solving the partial differential equations governing the electrical and thermal behavior of AlGaN/GaN HEMTs. FEM provides a versatile framework for analyzing and optimizing device performance by modeling the distribution of electric fields, carrier concentrations, and temperature profiles. It enables the exploration of design parameters and their impact on device characteristics, allowing for efficient device optimization.

Monte Carlo simulation is discussed as a statistical method that models electron transport in AlGaN/GaN HEMTs at the nanoscale. By considering scattering mechanisms, carrier dynamics, and mobility characteristics, Monte Carlo simulation offers insights into device performance metrics such as current–voltage characteristics, electron velocity, and electron mobility. This technique facilitates the identification of key performance limitations and aids in the design of strategies for enhanced device performance.

Furthermore, the chapter highlights the utilization of commercially available software packages, such as COMSOL and Sentaurus TCAD, in the numerical modeling of AlGaN/GaN HEMTs. These software packages provide comprehensive simulation environments with specialized modules tailored for semiconductor device analysis. The chapter discusses the advantages of using these tools, including their user-friendly interfaces, scalability, and ability to capture complex device physics.

In conclusion, the introduction establishes the significance of numerical modeling techniques in understanding and optimizing AlGaN/GaN HEMTs. By utilizing approaches such as density functional theory, finite element method, Monte Carlo simulation, and software packages like COMSOL and Sentaurus TCAD, researchers and engineers can gain invaluable insights into device behavior, accelerate the design process, and enhance the performance and reliability of AlGaN/GaN HEMTs for a wide range of applications.

2 Density Functional Theory

Density Functional Theory (DFT) is a powerful quantum mechanical method that allows for the accurate calculation of electronic properties based on the principles of quantum mechanics. It provides a comprehensive understanding of the material properties, electronic structure, and charge carrier transport within AlGaN/GaN HEMTs. By solving the Schrödinger equation self-consistently, DFT can predict the energy band structures, density of states, and charge carrier distributions in the device. One of the key advantages of DFT is its ability to capture the effects of polarization, strain, and alloy composition in AlGaN/GaN heterostructures. These factors have a significant impact on the band alignment, resulting in charge redistribution and modification of device characteristics. DFT calculations provide insights into the band offsets, polarization charges, and interface properties, aiding in the optimization of the device structure. Moreover, DFT enables the investigation of carrier transport mechanisms and scattering processes in AlGaN/GaN HEMTs. By calculating the electron and hole effective masses and the scattering rates due to various scattering mechanisms, such as phonons, impurities, and defects, DFT can provide valuable information on carrier mobility and transport characteristics. This information is crucial for assessing the performance and reliability of the device. DFT also plays a vital role in analyzing the impact of defects and interfaces on device behavior. By simulating the formation and migration of defects, as well as their impact on electronic properties, DFT can guide defect engineering strategies for improving device performance and reliability. Additionally, DFT calculations can provide insights into the interaction of charge carriers with surfaces, interfaces, and metal contacts, which is crucial for optimizing device operation.

Ambacher et al. (2002) delves into the material properties and characteristics of these structures using experimental techniques. The authors aim to understand the pyroelectric properties, which refer to the generation of electric charge due to temperature changes, in Al (In) GaN/GaN structures. These structures are of significant interest in semiconductor device technology, particularly in the development of optoelectronic devices. The article presents experimental results regarding the pyroelectric coefficient of Al (In) GaN/GaN heterostructures and quantum well structures. The authors investigate the influence of various parameters such as composition, strain, and doping on the pyroelectric properties. They provide a comprehensive analysis of the experimental data obtained from temperature-dependent measurements. Koleske et al. (2005) discusses the application of DFT to investigate the material properties of GaN, including its bandgap, band alignment, and charge carrier behavior. By employing DFT calculations, researchers were able to gain a fundamental understanding of the electronic structure of GaN, which directly impacts the device performance. For the initial stages, the polarization charges present in the channel region can be numerically simulated using this model (Ashok et al. 2009). They also determine vacancies and interstitials present in the Ga-rich content of GaN/AlGaN type HEMTs through which the release of hydrogens from hot-electrons can be found (Roy et al. 2010a, 2010b). Hu et al. in 2010 deduced a Guass bias model for solving the Schrodinger and Poisson equations in AlGaN/GaN HEMT device using the density functional theory (Hu et al. 2010). Li et al. (2011) worked on a Generalized Gradient Approximation (GGA) method of the DFT method for studying and evaluating the electronic structures of un-doped and Al-doped GaN HEMTs. One other technique in DFT is the Local Density Approximation (LDA) that uses the Exchange Correlation in DFT depending on energy value (Puzyrev et al. 2011a). The probability of an incoming particle interacting with a defect complex can match or even exceed the probability of a high-energy transfer collision event producing a defect in a previously defect-free region once a sufficient number of defects are produced, which is initially modeled using DFT (Roy 2011; Puzyrev et al. 2011b). Figure 1 shows a schematic illustration of a conventional AlGaN/GaN HEMT.

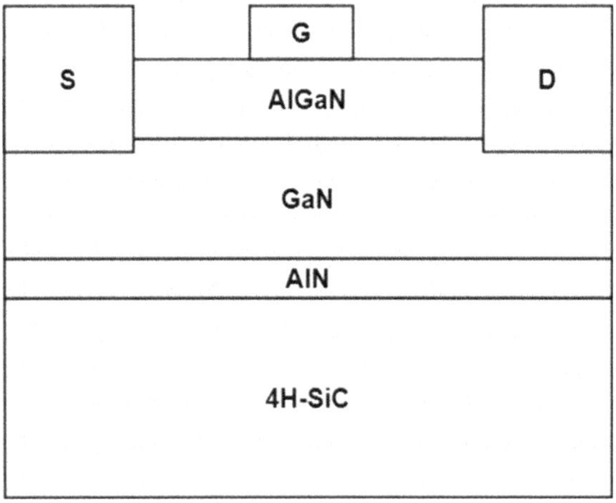

Fig. 1 Diagrammatic illustration of a conventional AlGaN/GaN HEMT structure

The AlGaN exhibits threshold voltage shift deterioration, which is consistent with the HEMT's cross-section in Fig. 1. In this regard, DFT can be used for developing surface-charge-controlled structure for these devices. One significant software package is Quantum Espresso which could be used for materials calculations (Kikkawa et al. 2012). Quantum ESPRESSO is an open-source software package widely used in materials science and computational physics for performing quantum mechanical calculations based on density functional theory (DFT). It provides a suite of tools for simulating and analyzing the properties of materials at the atomic and electronic levels. The software is designed to be modular and flexible, allowing researchers to perform various types of calculations for different materials and systems. Quantum ESPRESSO employs DFT, which approximates the quantum mechanical behavior of electrons in a solid or molecular system. By solving the Kohn–Sham equations, it provides information about the electronic structure, energetics, and other properties of materials.

$$H_S \Psi_i = (T_S + V_{ext} + V_H + V_{XC})\Psi_i \tag{1}$$

where H_S is the single-particle Hamiltonian for the non-interacting electrons; T_S is the kinetic energy operator; V_{ext} is the external potential, which includes the potential of the nuclei and any external fields; V_H is the Hartree potential, which is the potential energy due to the electron–electron interaction; V_{XC} is the exchange–correlation potential, which is the potential energy due to the exchange and correlation effects between the electrons. DFT calculations were used as the potential profiles and energy band structure for a GaN/AlN/GaN HEMT (Sołtys et al. 2014). Even while inspecting the structure's radiation hardness of the structure, the fall in energy levels can be combined with the dehydrogenation of O_N-H defects (Chen 2013; Chen et al. 2013).

2.1 *Heyd-Scuseria-Ernzerhof Hybrid DFT*

Heyd-Scuseria-Ernzerhof Calculations using a hybrid DFT model and plane wave supercells in wurtzite aluminium nitride was done for describing the nitrogen split interstitial defect's shape, formation energies, transition levels, and hyperfine tensors in a GaN HEMT (Szállás et al. 2014). The Heyd-Scuseria-Ernzerhof (HSE) hybrid density functional theory is a computational approach used in electronic structure calculations to improve the accuracy of predictions for materials and molecules. It is a variant of density functional theory (DFT) that incorporates a fraction of non-local exchange into the standard DFT exchange–correlation functional. The HSE hybrid functional seeks to improve the description of electronic properties by combining a portion of the exact Hartree–Fock (HF) exchange with the DFT exchange–correlation functional. This combination of HF exchange and DFT exchange–correlation introduces non-local effects that can better capture long-range electronic interactions. The HSE hybrid functional does come with increased computational cost

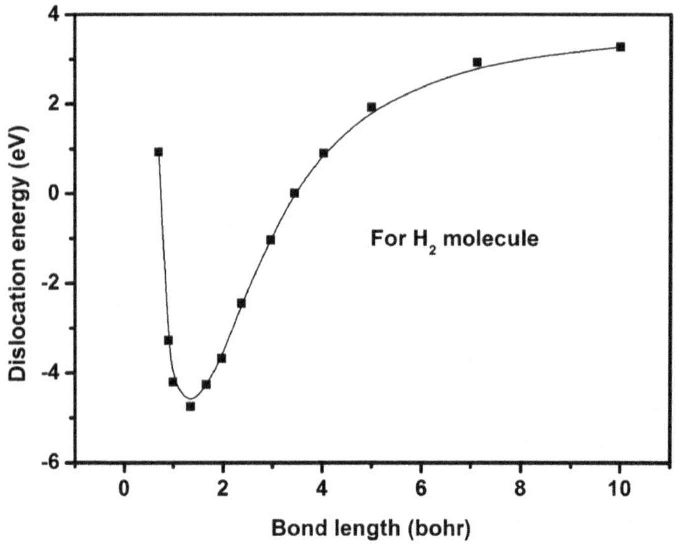

Fig. 2 Dissociation energy of H_2 molecule determined using HSE DFT calculation (Flores et al. 2018)

compared to standard DFT due to the inclusion of the Hartree–Fock exchange. However, it offers an improved description of electronic properties and is widely used in materials science, solid-state physics, and computational chemistry research. Figure 2 displays the findings for the H_2 dissociation curve produced using the HSE exchange–correlation functionals of the DFT approximation.

2.2 Generalized Gradient Approximation (GGA) DFT

Generalized Gradient Approximation (GGA) is a type of exchange–correlation functional used in density functional theory (DFT) calculations. In DFT, the exchange–correlation functional approximates the unknown exchange and correlation effects between electrons in a material. GGA functionals improve upon the local density approximation (LDA) by including the gradient of the electron density in the exchange–correlation functional. The gradient provides information about the spatial variations in electron density, which is important for capturing certain electronic and chemical properties. The GGA exchange–correlation functional takes the form:

$$E_{XC} = E_X + E_C \tag{2}$$

where E_X is the exchange contribution and E_C is the correlation contribution. The exchange term in GGA incorporates both the local density and the gradient of the electron density, while the correlation term includes similar contributions. The extreme underestimating of the semiconductor band gap, as shown in Table 1, results in a significant amount of uncertainty in the position of defect levels (and therefore, formation energies), which is a key flaw in LDA and GGA calculations.

DFT can also decrease the (1/f) noise and n-vacancy defects in GaN HEMTs when affected by radiation (Fleetwood 2015; Chen et al. 2015; Shen et al. 2015). A shift in E_2 modes from high to low due to an electric field component along the c-axis can also be determined (Bagnall et al. 2016; Chen et al. 2016). Using DFT calculations the significant defects present in substitutional complex degradations are studied (Mukherjee et al. 2016).

DFT is also known to determine excitation energies in GaN apart from the electron exchange–correlation discontinuity (Lev et al. 2018) and finding the O atom in an n-GaN (Lu et al. 2018). Calculations based on DFT and measurements of low frequency 1/f noise are used to detect potential flaws that might be causing radiation impacts and electrical stress-induced deterioration (Jiang 2018). Wang (2019, Elloh et al. (2019) talks about evaluating the thermal and electronic properties of GaN-based semiconductors using this method. The band offsets for different alloys in group III-nitrides are evaluated by use of a hybrid DFT model (Tsai and Bayram 2020) and produced the following energy values for binary and ternary nitrides using wurtzite (wz) and zincblende (zb) alloys (Fig. 3).

An increase in Gallium vacancies was observed through Deep Level Transient Spectroscopy (DLTS) measurements using DFT method while being affected by displacement damage effects (Wan et al. 2021). With regard to slab thickness, the preferred technique to improve the convergence of density-functional theory (DFT) surface calculations is to remove fake bands from the underside of surface slabs using pseudohydrogen atoms (Yoo et al. 2021). Niu et al. observed that the hot electron concentration will be the tradeoff between the density of electrons and the temperature during stress (Niu et al. 2021; Kamalian 2021). Supercell density functional theory simulations show that Si atoms are preferentially located where Ga atoms are not. Doped electrons penetrate the GaN conduction bands instead of forming distinct impurity states as they would in the case of Si interstitials (Lev et al. 2022). Effects of size, hydrogen passivation, and dipole corrections in a hybrid DFT analysis of zinc blende GaN and diamond surfaces and interfaces (Welch and Scolfaro 2022). Deep learning and density functional theory are used to create stable semiconductor materials (Siriwardane et al. 2022). In Wang and Li (2023), DFT is

Table 1 The formation enthalpy, bandgap and lattice parameters for GaN, AlN and InN materials evaluated using GGA and HSE DFT models (Lyons et al. 2014)

Method	Material											
	GaN properties (Å)				AlN properties (Å)				InN properties (Å)			
	A	c	E_g	ΔH_f	a	C	E_g	ΔH_f	a	c	E_g	ΔH_f
GGA	3.22	5.22	1.57	−1.02	3.12	5.02	4.11	−2.87	3.58	5.8	0	0.04
HSE	3.19	5.17	3.51	−1.33	3.1	4.96	6.19	−3.29	3.59	5.77	0.65	−0.2

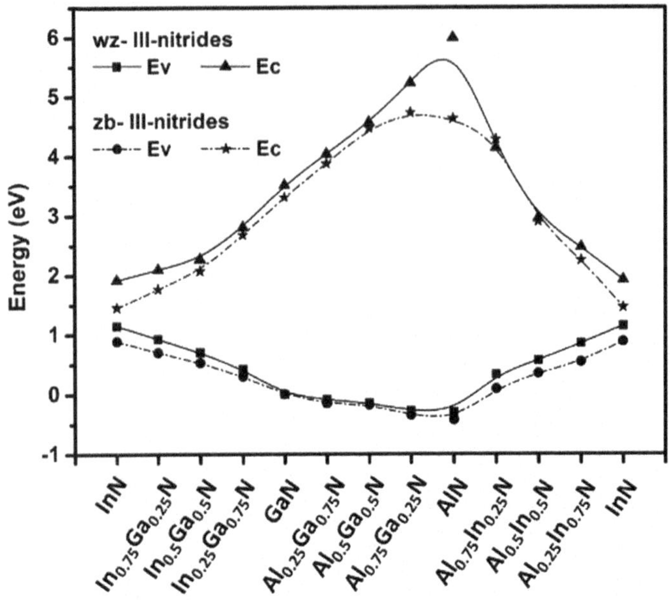

Fig. 3 Conduction and valence band edges in binary and ternary wz- and zb- III-nitrides. The energies are moved in relation to the wz- and zb- GaN valence band maximum (Tsai and Bayram 2020)

used to demonstrate that the electron transfer and redistribution between Al2O3 and GaN is what causes the dipole to develop at the Al2O3/GaN interface.

3 Finite Element Technique

HEMTs are crucial components in modern electronic devices, particularly in high-frequency and high-power applications. The Finite Element Method divides a complex geometry into smaller, simpler subdomains known as finite elements. Each element represents a portion of the original domain and is described by a set of mathematical equations based on fundamental physical principles. By assembling these equations for all elements, the FEM formulates a system of equations that can be solved numerically to obtain approximate solutions to the device behavior. The application of FEM to HEMT analysis involves several steps:

- The HEMT structure is divided into small geometric elements, such as triangles or quadrilaterals for 2D simulations and tetrahedrons or hexahedrons for 3D simulations. This process allows representing the complex device geometry in a more manageable way.

- Material properties are assigned to each element based on the characteristics of the materials used in the HEMT, such as semiconductor layers, metal contacts, and insulators. These properties include electron mobility, carrier concentration, dielectric constants, etc.
- The governing physical equations that describe the behavior of electrons and electromagnetic fields within the HEMT are formulated for each element. These equations usually involve the Poisson equation for the electric potential, the carrier transport equation, and other relevant equations based on the device physics.
- Boundary conditions are applied to the outer surfaces of the device to represent its operating conditions. These conditions can include voltage or current applied to specific terminals, insulating boundaries, and symmetry planes.

FEM technique was used initially in 2010 for AlGaN/GaN HEMTs by Das et al. for developing an optimized RF system design (Das et al. 2006). It was later used to extract the behavioral thermal model from a full-scale, time-domain, 3-D solution of the heat equation using the finite element method (Camarchia et al. 2007). In Green et al. (2008), finite element modeling outcomes are compared to measurements made using infrared and micro-Raman imaging. That paper compares thermal performance as determined by two optical thermometry methods: infrared and micro-Raman spectroscopy, with results from finite element modeling. In Dahmani et al. (2008), a precise method for extracting the thermal profile of large-size AlGaN/GaN HEMTs is proposed by utilizing modeling and measurement methods based on the Finite Element Method (FEM). Finite-element modeling has the drawback of being theoretical and time-consuming; nematic liquid crystals have the drawback of being intrusive and providing an average value over a large area; the micro-Raman technique has the drawback of being pricey and requiring special sample preparation; and the IR approach has the drawback of having relatively low resolution (Darwish et al. 2008). A finite element technique simulation was carried out later in in 2009 using FEMLAB to examine the impact of the slant sidewall angle on the parasitic capacitance (Pei et al. 2009).

FEMis also used to simulate the electron channel temperature for both power and temperature (Cuerdo et al. 2009). When compared to finite-element simulations, the results from the lumped-element thermal network are demonstrated to offer useful predictions of the thermal behavior of very large 2-D structures (Bertoluzza et al. 2009a). In Bertoluzza et al. (2009b), the findings of a three-dimensional finite-element thermal simulation of GaN HEMT devices are presented and discussed. Accurate channel to case temperature differentials is created through the use of IR microscopy and finite element analysis (FEA), from which a JC (junction to case thermal resistance) can be determined (Prejs et al. 2009). Finite element thermal simulations were used to examine the impact of device design on the maximum channel temperature for AlGaN/GaN HEMTs at dissipated power densities up to 5 W mm^{-1} (Douglas et al. 2011).

For instance, the temperature difference in the first device between the channel and the chuck can be predicted from a finite element simulation of the construction to be 26 C, with temperature contours (Darwish et al. 2011). A physics-based finite element model that simulates the operation of an AlGaN/GaN HEMT is presented using inputs from transmission electron microscope cross sections for device geometry. The model shown in Heller et al. (2011) is calibrated by comparison with measured electrical data that includes both well-known field-effect transistor metrics and lesser-known model parameters. The current study examines the effects of device design parameters on the junction temperature using finite element thermal simulations of gallium nitride high electron mobility transistors (GaN HEMTs) (Nochetto et al. 2011). FLOODS is the first fully-coupled solver in all three domains, using finite-element discretization techniques (Patrick et al. 2012). An approach for dynamic electro-thermal simulation of GaN microwave and power devices is provided in Ono et al. (2012). It is based on the extraction of compact thermal models through three-dimensional finite element simulation (Ono et al. 2012).

Carter et al. (2012), Sommet et al. (2012) provides real-world examples from infrared thermography, micro-Raman thermography, device transient electro-thermal response analysis on GaN HEMT devices, and finite element analysis (FEA) to show the effects of these variables. The results were validated using three-dimensional finite element (3D-FE) simulations. The relationship with 3D finite element simulations to determine the optimal combination of junction-level and heat-sink level cooling is quantified in Won et al. (2013). Later to measure the channel temperature variation beneath the source linked field plate, a physics-based electro-thermal finite element model was created (Heller et al. 2013). For AlGaN/GaN high electron mobility transistors, a finite-element model was created that combined the 2-D electron gas (2-DEG) density, piezoelectric polarization charge Q_P, and intrinsic stress caused by a nanocrystalline diamond capping layer (Wang et al. 2013a). Figure 4 shows a comparison of the channel temperatures with respect to the power dissipated evaluated using Finite element techniques.

The validity of this technique is then verified by comparing the temperature and stress values acquired to those derived from several Raman thermometry techniques and a three-dimensional finite element thermomechanical simulation (Choi et al. 2013). ANSYS 14.5 is used to compare the thermal performance of our unique GaN-on-Diamond method to that of commercially available GaN-on-SiC technology (Chu et al. 2013). For the thermo-mechanical co-simulation of such a 3-D structure, a self-developed technique based on a hybrid time-domain finite element method (TD-FEM) is used in place of commercial software (Zhang et al. 2014). In this study, the effects of a diamond heat spreader on the static and transient thermo-mechanical responses of an AlGaN/GaN HEMT are investigated. For the thermo-mechanical analysis, an effective hybrid time-domain finite element method (TD-FEM) approach is used (Zhang et al. 2015). The thermal and electrical properties of the devices under dc and pulsed operation circumstances were simulated using the finite element approach (Wang et al. 2013b).

The numerical finite-element approach is used to obtain the thermal parameters. The multi-layer power loop design can effectively lower the gate loop inductance and voltage overshoot on the devices, according to FEA and switching circuit simulation (Wang et al. 2014; Luo et al. 2014); whose approach was also used later for a three-heat system contrasted using finite element thermal analysis (Hou et al. 2014). The

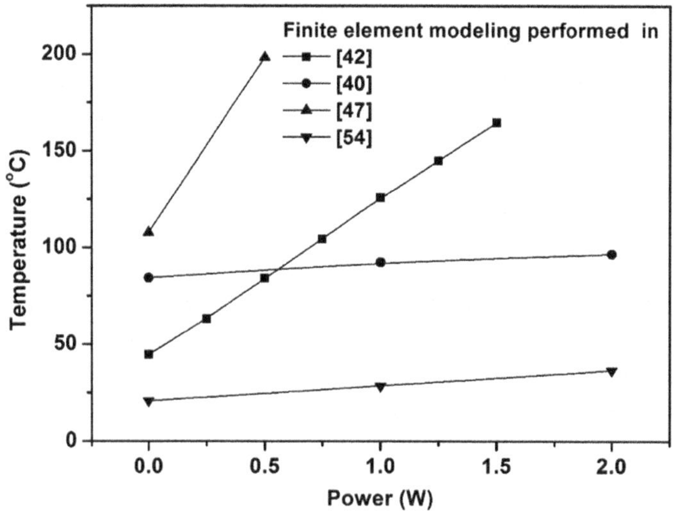

Fig. 4 Channel temperature for different GaN HEMT structures evaluated using finite element techniques for various power dissipation values

device's thermal location-dependent time constants are clearly visible at a temporal resolution of 50 ns, and this information might be used to further analyze the thermal structure. A Monolithic Microwave Integrated Circuit (MMIC) with an array of GaAs HEMT devices is additionally analyzed to comprehend the local thermal resistance distribution in comparison to a finite element analysis (Yazawa et al. 2014). FEM is also effective in determining the impact of diamond film thickness (between 1 and 50 m) (Dzuba et al. 2015; Zhao et al. 2015).

Additionally, the model's outcomes outperform those of finite-element numerical simulations for a variety of device geometry and material characteristics (Darwish et al. 2015). A model is further created and utilized to characterize test structures in the saturation regime for power densities up to 12 W/mm using linked technology computer aided design simulations (Oprins et al. 2015). A carbon-doped GaN layer with a doping concentration of 10^{17} cm^{-3} was given a thermal conductivity of $200 \pm 20 Wm^{-1}$ K^{-1} by comparing observed temperatures to finite element thermal modeling (Power et al. 2015);channel temperature can also be determined using FE thermal modeling (Pomeroy et al. 2015). In Jones et al. (2015), the findings of a finite element simulation of the transient stress response of an AlGaN/GaN HEMTare described. An internal parameterized tool that was previously developed to quickly build the thermal model of a transistor has been used to make a finite-element model of the eight gates HEMT using the ANSYS Professional program (Baczkowski et al. 2015). The Finite-Element-Analysis and Spice simulation demonstrate that our current design may successfully control the voltage overshoot and gate-drive ringing on HEMTs (Lu et al. 2016). Also, by employing FEM, the mechanical stress brought on by the creation and use of AlGaN/GaN transistors has been modeled in this article

(Joglekar et al. 2016). The finite element method is used to offer a three-dimensional thermal simulation for investigation of heat dissipation of AlGaN/GaN HEMTs with diamond substrates (Guo et al. 2017).

A thermal finite-element model has been created to simulate the heat dissipation throughout the devices in order to further validate this method (Pavlidis et al. 2016). Power device model for finite element method (FEM) simulation is proposedby Hariya et al. in Hariya et al. (2016) to investigate the impacts of magnetic flux on a GaN-HEMT. According to 2D finite element calculations, the dissipated power density in this area increases significantly and may even push the local temperature over the GaN/AlGaN structure's maximum operating temperature (Abbate et al. 2018). They can also determine the thermal resistances of these devices (Wu et al. 2017a). The derived capacitance model has a significantly lower level of complexity than the physics-based models used in Finite Element Analysis tools, making it more appropriate for inclusion in iterative design optimization procedures (Čučak et al. 2016). For the purpose of studying HEMTs, a two-dimensional finite-element model utilizing scanning electron microscopy (SEM) and infrared data was developed and validated by Liao et al. (2017), as well as a three-dimensional thermal study of GaN HEMTs on diamond substrate (Chen et al. 2018a). A finite element thermal simulation method is used to calculate the nonlinear thermal resistance with variable power dissipation and device dimensions in order to determine the channel temperatures as well as the access resistances under specific bias conditions (Chen et al. 2018b). The electrothermal finite-element technique simulations, which are scaled with the gate width W and power dissipations P_{diss} of the device, identify the geometry-dependent thermal resistance R_{ths} (Wu et al. 2017b). By contrasting the results with experimental data and finite-element simulations, the results are confirmed (Jia et al. 2018a).

With the use of finite-element analysis, parasitic inductance and resistance of the suggested design are extracted and explained (Gurpinar et al. 2017). Three different low inductance loop design approaches are examined and contrasted in-depth in this research. It is done to compare numbers using finite element analysis (FEA) (Sun et al. 2018). The finite element approach in three dimensions is used to systematically explore the transient thermal properties of pulse-operated AlGaN/GaN high electron mobility transistors (HEMT) utilized in high power amplifiers (Guo et al. 2020). Each design is assessed using thermal and flow finite element simulations in an effort to find the one that satisfies the steady-state temperature and pressure drop restrictions while delivering the highest current density in the smallest volume (Gurpinar et al. 2020). The distribution of heat generation and the maximum device temperature of high-electron-mobility transistors (HEMTs) made of gallium nitride (GaN) are examined in this paper using 2-D electrothermal and finite-element method (FEM) simulations (Odabaşı et al. 2020).

An effective model (for the electrical portion) and a 3-D finite element model (for the thermal part) are used to solve the problem in order to mimic the electrothermal behavior of gallium nitride (GaN) HEMTs (Valletta et al. 2021). The creation and linking of two models—a finite element model using COMSOL multiphysics® software (which will be explained in detail later in the chapter) and an RBDO model using MATLAB® software—is necessary for the execution of RBDO operations (Amar et al. 2021). To simulate the pulse behavior, a finite element-based 3-D transient thermal model is created. Accurate modeling is made significantly more difficult and challenging by the pronounced temperature variability that occurs in the periodic pulse mode from the microsecond to millisecond time scales (Zhang and Guo 2021). In a GaN HEMT device, it is shown that T_{max} can be calculated rapidly and accurately using a two-heat source technique and a finite element analysis (FEA) hybrid interaction with regard to different bias conditions (Kocer et al. 2022).

4 Monte Carlo Simulation

The Monte Carlo method takes its name from the famous Monte Carlo Casino, known for games of chance. This simulation technique relies on random sampling and statistical analysis to approximate the behavior of a complex system, which is difficult or impossible to solve analytically. In the context of AlGaN/GaN HEMTs, the method is used to model the movement of electrons through the semiconductor structure, accounting for scattering events, collisions, and interactions with the lattice. Here's a brief overview of the steps involved in performing Monte Carlo simulation for AlGaN/GaN HEMTs:

- Electron Trajectory Generation: The simulation starts by generating initial electron trajectories. Electrons are given random initial positions, velocities, and directions within the device structure. These electrons represent the charge carriers in the semiconductor.
- Time Step Advancement: The simulation advances in small time steps. During each time step, the electrons' positions and velocities are updated based on their interactions with the electric field, lattice, and other scattering mechanisms.
- Scattering Mechanisms: Various scattering mechanisms are considered in the simulation to account for electron–phonon scattering, impurity scattering, and other scattering events that influence electron mobility and energy dissipation.
- Collision Events: At each time step, the electrons may undergo scattering events due to collisions with lattice atoms or impurities. The probabilities of different scattering events are determined based on physical models and material properties
- Statistical Sampling: The Monte Carlo method relies on statistical sampling to approximate the electron behavior. Instead of tracking each electron individually, a large number of electrons are simulated, and their statistical properties, such as average velocity, mobility, and current, are analyzed.
- Device Characterization: The simulation results provide valuable insights into the device's electrical characteristics, such as current–voltage (I-V) characteristics, electron velocity distribution, and electron energy distribution. These results help in understanding the device's performance and limitations.

Monte Carlo simulation is especially useful for AlGaN/GaN HEMTs, as these devices often operate at high electron energies and encounter significant scattering effects. By considering statistical variations and averaging over a large number of simulated electrons, Monte Carlo simulation enables researchers and engineers to gain a deeper understanding of device behavior and optimize HEMT designs for specific applications.

The impacts of the two-dimensional electric field profile, the entire features of the two-dimensional electron gas, velocity overshoot, and ballistic transport, and the unique inclusion of actual space transfer in an ensemble Monte Carlo computation in Park et al. (1989). MC was later used in 1996 for evaluating the drain current and external resistances in MESFET and HEMT devices (Babiker et al. 1996). Using Monte Carlo simulation, (Zandler et al. 1999) described the dynamic behaviour of breakdown effects and electroluminescence associated with impact ionisation in 0.25 lm double heterojunction HEMTs. They include Real-space transfer phenomena for precisely determining the electron transport behavior (Li 2001). The Poisson's equation and a typical Monte Carlo (MC) approach are used to model a particle transport (Kalna and Asenov 2003). The compilation of statistical averages is the simulation's final stage. The time average is carried out in the ensemble MC simulation of the steady-state transport as follows. If N is the total number of simulated particles, we can calculate the overall average value of the quantity Q(k) for a single history of time t as follows:

$$Q(k) = \frac{1}{t}\int_{0}^{t} dt' \frac{1}{N}\sum_{j=1}^{N} Q[k_j(t')] = \frac{1}{t}\frac{1}{N}\sum_{i}\int_{o}^{t_i} dt' \sum_{j=1}^{N} Q[k_j(t')] \quad (3)$$

Here j is the particle index; the total time t is segregated into sum of integrals given as t_i.

5 Cellular Monte Carlo Approach

The Cellular Monte Carlo Approach (CMCA) is a computational technique used to simulate and analyze complex systems, particularly in the field of statistical physics. It is an extension of the traditional Monte Carlo method that incorporates spatial and lattice-based structures, making it particularly suitable for simulating systems with spatial correlations and interactions. The basic idea behind the Cellular Monte Carlo Approach is to divide the system of interest into discrete cells or lattice sites, forming a spatial grid. Each cell represents a specific state or configuration of the system, and the state of each cell can evolve over time based on probabilistic rules. These rules are often derived from thermodynamic principles and microscopic interactions between neighboring cells. Yamakawa et al. (2004) used the AlGaN/GaN HEMT structure with the particle-based CMC model. The effective potential technique is used to take quantum–mechanical phenomena brought on by electron confinement into account. The electron distribution of the Schrodinger–Poisson (S-P) self-consistent solution and the effective potential method are compared. Figure 5

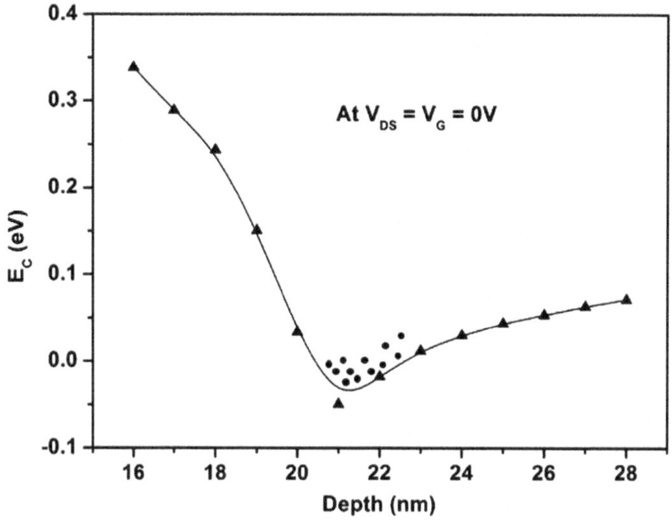

Fig. 5 Distribution of electrons depicted using the conduction band of AlGaN/GaN HEMT simulated using Cellular Monte Carlo approach (Yamakawa et al. 2004)

shows the electron distribution simulated using Cellular Monte Carlo approach with quantum correction.

Russo (2009) uses a self-consistent Monte Carlo code in two dimensions that takes into consideration three valence bands and four conduction band valleys, as shown in Fig. 6.

Using a full band Cellular Monte Carlo (CMC) simulator, (Marino et al. 2010) presented simulation findings for high-frequency, high-power state-of-the-art GaN HEMTs, which contain all of the specifics of the spectra of the phonons and band

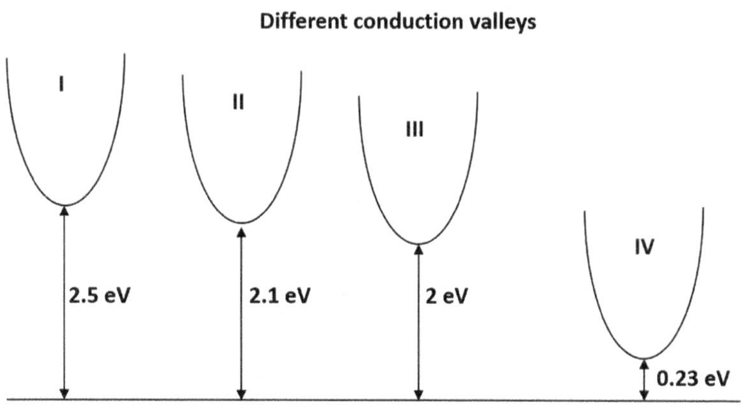

Fig. 6 Conduction band structure of four different conduction band valleys used for two-dimensional self-consistent Monte Carlo simulation of GaN HEMT in Russo (2009)

structure. For an accurate modelling of the nanoscale physics of GaN HEMTs and the determination of the corresponding intrinsic small-signal characteristics under various bias values, the Cellular Monte Carlo approach can be a useful tool (Guerra et al. 2011). Guerra et al. carried out particle-based device simulations on GaN HEMT for sub-millimeter wave power amplifiers (Guerra 2011). By segmenting the channel into several meshes and analysing the scattering rates of the electronic states in each mesh, Monte Carlo simulation is carried out. Next, taking into account the probable spread of the specified device from the stages are continued until a steady state is reached in the two-dimensional Poisson equation (Aadit et al. 2017).

5.1 Phonon Temperature and Thermal Calculations Using Monte Carlo Simulations

The electron MC simulation is performed first in each cycle. There are several spatial divisions known as "subcells" that make up the whole computational domain. The phonon emission/absorption by electrons and the related electron–phonon energy exchange inside each subcell are tracked by observing the mobility and scattering of individual electrons. The linked Fourier's law study and the phonon MC simulations will both utilize this knowledge (Hao et al. 2018). Figure 7 depicts the interaction of the hybrid simulation's three kernels, namely the Fourier's law analysis, the electron MC simulation, and the phonon MC simulation.

Li et al. (2018) creates a novel hybrid phonon Monte Carlo-diffusion technique that precisely captures tiny characteristics for ballistic-diffusive heat conduction while effectively simulating large-scale structures. The MC model employed in Shi et al. (2013) is a 2-D Poisson solver self-consistently linked with a semi-classical ensemble simulator. An analytical non-parabolic model for the six ellipsoidal delta valleys along the [1 0 0] directions at 85% of the Brillouin zone border is used to illustrate the conduction band for silicon. Also later in 2013, utilizing Boltzmann-Equation Monte Carlo Simulations and First-Principles Calculations, Electron Transport Properties

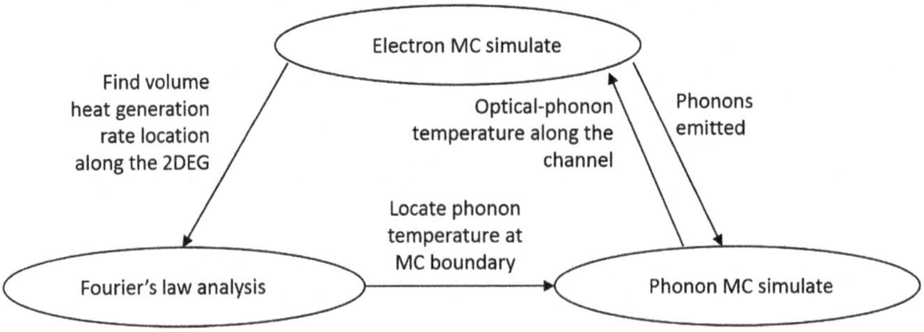

Fig. 7 The interactions between Fourier's law analysis, phonon MC simulations, and electron MC simulations, limited to area of the transistor as given in Hao et al. (2018)

of $Al_xGa_{1-x}N/Ga_{1-x}N$ transistors were carried out (Fang et al. 2019). In the hybrid method proposed in Li et al. (2022a), phono MC simulation and the Fourier-based finite element method (FEM) are combined by an alternating process. Figure 8 shows the flowchart for the hybrid technique (Li et al. 2022a).

In Sánchez-Martín et al. (2022) MC calculations has been carried out using two different thermal models, the electro-thermal model (ETM) and the Thermal Resistance Model (TRM) which is validated with DC I-V curves and different elements of Small-Signal Equivalent Circuit (SSEC) model. (Shen et al. 2023) has explored the thermal spreading resistance in GaN HEMTs using first-principle-based steady-state full-band phonon tracing MC simulations.

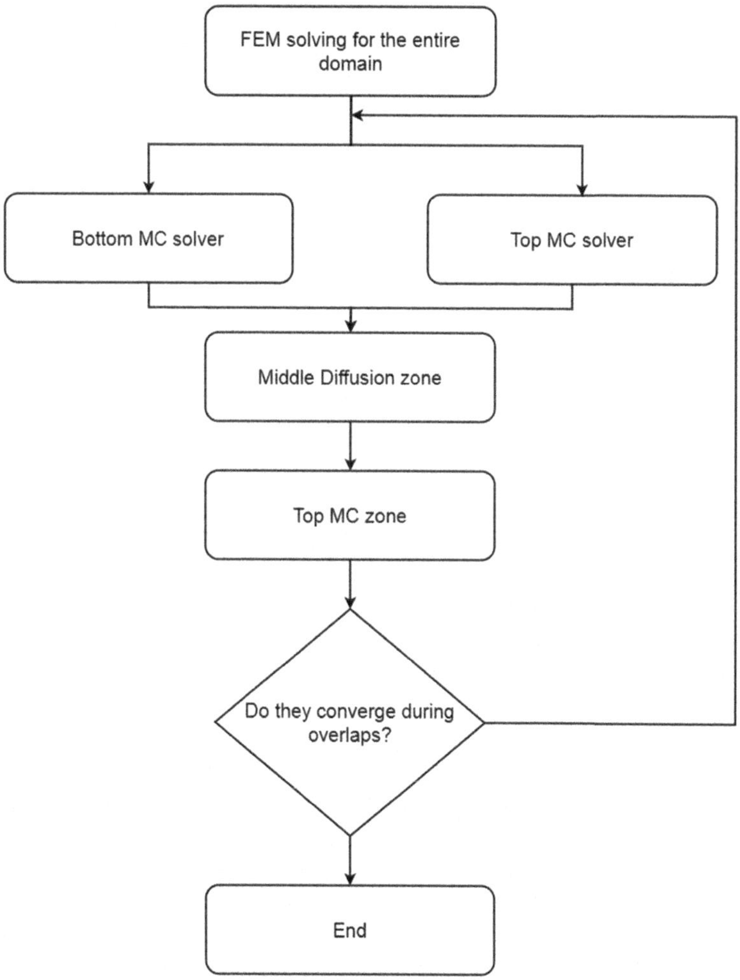

Fig. 8 Flow for the hybrid Monte Carlo Diffusion technique defined in Li et al. (2022a)

6 Numerical Modeling Using Ansys

ANSYS is a powerful software suite widely used for numerical modeling and simulation in various engineering fields. It allows engineers and researchers to analyze complex systems and predict their behavior under different conditions. ANSYS provides a vast array of simulation tools and capabilities, making it suitable for diverse applications such as structural analysis, fluid dynamics, electromagnetic simulations, and more. In this introduction to numerical modeling using ANSYS, we'll cover the basic steps and concepts involved in creating and running simulations.

- Geometry Creation: The first step in numerical modeling using ANSYS is to create or import the geometry of the system or structure you want to analyze. ANSYS supports various ways to build geometries, including 2D sketches, 3D modeling, and importing CAD models from other software.
- Mesh Generation: After creating the geometry, the next step is to generate a mesh. Meshing involves dividing the geometry into small elements (such as triangles or tetrahedra in 2D or 3D, respectively) to represent the shape for numerical calculations. The mesh quality significantly influences the accuracy and efficiency of the simulation.
- Material Properties: Assign appropriate material properties to the model. These properties could include mechanical properties like Young's modulus and Poisson's ratio for solids, thermal properties for heat transfer simulations, or electrical properties for electromagnetic simulations.
- Boundary Conditions: Specify the boundary conditions that represent how the system interacts with its environment. Boundary conditions can include applied loads, constraints, temperature conditions, and other external influences.

Ansys tool has been used to create a 3D finite element model (FEM) in order to more precisely describe the mechanical structure. The polarization may be determined immediately from this model's piezoelectric field distribution on the device, as given in Silvaco TCAD for GaN Technology manual (n.d.), Chen et al. (2014). GaN HEMT device upon SiC are lateral so used in two-phase simulations in microchannel utilizing the Volume-of-Fluid (VOF) approach done using ANSYS Fluent to determine two-phase flow patterns of the microcooler built using GaN HEMT, flow boiling, and temperature distributions (Lee et al. 2015).

6.1 *Ansys Icepak*

A new model for Ansys solver includes the Icepak feature. ANSYS Icepak is a powerful software tool developed by ANSYS, Inc. for electronic thermal analysis and electronics cooling simulations. It is a part of the ANSYS Electronics Suite and is widely used in various industries, including electronics, aerospace, automotive, and telecommunications, to predict and optimize the thermal performance of electronic devices and systems (Chou and Cheng 2017). It uses Computational Fluid Dynamics (CFD) techniques to simulate heat transfer and fluid flow, enabling engineers to assess and improve the thermal management of electronic products.

6.2 Ansys HFSS

In Jia et al. (2018b) another feature of Ansys tool called the High Frequency Structure Simulator (HFSS) was used to evaluate the S and Y-parameters of an AlGaN/GaN HEMT. HFSS is widely used in the field of RF and microwave engineering, antenna design, signal integrity analysis, and other high-frequency applications. It employs finite element method (FEM) and finite element–boundary integral (FEBI) techniques to accurately model complex electromagnetic phenomena.

6.3 Ansys Maxwell

Ansys Maxwell is also a powerful tool for calculating the DC and breakdown characteristics of a GaN device (Wang et al. 2018). Maxwell uses finite element analysis (FEA) techniques to solve a wide range of electromagnetic problems, including static, low-frequency, and high-frequency applications. It is widely used in various industries, such as electrical engineering, power electronics, motors and generators, transformers, sensors, and more. Key features and capabilities of ANSYS Maxwell include:

- It can simulate and analyze electromagnetic fields in both 2D and 3D domains, making it suitable for a wide range of applications.
- It is capable of analyzing static and low-frequency electromagnetic phenomena, such as magnetic fields, electric fields, and eddy currents.
- It can handle high-frequency electromagnetic problems, including radio-frequency (RF) and microwave applications, such as antennas, waveguides, and RF circuits.
- It also allows for time-domain simulations, enabling the study of transient electromagnetic phenomena and transient responses of devices.

The temperature dependent thermal resistance is extracted using a 3D simulation procedure in the ANSYS (v.17.0) software in Chen et al. (2020).

7 Numerical Modeling of AlGaN/GaN Hemts Using Comsol Multiphysics

COMSOL Multiphysics is a powerful and versatile simulation software developed by COMSOL Inc. It is widely used in various fields of science and engineering to model and solve complex physical phenomena through numerical simulations. The software uses the finite element method (FEM) as the primary numerical technique to solve partial differential equations governing various physical processes. COMSOL Multiphysics allows users to create and analyze multi-physics simulations, combining multiple physical phenomena that may interact with each other. In this study, we aim to explore the numerical modeling of AlGaN/GaN HEMTs using COMSOL Multiphysics. The AlGaN/GaN heterostructure presents unique challenges and opportunities, such as carrier transport, polarization effects, and impact ionization, which significantly influence the device performance.

COMSOL has been initially used by Djelti et al. in 2010 for finding the heat conduction and thermal effects in NMOSFETs (Djelti et al. 2010). In order to examine the effects of the thermal boundary resistance (TBR) between the GaN layer and 6H-SiC substrate, a supplementary "transition" layer with thickness t_{trl} and thermal conductivity k_{trl}—absent in actual transistors—is introduced in Russo et al. (2012) while evaluating the thermal behaviour of AlGaN/GaN HEMTs. Later, the mechanical response of the modelled device was directly calculated using COMSOL, and the electrical potential and thermal heat production distributions were transferred to a large-scale model created to determine the stress and strain profiles for certain bias conditions (Jones et al. 2014). Also, (Agarwal et al. 2017) describes the thermal model and numerical analysis for a multifinger HEMT that is cooled by microfluidic channels. FEA was carried out utilizing a mixed fluid-thermal framework and the modelling program COMSOL MULTIPHYSICSTM (version 4.4).

Further the COMSOL tool has been enhanced to be developed for GaN-On-Diamond HEMT technology for performing two-dimensional electrothermal simulation (Tadjer et al. 2019; Li et al. 2021). It can also corroborate the outcome temperature of multifinger devices and accurately measure the thermal coupling among each gate fingers (Li et al. 2022b).

8 Numerical Modeling and Simulation of AlGaN/GaN HEMT Using Sentaurus TCAD

Sentaurus Technology Computer-Aided Design (TCAD) is a suite of semiconductor device simulation tools used in the semiconductor industry for the design and analysis of various semiconductor devices and processes. TCAD tools are essential for understanding and optimizing the behavior of semiconductor devices, such as transistors, diodes, and integrated circuits, at the process level. These tools simulate the physical and electrical behavior of devices based on fundamental physics equations, enabling engineers to gain insights into device performance, reliability, and characteristics before actual fabrication. The following details show how numerical modeling and simulation can be performed for GaN devices using TCAD.

- Material Modeling: Sentaurus TCAD includes models for the electrical and thermal properties of AlGaN/GaN materials, including the wide bandgap energy characteristics of these materials.
- Device Structure: The first step in the modeling process is to define the AlGaN/GaN HEMT's physical structure, including the material layers, gate length, gate material, and source-drain regions.
- Process Simulation: This includes simulating growth processes, doping profiles, and material interfaces.
- Device Simulation: After defining the device structure and material properties, device simulation is performed to study the electrical characteristics of the AlGaN/GaN HEMT. The simulation considers the charge transport, electron mobility, and other relevant physical phenomena.
- Gate and Bias Conditions
- Output Analysis: Sentaurus TCAD provides detailed output analysis, such as current–voltage (I-V) characteristics, transconductance, output conductance, and capacitances. These results help in understanding the device's behavior and identifying potential areas for improvement.
- Parameter Extraction: The TCAD simulations can be used to extract important device parameters, such as threshold voltage, breakdown voltage, and cutoff frequency.
- Optimization: Based on the simulation results and parameter extraction, designers can make informed decisions to optimize the device's performance, enhance its efficiency, and improve its reliability.

There are several numerical models available in TCAD such as Drift-Diffusion (DD), SRH Generation-Recombination, Polarization, impact ionization, thermodynamic and hydrodynamic models.

8.1 Drift–Diffusion Model

The drift–diffusion model (DDM) is a simplified model of carrier transport in semiconductor devices. It is based on the following assumptions:

- Carriers move in a deterministic manner under the influence of an electric field.
- They are scattered by impurities and defects in the semiconductor material.
- They are generated and recombined at random locations in the semiconductor material.

The DDM can be used to calculate the carrier concentration, current density, and electric field in semiconductor devices. It is a widely used model in technology computer-aided design (TCAD) tools for simulating the performance of AlGaN/GaN

HEMTs. However, this is not a perfect model of carrier transport in semiconductor devices. It does not take into account the quantum mechanical effects of carrier transport, such as wave-like behavior and tunneling. It is a good approximation for many semiconductor devices, and it is much simpler to use than more accurate models. The analytical formulation for the model is given as:

8.2 Continuity Equation

$$\frac{dn}{dt} = G - R + (D_n * \nabla_n) \tag{4}$$

where n is the carrier concentration, G is the generation rate, R is the recombination rate, D_n is the carrier diffusion constant, and ∇_n is the gradient of the carrier concentration.

8.3 Transport Equation

$$J_n = \mu_n * E + (D_n * \nabla_n) \tag{5}$$

where J_n is the carrier current density, μ_n is the carrier mobility, E is the electric field, and ∇_n is the gradient of the carrier concentration.

8.4 Poisson Equation

$$\nabla^2 \varphi = -q * (n - p) \tag{6}$$

where φ is the electrostatic potential, q is the electron charge, n is the electron concentration, and p is the hole concentration.

The DDM can be solved using a variety of numerical methods, such as finite difference, finite element, and Monte Carlo methods. The choice of numerical method depends on the complexity of the device being simulated. The DDM has been used to simulate a wide variety of semiconductor devices, including transistors, diodes, and solar cells. It has also been used to study the effects of doping, temperature, and bias on carrier transport in semiconductor devices.

9 SRH Generation/Recombination Model

9.1 Generation

The SRH generation model is a model that describes the generation of carriers in a semiconductor material, namely an AlGaN/GaN HEMT, through the recombination of electrons and holes at deep levels (Mukherjee et al. 2017; Catoggio et al. 2023). The model is based on the following assumptions:

- There are deep levels in the semiconductor material that can trap electrons or holes.
- The deep levels are occupied by electrons or holes with a certain probability.
- When an electron and a hole are trapped at the same deep level, they can recombine and generate a new electron and hole pair.

The SRH generation rate is given by the following equation:

$$J_{gen} = A * N * T * e^{\left(-\frac{E_G}{kT}\right)} \tag{7}$$

Here J_{gen} is the generation rate per unit volume; A is a constant; N is the density of deep levels; T is the temperature; E_G is the energy gap of the semiconductor material; k is Boltzmann's constant.

The SRH generation model is used in TCAD simulations to calculate the generation current. The generation current is a major contributor to the dark current in GaN devices (Mukherjee et al. 2017). Here are the steps on how to use the SRH generation model in TCAD:

- Create a model for the deep levels in the semiconductor material. The model should include the energy level of the deep levels, the density of deep levels, and the capture cross-sections for electrons and holes.
- Set the temperature in the TCAD simulation.
- Calculate the SRH generation rate using the equation above.
- Add the SRH generation current to the current density equation in the TCAD simulation.

9.2 Recombination

The SRH recombination model is a semiconductor device physics model that describes the recombination of electrons and holes in a semiconductor material. The model is based on the following assumptions:

- Electrons and holes can recombine through two mechanisms: direct recombination and indirect recombination.

- Direct recombination occurs when an electron and hole meet and annihilate each other.
- Indirect recombination occurs when an electron and hole meet and create a third particle, such as a phonon or a photon.
- The rate of recombination is proportional to the product of the electron and hole concentrations.

The model can be used to calculate the carrier lifetime in a semiconductor material, which is the average time an electron or hole can exist before it recombines with a carrier of the opposite type. The carrier lifetime is an important parameter for determining the performance of HEMT. The SRH recombination model can be expressed mathematically as follows:

$$R = (\alpha_n * p) + (\beta_n * p * \tau)$$

Here R is the recombination rate; α is the direct recombination coefficient; β is the indirect recombination coefficient; n is the electron concentration; p is the hole concentration; τ is the carrier lifetime. The direct recombination coefficient and the indirect recombination coefficient are material-dependent parameters (Murugapandiyan et al. 2017). The carrier lifetime is a function of the temperature and doping concentration of the GaN HEMT.

10 Polarization Model

A polarization model in TCAD is a mathematical model that is used to describe the polarization of a semiconductor material. Polarization is the separation of charge in a material due to the presence of electric fields or strain. In TCAD, polarization models are used to calculate the electrostatic potential and electric field distributions in a semiconductor device. There are two main types of polarization models in TCAD:

- Spontaneous polarization: This is the polarization that exists in a material in the absence of any external fields. It is caused by the intrinsic properties of the material, such as the atomic structure and the presence of defects.
- Piezoelectric polarization: This is the polarization that is induced in a material by the application of a strain. It is caused by the realignment of the atoms in the material in response to the strain (Raja et al. 2021).

TCAD polarization models can be used to calculate the threshold voltage, 2DEG density, Drain-Induced Barrier Lowering (DIBL), subthreshold swing, gate leakage etc. of AlGaN/GaN HEMTs:

Polarization models are an essential tool for the design and optimization of semiconductor devices. They can be used to predict the performance of devices under different operating conditions and to identify the factors that limit device performance (Anju et al. 2023). Here are some examples of how polarization models are used in TCAD:

- For calculating the threshold voltage of anAlGaN/GaN HEMT, the polarization model is used to calculate the built-in potential at the gate-oxide interface. The built-in potential is the difference in electrostatic potential between the semiconductor and the gate oxide. The threshold voltage is the gate voltage at which the built-in potential is equal to the difference in Fermi levels between the semiconductor and the gate oxide.
- To evaluate the subthreshold slope of a GaN HEMT, the polarization model is used to calculate the electric field distribution in the semiconductor under different gate voltages. The subthreshold slope is the slope of the I-V curve at low gate voltages. It is a measure of the sensitivity of the device to gate voltage changes.
- For finding the 2DEG density in a GaN HEMT, the polarization model is used to calculate the spontaneous polarization in the AlGaN layer. The 2DEG density is the density of electrons in the GaN layer that is confined to the interface between the AlGaN and GaN layers. The 2DEG density is a critical parameter for the performance of GaN HEMTs (Wang 2014).
- In the case of DIBL, the polarization model is used to calculate the electric field distribution in the semiconductor under different drain voltages. The DIBL is the change in threshold voltage with drain voltage. It is a measure of the sensitivity of the device to drain voltage changes.
- For leakage current, the polarization model is used to calculate the electric field distribution in the gate oxide. The gate leakage current is the current that flows between the gate and the semiconductor through the gate oxide. It is a critical parameter for the reliability of HEMTs.

Polarization models are a powerful tool for the design and optimization of semiconductor devices (Garcia 2018). They can be used to predict the performance of devices under different operating conditions and to identify the factors that limit device performance.

11 Impact Ionization Model

Impact ionization is a process in which an electron or hole, moving through a semiconductor material, collides with another carrier and knocks it out of its orbit. This creates two new carriers, one of which is the original carrier (now with a higher energy) and the other is the newly created carrier (Subramani 2017; Reddy et al. 2019). The impact ionization model in TCAD for AlGaN/GaN HEMT is based on the following equations:

$$\alpha e(E) = \alpha e(E) e^{\left(\frac{-E}{E_0}\right)} \tag{8}$$

$$\alpha h(E) = \alpha h(E)e^{\left(\frac{-E}{E_0}\right)} \quad (9)$$

where $\alpha e(E)$ and $\alpha h(E)$ are the impact ionization coefficients for electrons and holes, respectively, E is the energy of the carrier, αe0 and αh0 are the impact ionization coefficients at low energies, and E0 is the critical energy for impact ionization. The critical energy is typically around 3 eV for silicon and 4 eV for GaN. The avalanche multiplication factor is used to calculate the breakdown voltage of the device.

12 Thermodynamic Model

A thermodynamic model in TCAD is a model that describes the transport of charge carriers in a semiconductor device by considering the interactions of these carriers with the lattice of the semiconductor. This model takes into account the temperature of the semiconductor, which affects the mobility of the charge carriers and the band structure of the semiconductor. The thermodynamic model is more accurate than the drift–diffusion model, which is the other main model used in TCAD (Joshi 2018). The drift–diffusion model does not consider the temperature of the semiconductor, and therefore can underestimate the effects of self-heating on device performance.

13 Hydrodynamic Model

The hydrodynamic model, also known as the hydrodynamic transport model or the hydrodynamic drift–diffusion model, is a more advanced transport model used in Sentaurus TCAD to simulate the behavior of charge carriers in semiconductor devices. It is an extension of the traditional drift–diffusion model that includes additional physics, accounting for carrier momentum and energy in addition to carrier concentrations. The hydrodynamic model is particularly useful for simulating devices at high electric fields, high carrier velocities, and in situations where momentum and energy transport become significant (Pardeshi 2015). Key features of the hydrodynamic model in TCAD include:

- Momentum Transport: The hydrodynamic model considers the momentum of charge carriers, taking into account their velocity and acceleration in response to the electric field. This enables the simulation of carrier transport in devices where carrier velocities are significant, such as in high-electron-mobility transistors (HEMTs) and advanced nanoscale devices.
- Energy Transport: In addition to momentum, the hydrodynamic model includes the transport of carrier energy. It considers the energy distribution and energy transfer mechanisms, allowing the simulation of devices where carrier energy becomes important, such as in hot-carrier effects.

- High-Frequency Effects: The hydrodynamic model is well-suited for simulating devices operating at high frequencies, where conventional drift–diffusion models may not accurately capture the carrier behavior.
- Ballistic Transport: At high electric fields and short channel lengths, the carrier transport can become ballistic, meaning carriers move without scattering. The hydrodynamic model can handle both diffusive and ballistic transport regimes.

The hydrodynamic model is computationally more intensive than the traditional drift–diffusion model due to the additional physics it considers. However, it provides more accurate results in scenarios where conventional models might not be sufficient.

14 Conclusion

In conclusion, this chapter delves into the intriguing field of numerical modeling applied to AlGaN/GaN High-Electron-Mobility Transistors (HEMTs), shedding light on the critical role it plays in understanding, optimizing, and designing these advanced semiconductor devices. Throughout this research, several key takeaways emerge: (i) Numerical modeling provides valuable insights into the intricate physics and behaviors governing AlGaN/GaN HEMTs. By simulating the underlying processes at the microscopic level, researchers and engineers gain a deeper understanding of carrier transport, electric field distribution, and device performance under varying conditions; (ii) Through simulations, researchers can fine-tune the device parameters and geometries to optimize performance metrics like carrier mobility, breakdown voltage, and current density. These optimizations lead to HEMTs with enhanced efficiency, power handling capabilities, and reliability; (iii) As semiconductor technology continues to evolve, numerical modeling serves as a fundamental tool in exploring novel device architectures and materials. It empowers researchers to push the boundaries of AlGaN/GaN HEMT technology and develop innovative solutions for emerging challenges in the field of high-power electronics.

However, it is essential to acknowledge that numerical modeling has its limitations. While it offers invaluable insights, its accuracy depends on the quality of the physical models, material parameters, and boundary conditions used. Therefore, experimental validation remains crucial to ensure the reliability of the simulation results. In conclusion, numerical modeling of AlGaN/GaN HEMTs is a powerful and indispensable tool in thc quest for better-performing and more efficient high-power electronic devices. Its integration with experimental research paves the way for significant advancements in this exciting field, holding the potential to revolutionize various applications, from wireless communication systems to power amplifiers and beyond.

Acknowledgements The authors are grateful to the Centre for VLSI Design, Department of Electronics and Communication Engineering, Kalasalingam Academy of Research and Education, as well as the Department of ECE, Mangalore Institute of Technology & Engineering, for providing your support during this review work for completing the chapter.

References

Aadit MNA, Kirtania SG, Afrin F, Alam MK, Khosru QDM (2017) High electron mobility transistors: performance analysis, research trend and applications. Different types of field-effect transistors-theory and applications, pp 45–64

Abbate C, Busatto G, Sanseverino A, Tedesco D, Velardi F (2018) Failure analysis of 650 V enhancement mode GaN HEMT after short circuit tests. Microelectron Reliab 88:677–683

Agarwal G, Kazior T, Kenny T, Weinstein D (2017) Modeling and analysis for thermal management in gallium nitride HEMTs using microfluidic cooling. J Electron Packag 139(1):011001

Amar A, Radi B, El Hami A (2021) Reliability based design optimization applied to the high electron mobility transistor (HEMT). Microelectron Reliab 124:114299

Ambacher O, Majewski J, Miskys C, Link A, Hermann M, Eickhoff M, Stutzmann M, Bernardini F, Fiorentini V, Tilak V, Schaff B (2002) Pyroelectric properties of Al (In) GaN/GaN hetero-and quantum well structures. J Phys: Condens Matter 14(13):3399

Anju S, Babu VS, Jacob B (2023) A nanoscale AlGaN/GaN HEMT on BGO substrate with recessed T gate for high frequency applications. Materials Today: Proc 80:2076–2079

Ashok A, Vasileska D, Goodnick SM, Hartin OL (2009) Importance of the gate-dependent polarization charge on the operation of GaN HEMTs. IEEE Trans Electron Dev 56(5):998–1006

Babiker S, Asenov A, Cameron N, Beaumont SP (1996) Simple approach to include external resistances in the Monte Carlo simulation of MESFETs and HEMTs. IEEE Trans Electron Dev 43(11):2032–2034

Baczkowski L, Jacquet JC, Jardel O, Gaquiere C, Moreau M, Carisetti D, Brunel L, Vouzelaud F, Mancuso Y (2015) Thermal characterization using optical methods of AlGaN/GaN HEMTs on SiC substrate in RF operating conditions. IEEE Trans Electron Devices 62(12):3992–3998

Bagnall KR, Dreyer CE, Vanderbilt D, Wang EN (2016) Electric field dependence of optical phonon frequencies in wurtzite GaN observed in GaN high electron mobility transistors. J Appl Phys 120(15):155104

Bertoluzza F, Delmonte N, Menozzi R (2009b) Three-dimensional finite-element thermal simulation of GaN-based HEMTs. Microelectron Reliab 49(5):468–473

Bertoluzza F, Sozzi G, Delmonte N, Menozzi R (2009a) Hybrid large-signal/lumped-element electro-thermal modeling of GaN-HEMTs. IEEE Trans Microw Theory Tech 57(12):3163–3170

Camarchia V, Cappelluti F, Pirola M, Guerrieri SD, Ghione G (2007) Self-consistent electrothermal modeling of class A, AB, and B power GaN HEMTs under modulated RF excitation. IEEE Trans Microw Theory Tech 55(9):1824–1831

Carter JA, Acord J, Hoffmann D, Trageser A, Pagel C (2012) Thermal factors influencing the reliability of GaN HEMTs. In: 2012 28th Annual IEEE semiconductor thermal measurement and management symposium (SEMI-THERM) (pp 182–188). IEEE

Catoggio E, Donati Guerrieri S, Bonani F (2023) TCAD Modeling of GaN HEMT output admittance dispersion through trap rate equation green's functions. Electronics 12(11):2457

Chen X, Donmezer FN, Kumar S, Graham S (2014) A numerical study on comparing the active and passive cooling of AlGaN/GaN HEMTs. IEEE Trans Electron Dev 61(12):4056–4061

Chen J, Puzyrev YS, Jiang R, Zhang EX, McCurdy MW, Fleetwood DM, Schrimpf RD, Pantelides ST, Arehart AR, Ringel SA, Saunier P (2015) Effects of applied bias and high field stress on the radiation response of GaN/AlGaN HEMTs. IEEE Trans Nucl Sci 62(6):2423–2430

Chen J, Puzyrev YS, Zhang EX, Fleetwood DM, Schrimpf RD, Arehart AR, Ringel SA, Kaun SW, Kyle EC, Speck JS, Saunier P (2016) High-field stress, low-frequency noise, and long-term reliability of AlGaN/GaN HEMTs. IEEE Trans Dev Mater Reliab 16(3):282–289

Chen J, Puzyrev YS, Zhang CX, Zhang EX, McCurdy MW, Fleetwood DM, Schrimpf RD, Pantelides ST, Kaun SW, Kyle EC, Speck JS (2013) Proton-induced dehydrogenation of defects in AlGaN/GaN HEMTs. IEEE Trans Nucl Sci 60(6):4080–4086

Chen Y, Xu Y, Wang F, Wang C, Wu Q, Qiao S, Yan B, Xu R (2018b) A scalable and multibias parameter extraction method for a small-signal GaN HEMT model. Int J Numer Model Electron Networks Devices Fields 31(5):e2347

Chen Y, Xu Y, Wang F, Wang C, Zhang Y, Yan B, Xu R (2020) Improved quasi-physical zone division model with analytical electrothermal Ids model for AlGaN/GaN heterojunction high electron mobility transistors. Int J Numer Model Electron Networks Dev Fields 33(3):e2630

Chen X, Zhai W, Zhang J, Bu R, Wang H, Hou X (2018a) FEM thermal analysis of high power GaN-on-diamond HEMTs. J Semicond 39(10):104005

Chen J (2013) Radiation response and reliability of AlGaN/GaN HEMTS (Doctoral dissertation)

Choi S, Heller ER, Dorsey D, Vetury R, Graham S (2013) Thermometry of AlGaN/GaN HEMTs using multispectral raman features. IEEE Trans Electron Devices 60(6):1898–1904

Chou HP, Cheng S (2017) Thermal performance evaluation of cascode Paralleled-GaN-HEMTs packaging for high power switching applications. J Thermal Sci Technol 12(2):JTST0022-JTST0022

Chu KK, Yurovchak T, Chao PC, Creamer CT (2013) Thermal modeling of high power GaN-on-diamond HEMTs fabricated by low-temperature device transfer process. In: 2013 IEEE compound semiconductor integrated circuit symposium (CSICS) (pp 1–4). IEEE

Cuerdo R, Sillero E, Romero MF, Uren MJ, di Forte Poisson MA, Muñoz E, Calle F (2009) High-temperature microwave performance of submicron AlGaN/GaN HEMTs on SiC. IEEE Electron Dev Lett 30(8):808–810

Čučak D, Vasić M, García O, Oliver JA, Alou P, Cobos JA, Wang A, Martin-Horcajo S, Romero MF, Calle F (2016) Physics-based analytical model for input, output, and reverse capacitance of a GaN HEMT with the field-plate structure. IEEE Trans Power Electron 32(3):2189–2202

Dahmani S, Mengistu ES, Kompa G (2008) Thermal model extraction of GaN HEMTs for large-signal modeling. In 2008 European microwave integrated circuit conference, pp 226–229. IEEE

Darwish AM, Bayba AJ, Hung HA (2008) Utilizing diode characteristics for GaN HEMT channel temperature prediction. IEEE Trans Microw Theory Tech 56(12):3188–3192

Darwish A, Bayba AJ, Hung HA (2015) Channel temperature analysis of GaN HEMTs with nonlinear thermal conductivity. IEEE Trans Electron Dev 62(3):840–846

Darwish AM, Ibrahim AA, Hung HA (2011) Temperature dependence of GaN HEMT small signal parameters. Int J Microwave Sci Technol

Das J, Oprins H, Ji H, Sarua A, Ruythooren W, Derluyn J, Kuball M, Germain M, Borghs G (2006) Improved thermal performance of AlGaN/GaN HEMTs by an optimized flip-chip design. IEEE Trans Electron Dev 53(11):2696–2702

Djelti H, Ouslimani A, Feham M, Kasbari A (2010) NMOSFET numerical model to investigate the heat conduction effects. In: 2010 International symposium on signals, systems and electronics (Vol 1, pp 1–4). IEEE

Douglas EA, Ren F, Pearton SJ (2011) Finite-element simulations of the effect of device design on channel temperature for AlGaN/GaN high electron mobility transistors. J Vac Sci Technol, b: Nanotechnol Microelectron: Mater, Process, Meas, Phenom 29(2):020603

Dzuba J, Vanko G, Vojs M, Rýger I, Ižák T, Jirásek V, Kutiš V, Lalinský T (2015) Finite element analysis of AlGaN/GaN micro-diaphragms with diamond coating. In: Smart sensors, actuators, and MEMS VII; and cyber physical systems (Vol 9517, pp 371–377). SPIE

Elloh VW, Yaya A, Gebreyesus G, Dua P, Mishra AK (2019) New 2D structural materials: carbon–gallium nitride (CC–GaN) and boron–gallium nitride (BN–GaN) heterostructures—materials design through density functional theory. ACS Omega 4(1):1722–1728

Fang J, Fischetti MV, Schrimpf RD, Reed RA, Bellotti E, Pantelides ST (2019) Electron transport properties of Al x Ga 1–x N/Ga N transistors based on first-principles calculations and boltzmann-equation monte carlo simulations. Phys Rev Appl 11(4):044045

Fleetwood DM (2015) $1/f $ noise and defects in microelectronic materials and devices. IEEE Trans Nucl Sci 62(4):1462–1486

Flores MA, Orellana W, Menéndez-Proupin E (2018) Accuracy of the Heyd-Scuseria-Ernzerhof hybrid functional to describe many-electron interactions and charge localization in semiconductors. Phys Rev B 98(15):155131

Garcia FD (2018) A physics-based analytical compact model, TCAD simulation, and empirical spice models of GaN devices for power applications

Green DS, Vembu B, Hepper D, Gibb SR, Jin D, Vetury R, Shealy JB, Beechem LT, Graham S (2008) GaN HEMT thermal behavior and implications for reliability testing and analysis. Physica Status Solidi c 5(6):2026–2029

Guerra D, Marino FA, Goodnick S, Ferry D, Saraniti M (2011) Extraction of gate capacitance of high-frequency and high-power gan hemts by means of cellular monte carlo simulations. Int J High Speed Electron Syst 20(03):423–430

Guerra D (2011) Gallium nitride HEMT modeling and design for millimeter and sub-millimeter wave power amplifiers through Monte Carlo particle-based device simulations. Arizona State University

Guo H, Chen T, Shi S (2020) Transient simulation for the thermal design optimization of pulse operated AlGaN/GaN HEMTs. Micromachines 11(1):76

Guo H, Kong Y, Chen T (2017) Thermal simulation of high power GaN-on-diamond substrates for HEMT applications. Diam Relat Mater 73:260–266

Gurpinar E, Iannuzzo F, Yang Y, Castellazzi A, Blaabjerg F (2017) Design of low-inductance switching power cell for GaN HEMT based inverter. IEEE Trans Ind Appl 54(2):1592–1601

Gurpinar E, Sahu R, Ozpineci B, DeVoto D (2020) Analysis and optimization of a multi-layer integrated organic substrate for high current GaN HEMT-based power module. In: 2020 IEEE workshop on wide bandgap power devices and applications in Asia (WiPDA Asia) (pp 1–6). IEEE

Hao Q, Zhao H, Xiao Y, Kronenfeld MB (2018) Electrothermal studies of GaN-based high electron mobility transistors with improved thermal designs. Int J Heat Mass Transf 116:496–506

Hariya A, Koga T, Matsuura K, Yanagi H, Tomioka S, Ishizuka Y, Ninomiya T (2016) Circuit design techniques for reducing the effects of magnetic flux on GaN-HEMTs in 5-MHz 100-W high power-density LLC resonant DC–DC converters. IEEE Trans Power Electron 32(8):5953–5963

Heller E, Choi S, Dorsey D, Vetury R, Graham S (2013) Electrical and structural dependence of operating temperature of AlGaN/GaN HEMTs. Microelectron Reliab 53(6):872–877

Heller ER, Vetury R, Green DS (2011) Development of a versatile physics-based finite-element model of an AlGaN/GaN HEMT capable of accommodating process and epitaxy variations and calibrated using multiple DC parameters. IEEE Trans Electron Dev 58(4):1091–1095

Hou M, Pan CC, Asheghi M, Senesky DG (2014) Finite element thermal analysis of localized heating in AlGaN/GaN HEMT based sensors. In: Fourteenth intersociety conference on thermal and thermomechanical phenomena in electronic systems (ITherm) (pp 25–30). IEEE

Hu Y, Wang J, Wen CP, Yu M, Mao L, Hao Y (2010) Heterostructure HEMT electronic device simulation with density functional theory. ECS Trans 27(1):79

Jia Y, Xu Y, Guo Y (2018a) A universal scalable thermal resistance model for compact large-signal model of AlGaN/GaN HEMTs. IEEE Trans Microw Theory Tech 66(10):4419–4429

Jia Y, Xu Y, Xu R, Li Y (2018b) An accurate parasitic parameters extraction method based on FW-EM for AlGaN/GaN HEMT up to 110 GH z. Int J Numer Model Electron Networks Dev Fields 31(1):e2270

Jiang R (2018) Bias dependence of radiation response and reliability of AlGaN/GaN HEMTs. Vanderbilt University

Joglekar S, Lian C, Baskaran R, Zhang Y, Palacios T, Hanson A (2016) Finite element analysis of fabrication-and operation-induced mechanical stress in AlGaN/GaN transistors. IEEE Trans Semicond Manuf 29(4):349–354

Jones JP, Heller E, Dorsey D, Graham S (2015) Transient stress characterization of AlGaN/GaN HEMTs due to electrical and thermal effects. Microelectron Reliab 55(12):2634–2639

Jones JP, Rosenberger MR, King WP, Vetury R, Heller E, Dorsey D, Graham S (2014) Electro-thermo-mechanical transient modeling of stress development in AlGaN/GaN high electron mobility transistors (HEMTs). In: Fourteenth intersociety conference on thermal and thermomechanical phenomena in electronic systems (ITherm) (pp 959–965). IEEE

Joshi V (2018) Modeling and simulation of gallium nitride high electron mobility transistors and optimization of buffer layer (Doctoral dissertation, Indian Institute of Technology Jodhpur)

Kalna K, Asenov A (2003) Nonequilibrium and ballistic transport, and backscattering in decanano HEMTs: a Monte Carlo simulation study. Math Comput Simul 62(3–6):357–366

Kamalian M (2021) Investigation of structural, electronic and transport properties of III-V nanotubes using by density functional theory. J Res Many-Body Syst 11(4):110–123

Kikkawa T, Joshin K, Kanamura M (2012) GaN device for highly efficient power amplifiers. Fujitsu Sci Tech J 48(1):40–46

Kocer H, Durna Y, Gunes B, Tendurus G, Butun B, Ozbay E (2022) Fast unveiling of T max in GaN HEMT devices via the electrical measurement-assisted two-heat source model. IEEE Trans Electron Dev 69(5):2319–2324

Koleske DD, Allerman AA, Kurtz SR, Lee SR, Wright AF, Seager CH, Missert NA, Baca AG, Briggs RD, Marsh PF, Tigges CP (2005) Materials physics and device development for improved efficiency of GaN HEMT high power amplifiers

Lee H, Won Y, Houshmand F, Gorle C, Asheghi M, Goodson KE (2015) Computational modeling of extreme heat flux microcooler for GaN-based HEMT. In: International electronic packaging technical conference and exhibition (Vol 56895, p V002T06A018). American Society of Mechanical Engineers

Lev LL, Maiboroda IO, Grichuk ES, Chumakov NK, Schröter NBM, Husanu MA, Schmitt T, Aeppli G, Zanaveskin ML, Valeyev VG, Strocov VN (2022) Impact of band-bending on the k-resolved electronic structure of Si-doped GaN. Phys Rev Res 4(1):013183

Lev LL, Maiboroda IO, Husanu MA, Grichuk ES, Chumakov NK, Ezubchenko IS, Chernykh IA, Wang X, Tobler B, Schmitt T, Zanaveskin ML (2018) K-space imaging of anisotropic 2D electron gas in GaN/GaAlN high-electron-mobility transistor heterostructures. Nat Commun 9(1):2653

Li HL, Hua YC, Cao BY (2018) A hybrid phonon Monte Carlo-diffusion method for ballistic-diffusive heat conduction in nano-and micro-structures. Int J Heat Mass Transf 127:1014–1022

Li Y, Mao S, Fu Y, Xu R, Yan B, Zhang Y, Xu Y (2021) A scalable electrothermal model using a three-dimensional thermal analysis model for GaN-on-diamond high-electron-mobility transistors. Int J Numer Model Electron Networks Dev Fields 34(5):e2875

Li J, Tang M, Mao J (2022b) Analytical thermal model for AlGaN/GaN HEMTs using conformal mapping method. IEEE Trans Electron Dev 69(5):2313–2318

Li T (2001) Ensemble Monte Carlo based simulation analysis of gallium nitride HEMTs for high-power microwave device applications. Old Dominion University

Li HL, Shen Y, Hua YC, Sobolev SL, Cao BY (2022) Thermal spreading resistance of GaN HEMTs with heat source heating studied by hybrid Monte Carlo-diffusion simulations. arXiv preprint arXiv:2202.13770

Li E, Hou L, Li L, Liu M, Xi M, Wang X, Dai Y (2011) The study of electronic structures and optical properties of Al-doped GaN. In: Journal of physics: conference series (Vol 276, No 1, p 012044). IOP Publishing

Liao Z, Guo C, Meng J, Jiang B, Gao L, Su Y, Wang R, Feng S (2017) Thermal evaluation of GaN-based HEMTs with various layer sizes and structural parameters using finite-element thermal simulation. Microelectron Reliab 74:52–57

Lu F, Liu L, Xia S, Diao Y, Feng S (2018) Electronic structure and optical properties of N vacancy and O filling on n-GaN (0001) surface. Superlattices Microstruct 118:160–169

Lu J, Bai H, Brown A, McAmmond M, Chen D, Styles J (2016) Design consideration of gate driver circuits and PCB parasitic parameters of paralleled E-mode GaN HEMTs in zero-voltage-switching applications. In: 2016 IEEE applied power electronics conference and exposition (APEC) (pp 529–535). IEEE

Luo F, Chen Z, Xue L, Mattavelli P, Boroyevich D, Hughes B (2014) Design considerations for GaN HEMT multichip halfbridge module for high-frequency power converters. In: 2014 IEEE applied power electronics conference and exposition-APEC 2014 (pp 537–544). IEEE

Lyons JL, Janotti A, Van de Walle CG (2014) Effects of carbon on the electrical and optical properties of InN, GaN, and AlN. Phys Rev B 89(3):035204

Marino FA, Guerra D, Goodnick S, Ferry D, Saraniti M (2010) RF and DC characterization of state-of-the-art GaN HEMT devices through cellular Monte Carlo simulations. Physica Status Solidi C 7(10):2445–2449

Mukherjee K, Darracq F, Curutchet A, Malbert N, Labat N (2017) TCAD simulation capabilities towards gate leakage current analysis of advanced AlGaN/GaN HEMT devices. Microelectron Reliab 76:350–356

Mukherjee S, Puzyrev Y, Chen J, Fleetwood DM, Schrimpf RD, Pantelides ST (2016) Hot-carrier degradation in GaN HEMTs due to substitutional iron and its complexes. IEEE Trans Electron Devices 63(4):1486–1494

Murugapandiyan P, Ravimaran S, William J, Sundaram KM (2017) Design and analysis of 30 nm T-gate InAlN/GaN HEMT with AlGaN back-barrier for high power microwave applications. Superlattices Microstruct 111:1050–1057

Niu X, Ma X, Hou B, Yang L, Lin YS, Zhu Q, Ciou FM, Chen KH, Chen Y, Du J, Wu M (2021) Electrical degradation of in situ SiN/AlGaN/GaN MIS-HEMTs caused by dehydrogenation and trap effect under hot carrier stress. IEEE Trans Electron Dev 68(9):4283–4288

Nochetto HC, Jankowski NR, Bar-Cohen A (2011) The impact of GaN/substrate thermal boundary resistance on a HEMT device. In: ASME international mechanical engineering congress and exposition (Vol 54969, pp 241–249)

Odabaşı O, Akar MÖ, Bütün B, Özbay E (2020) Improved T MAX estimation in GaN HEMTs using an equivalent hot point approximation. IEEE Trans Electron Dev 67(4):1553–1559

Ono S, Ciappa M, Hiura S, Fichtner W (2012) Electro-thermal simulation in the time domain of GaN HEMT for RF switch-mode amplifier. Microelectron Reliab 52(9):2224

Oprins H, Stoffels S, Baelmans M, De Wolf I (2015) Influence of field-plate configuration on power dissipation and temperature profiles in AlGaN/GaN on silicon HEMTs. IEEE Trans Electron Dev 62(8):2416–2422

Pardeshi H (2015) Analog/RF performance of AlInN/GaN underlap DG MOS-HEMT. Superlattices Microstruct 88:508–517

Park DH, Wang Y, Brennan KF (1989) Ensemble monte carlo simulation of a 0.35-mu m pseudomorphic HEMT. IEEE Electron Dev Lett 10(3):107–110

Patrick E, Horton D, Griglione M, Law ME (2012) A self-consistent electro-thermo-mechanical device simulator based on the finite-element method. SISPAD

Pavlidis G, Pavlidis S, Heller ER, Moore EA, Vetury R, Graham S (2016) Characterization of AlGaN/GaN HEMTs using gate resistance thermometry. IEEE Trans Electron Dev 64(1):78–83

Pei Y, Chen Z, Brown D, Keller S, Denbaars SP, Mishra UK (2009) Deep-submicrometer AlGaN/GaN HEMTs with slant field plates. IEEE Electron Dev Lett 30(4):328–330

Pomeroy JW, Uren MJ, Lambert B, Kuball M (2015) Operating channel temperature in GaN HEMTs: DC versus RF accelerated life testing. Microelectron Reliab 55(12):2505–2510

Power M, Pomeroy JW, Otoki Y, Tanaka T, Wada J, Kuzuhara M, Jantz W, Souzis A, Kuball M (2015) Measuring the thermal conductivity of the GaN buffer layer in AlGaN/GaN HEMTs. Physica Status Solidi (A) 212(8):1742–1745

Prejs A, Wood S, Pengelly R, Pribble W (2009) Thermal analysis and its application to high power GaN HEMT amplifiers. In 2009 IEEE MTT-S international microwave symposium digest, pp 917–920. IEEE

Puzyrev YS, Roy T, Beck M, Tuttle BR, Schrimpf RD, Fleetwood DM, Pantelides ST (2011a) Dehydrogenation of defects and hot-electron degradation in GaN high-electron-mobility transistors. J Appl Phys 109(3):034501

Puzyrev YS, Roy T, Zhang EX, Fleetwood DM, Schrimpf RD, Pantelides ST (2011b) Radiation-induced defect evolution and electrical degradation of AlGaN/GaN high-electron-mobility transistors. IEEE Trans Nucl Sci 58(6):2918–2924

Raja PV, Subramani NK, Gaillard F, Bouslama M, Sommet R, Nallatamby JC (2021) Identification of buffer and surface traps in Fe-doped AlGaN/GaN HEMTs using Y21 frequency dispersion properties. Electronics 10(24):3096

Reddy MK, Lakshmi J, Hemanth A, Kumar BH, Bandi L, Sheu G, Song YL, Chen PA, Chang LM (2019) Physics Based TCAD Simulation and Calibration of GaN/AlGaN/GaN HEMT Device. In: 2019 6th International conference on systems and informatics (ICSAI) (pp 253–256). IEEE

Roy T, Puzyrev YS, Tuttle BR, Fleetwood DM, Schrimpf RD, Brown DF, Mishra UK, Pantelides ST (2010a) Electrical-stress-induced degradation in AlGaN/GaN high electron mobility transistors grown under gallium-rich, nitrogen-rich, and ammonia-rich conditions. Appl Phys Lett 96(13):133503

Roy T, Zhang EX, Puzyrev YS, Fleetwood DM, Schrimpf RD, Choi BK, Hmelo AB, Pantelides ST (2010b) Process dependence of proton-induced degradation in GaN HEMTs. IEEE Trans Nucl Sci 57(6):3060–3065

Roy T (2011) Reliability-limiting defects in gallium nitride/aluminium gallium nitride high electron mobility transistors. Vanderbilt University

Russo S, d'Alessandro V, Costagliola M, Sasso G, Rinaldi N (2012) Analysis of the thermal behavior of AlGaN/GaN HEMTs. Mater Sci Eng B 177(15):1343–1351

Russo S (2009) Gallium nitride-based device simulation and development

Shen X, Puzyrev YS, Fleetwood DM, Schrimpf RD, Pantelides ST (2015) Quantum mechanical modeling of radiation-induced defect dynamics in electronic devices. IEEE Trans Nucl Sci 62(5):2169–2180

Shen Y, Yang HA, Cao BY (2023) Near-junction phonon thermal spreading in GaN HEMTs: a comparative study of simulation techniques by full-band phonon Monte Carlo method. Int J Heat Mass Transf 211:124284

Shi M, Saint-Martin J, Bournel A, Querlioz D, Wichmann N, Bollaert S, Danneville F, Dollfus P (2013) Monte Carlo analysis of the dynamic behavior of III–V MOSFETs for low-noise RF applications. Solid-State Electron 87:51–57

Silvaco TCAD for GaN Technology manual

Siriwardane EMD, Zhao Y, Perera I, Hu J (2022) Generative design of stable semiconductor materials using deep learning and density functional theory. npj Comput Mater 8(1):164

Sommet R, Mouginot G, Quéré R, Ouarch Z, Camiade M (2012) Thermal modeling and measurements of AlGaN/GaN HEMTs including thermal boundary resistance. Microelectron J 43(9):611–617

Sołtys J, Ptasinska M, Piechota J, Krukowski S (2014) Density functional study of GaN (0001)/ AlN (0001) high electron mobility transistor structures. J Cryst Growth 401:30–32

Subramani NK (2017) Physics-based TCAD device simulations and measurements of GaN HEMT technology for RF power amplifier applications (Doctoral dissertation, Université de Limoges)

Sun B, Zhang Z, Andersen MA (2018) Research of low inductance loop design in GaN HEMT application. In: IECON 2018–44th annual conference of the IEEE industrial electronics society (pp 1466–1470). IEEE

Szállás A, Szász K, Trinh XT, Son NT, Janzén E, Gali A (2014) Characterization of the nitrogen split interstitial defect in wurtzite aluminum nitride using density functional theory. J Appl Phys 116(11):113702

Sánchez-Martín H, Íñiguez-de-la-Torre I, García-Sánchez S, Mateos J, González T (2022) Monte Carlo analysis of thermal effects in the DC and AC performance of AlGaN/GaN HEMTs. Solid-State Electron 193:108289

Tadjer MJ, Anderson TJ, Ancona MG, Raad PE, Komarov P, Bai T, Gallagher JC, Koehler AD, Goorsky MS, Francis DA, Hobart KD (2019) GaN-on-diamond HEMT technology with T AVG= 176 C at P DC, max= 56 W/mm measured by transient thermoreflectance imaging. IEEE Electron Dev Lett 40(6):881–884

Tsai YC, Bayram C (2020) Band alignments of ternary wurtzite and zincblende III-nitrides investigated by hybrid density functional theory. ACS Omega 5(8):3917–3923

Valletta A, Mussi V, Rapisarda M, Lucibello A, Natali M, Peroni M, Lanzieri C, Fortunato G, Mariucci L (2021) Hybrid electrothermal simulations of GaN HEMT devices based on self-heating free virtual electrical characteristics. IEEE Trans Electron Dev 68(8):3740–3747

Wan P, Yang J, Lv G, Lv L, Dong S, Li W, Xu X, Peng C, Zhang Z, Li X (2021) Effect of hydrogen on radiation-induced displacement damage in AlGaN/GaN HEMTs. IEEE Trans Nucl Sci 68(6):1258–1264

Wang Z, Cao J, Wang F, Chen W, Zhang B, Guo S, Yao Y (2018) Proposal of a novel enhancement type AlGaN/GaN HEMT using recess-free field coupled gate. Superlattices Microstruct 122:343–348

Wang C, Li X (2023) Interface dipole induced threshold voltage shift in the Al_2O_3/GaN heterostructure. Appl Surf Sci 622:156954

Wang A, Tadjer MJ, Anderson TJ, Baranyai R, Pomeroy JW, Feygelson TI, Hobart KD, Pate BB, Calle F, Kuball M (2013a) Impact of intrinsic stress in diamond capping layers on the electrical behavior of AlGaN/GaN HEMTs. IEEE Trans Electron Devices 60(10):3149–3156

Wang A, Tadjer MJ, Calle F (2013b) Simulation of thermal management in AlGaN/GaN HEMTs with integrated diamond heat spreaders. Semicond Sci Technol 28(5):055010

Wang C, Xu Y, Yu X, Ren C, Wang Z, Lu H, Chen T, Zhang B, Xu R (2014) An electrothermal model for empirical large-signal modeling of AlGaN/GaN HEMTs including self-heating and ambient temperature effects. IEEE Trans Microw Theory Tech 62(12):2878–2887

Wang R (2014) AlGaN/GaN HEMT DC simulation (Doctoral dissertation, Auburn University)

Wang T (2019) Electronic and thermal properties of wide bandgap materials from density functional theory. University of Delaware

Welch E, Scolfaro L (2022) Hybrid density functional theory study on zinc blende GaN and diamond surfaces and interfaces: Effects of size, hydrogen passivation, and dipole corrections. Comput Condensed Matter 30:e00653

Won Y, Cho J, Agonafer D, Asheghi M, Goodson KE (2013) Cooling limits for GaN HEMT technology. In: 2013 IEEE compound semiconductor integrated circuit symposium (CSICS) (pp 1–5). IEEE

Wu Q, Xu Y, Chen Y, Wang Y, Fu W, Yan B, Xu R (2017b) A scalable multiharmonic surface-potential model of AlGaN/GaN HEMTs. IEEE Trans Microw Theory Tech 66(3):1192–1200

Wu Q, Xu Y, Zhou J, Kong Y, Chen T, Wang Y, Lin F, Fu Y, Jia Y, Zhao X, Yan B (2017a) Performance comparison of GaN HEMTs on diamond and SiC substrates based on surface potential model. ECS J Solid State Sci Technol 6(12):Q171

Yamakawa S, Goodnick S, Aboud S, Saraniti M (2004) Quantum corrected full-band cellular Monte Carlo simulation of AlGaN/GaN HEMTs. J Comput Electron 3:299–303

Yazawa K, Kendig D, Maize K, Shakouri A (2014) Transient thermal characterization of HEMT devices. In: 2014 IEEE MTT-S international microwave symposium (IMS2014) (pp 1–4). IEEE

Yoo SH, Lymperakis L, Neugebauer J (2021) Efficient electronic passivation scheme for computing low-symmetry compound semiconductor surfaces in density-functional theory slab calculations. Phys Rev Mater 5(4):044605

Zandler G, Rossi L, DiCarlo A, Tocca L, Bonfiglio A, Brunori M, Lugli P, Meneghesso G, Zanoni R (1999) Monte Carlo simulation of impact ionization and light emission in pseudomorphic HEMTs. Physica B 272(1–4):558–561

Zhang H, Guo Z (2021) Thickness dependence and anisotropy of capped diamond thermal conductivity on cooling of pulse-operated GaN HEMTs. IEEE Trans Components Packaging Manuf Technol 11(2):233–240

Zhang R, Zhao WS, Yin WY (2014) Investigation on thermo-mechanical responses in high power multi-finger AlGaN/GaN HEMTs. Microelectron Reliab 54(3):575–581

Zhang R, Zhao WS, Yin WY, Zhao ZG, Zhou HJ (2015) Impacts of diamond heat spreader on the thermo-mechanical characteristics of high-power AlGaN/GaN HEMTs. Diam Relat Mater 52:25–31

Zhao X, Xu R, Xu Y (2015) An improved nonlinear thermal resistance extraction method for AlGaN/GaN HEMTs. In: 2015 IEEE international conference on communication problem-solving (ICCP) (pp 261–263). IEEE

Dr. V. Sandeep carried out his Bachelor's degree (B.E.) in Electronics and Communication Engineering from Hindusthan College of Engineering and Technology, Coimbatore in the year 2012 and pursued his Master of Technology (M.Tech.) in VLSI Design and Embedded systems from Reva Institute of Technology and Management, Bengaluru. Following that, he started his Ph.D. in Kalasalingam Academy of Research and Education in the year 2017. He worked in the field of High electron mobility transistors (HEMTs), which is also his main field of interest as of now, solving interface state distributions and parasitic channel effects, the findings of which have been published in several reputed international journals during the tenure. He completed and was awarded his Ph.D. in December, 2021. He has around 16 international journal and conference publications, one among them being IEEE Transactions on Electron Devices. He later continued in Kalasalingam Academy of Research and Education as a Post-doctoral researcher from May 2022 till January 2024. The nature of work he does there includes collaborating with a US-based semiconductor organization and working on numerical simulation of Gallium Nitride (GaN)-based device design. From February 2024, he has been working as an Assistant Professor in Mangalore Institute of Technology & Engineering, Karnataka. His main research interests include semiconductor device modelling and simulation, GaN-based HEMTs, characterization and VLSI design.

Dr. J. Charles Pravin has obtained his B.E. degree in Electrical and Electronics Engineering from Anna University, and M.Tech in Applied Electronics from Jerusalem College of Engineering. He has completed Ph.D in Electronics and Communication Engineering from Karunya University in the year 2017. He is presently working as an Associate Professor in the department of Electronics and Communication engineering, Kalasalingam Academy of Research and Education. His research interests include Semiconductor Device Modelling and Simulation, Junctionless Nanowire Transistor, LED, Device Fabrication and Characteristics, Nano Material Synthesis for Electronic Devices. He has published more than 47 research articles in reputed journals and conferences. He has guided 3 full time research scholars and 2 Master's thesis in his field of research. He has received an overseas R&D grant from Tagore Technology (Chicago-USA) for conducting industrial research. He has also received a recognition award from IEEE Madras Section for publishing research article in IEEE Transactions on Electron Devices in the year 2020.

Evolution of GaN Based HEMTs Towards Achieving Enhancement Mode Operation

Ashutosh Chakrabarty and Raghunandan Swain

Abstract In power electronic circuits, enhancement mode (E-mode) devices play a vital role while realizing fail-safe operation. Presence of in-built two-dimensional electron gas (2DEG), near the heterointerface in High Electron Mobility Transistors (HEMTs), make these devices very difficult and challenging for the researchers to achieve E-mode of operation. In this chapter, a comprehensive study has been performed first on AlGaAs/GaAs based HEMTs and gradually the study continued up to AlGaN/GaN based tri-gate fin-shaped HEMTs (FinHEMTs). In all these devices under consideration, efforts have been made to investigate the device operation with respect to 2DEG formation near the heterointerface and how it can be depleted in order to realize E-mode operation with reference to the existing literature. Moreover, as gate leakage current becoming an issue in smooth device characteristics, one section is dedicated towards explaining the different mechanisms governing this current scenario with respect to different operating temperatures in GaN-based HEMTs. It is very fascinating to accept that the recent advances in HEMT technology such as tri-gate FinHEMT structures are promising candidates not only for realizing E-mode operation but also supressing short channel effects while downscaling in accordance with Moore's law.

Keywords E-mode · FinHEMT · HEMT · MOSHEMT · Tri-gate

A. Chakrabarty (✉)
Centre for Internet of Things, Department of Electronics and Communication Engineering, ITER, Siksha O Anusandhan University, Bhubaneswar, Odisha 751030, India
e-mail: ashutoshchakrabarty@soa.ac.in

R. Swain
Department of Electronics and Telecommunication, Parala Maharaja Engineering College, Berhampur, Odisha 761003, India
e-mail: raghu.etc@pmec.ac.in

D. Nirmal and J. Ajayan (eds.), *Modeling of AlGaN/GaN High Electron Mobility Transistors*, Springer Tracts in Electrical and Electronics Engineering,
https://doi.org/10.1007/978-981-97-7506-4_10

1 Introduction

HEMTs operate efficiently at high frequency and power domain. GaN is the most deserving material among the gamut of III-V compound semiconductor based state of the art technology since more than two decades. In this chapter, first advantages of GaN-HEMTs over GaAs-HEMTs are discussed. Then the operation of GaN-MOSHEMT is discussed in detail. As realization of enhancement mode (E-mode) operation is a major challenge in these devices, literature regarding existing analytical modeling and numerical simulation techniques for 2DEG density and gate leakage current phenomena in HEMTs are presented. Finally, the recent fin-shaped tri-gate HEMT (FinHEMT) technology is reviewed.

2 GaN Over GaAs Based HEMTs

For the first time Mimura et al. had reported AlGaAs/GaAs based HEMTs in 1980 (Mimura et al. 1980). The structure was consisting of alternate layers of AlGaAs doped with Si and GaAs layers grown by molecular beam epitaxy technique. The structure along with the band diagram at AlGaAs/GaAs heterojunction is demonstrated in Fig. 1. Due to band offset, there is formation of a triangular potential well at AlGaAs/GaAs heterojunction. Due to the higher work function of GaAs the electrons move from AlGaAs layer to GaAs layer leaving behind the positively charged donors at the heterointerface between the barrier and buffer layer and form 2DEG at the quantum well. High mobility of electrons inside the 2DEG layer is due to their spatial distribution from parent donors and better confinement inside the quantum well. A maximum carrier concentration of 2×10^{17} cm^{-3} is found at 77 K after measuring with a differential capacitance feedback profiler.

AlGaAs/GaAs based HEMTs were popular for more than a decade until AlGaN/GaN HEMT was proposed initially by Asif Khan et al. in 1993 (Asif Khan et al. 1993). In this device, the doped interlayer between the barrier and channel is not required. AlGaN is fabricated by the metal organic chemical vapor deposition (MOCVD) process with the source material as ammonia, trimethyl aluminium (TMAl) and trimethylgallium (TMGa). A schematic of the device along with the charges present in various layers is shown in Fig. 2. Due to polarity and unequal lattice constants of AlGaN and GaN, there exists spontaneous polarization for GaN and both piezoelectric and spontaneous polarization for AlGaN. These polarization charges create an electric field which drifts the electrons from trap states present in AlGaN to the 2DEG layer. This 2DEG behaves as a channel for electron movement responsible for current conduction. The gate bias controls the amount of electon density inside 2DEG. Since it is formed even in the absence of gate bias, these devices generally operate in normally-On mode. Hence, in order to realize E-mode, the 2DEG must be absent at zero gate bias and it should be formed at the quantum well only upon application of positive gate bias.

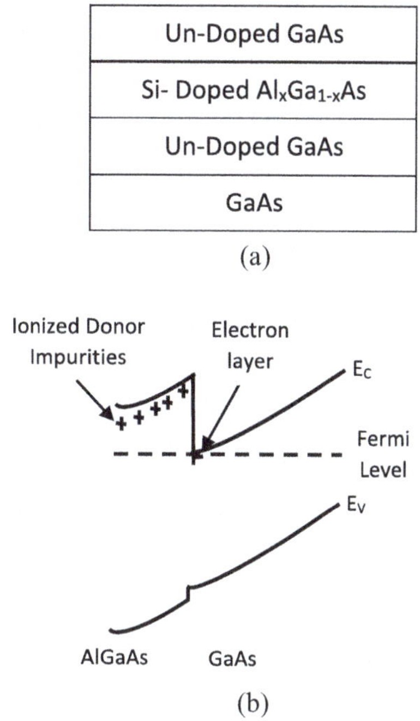

Fig. 1 **a** AlGaAs/GaAs Heterostructure **b** demonstration of formation of two-dimensional electron layer inside triangular quantum well through energy band diagram (Mimura et al. 1980)

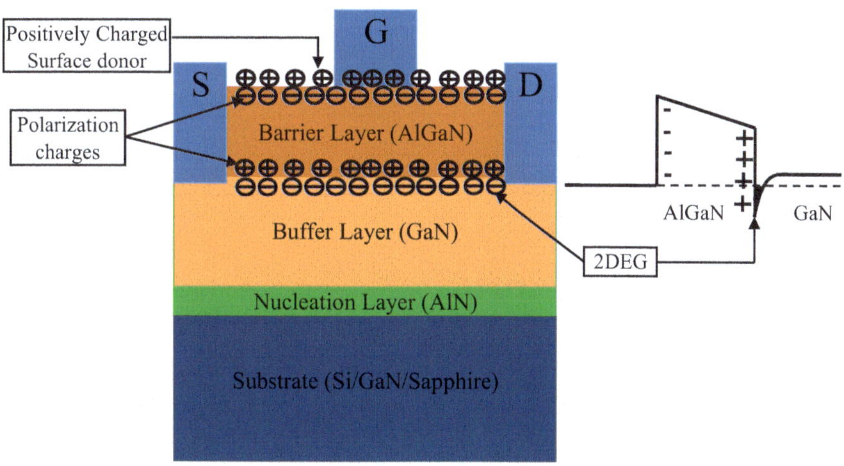

Fig. 2 AlGaN/GaN HEMT Structure and corresponding conduction band energy diagram across the junction showing the formation of charges at different edges and 2DEG inside the triangular quantum well (Asif Khan et al. 1993)

However, AlGaN/GaN HEMTs perform efficiently in RF power domain over AlGaAs/GaAs HEMTs due to favorable properties of AlN and GaN such as high saturation electron drift velocity greater than 2×10^7 cm/s, high mobility of nearly 1000 $cm^2/V.s$, large breakdown electric field strength of nearly 3×10^6 V/cm, high 2DEG density of nearly 2×10^{13} cm^{-2}, wide band gap of 6.2 eV in AlN and 3.4 eV in GaN, low relative dielectric constant of 9.5 for GaN and 10.8 for AlN (Dabiran et al. 2008). One of the AlGaN/GaN devices even possesses maximum oscillation frequency (f_{MAX}) of 91 GHz and output power density of 3.65 W/mm at 18 GHz (Feng et al. 2010). In contrast to AlGaAs/GaAs devices, where doping concentration in barrier layer decides the amount of electrons in quantum well, in AlGaN/GaN HEMT same is decided by the thickness and mole fraction of Al within AlGaN (Lenka and Panda 2011a, 2011b; Miyoshi et al. 2004). In AlGaN/GaN HEMT with T gate structure f_{MAX} and Johnson's figure of merit as high as 182 GHz and 6.43×10^{12} V/s respectively can be achieved (Huang et al. 2014a).

In case of AlN/GaN junction even with very low thickness a very high 2DEG can be achieved due to 100% Al content in the barrier layer (Hubbard et al. 2002). Also thin barrier facilitates the realization of E-mode operation due to lack of sufficient 2DEG at zero gate bias and current is formed only after applying a positive gate voltage (Zimmermann et al. 2011). Due to the difference in polarizations, dielectric constant and band gap, AlN/GaN junction show firm confinement of carriers at quantum well (Medjdoub et al. 2011; Medjdoub et al. 2011). A novel AlN/GaN HEMT on Silicon can possess f_T more than 100 GHz, drain current of 1.74 A/mm, maximum transconductance (g_m) of 530 mS/mm, and f_{max} of 162 GHz (Medjdoub et al. 2011). However, on sapphire substrate the same device with T-gate structure shows maximum drain current of 1.7 mA/mm, g_m of 500 mS/mm (Chabak et al. 2011). Till date the maximum f_T of 220 GHz, f_{MAX} of 400 GHz and higher g_m of 723 mS/mm are obtained in AlN/GaN/AlGaN HEMT structure (Shinohara et al. 2010).

Similarly, InAlN/GaN HEMTs with high mole fractionof Al up to 83% are also highly exploited because of its improved electrostatics even at thinner barrier due to high amount of strain induced polarization charges (Kuzmik 2001). The lattice mismatching problem in AlGaN/GaN interface responsible for 2DEG degradation is overcome by AlInN/GaN with Al mole fraction of 83%. So these devices are gaining popularity for high frequency applications with drain current more than 2 A/mm at room temperature for a gate length of 0.25 μm (Medjdoub et al. 2006).

3 GaN Based MOSHEMTs

Scaling down the device dimensions in order to meet the current technology challenges such as realizing normally-Off operation makes the device prone to high gate leakage current. Higher leakage current is responsible for degrading the reliability and device performance by affecting the noise figure (Sanabria et al. 2006), subthreshold characteristics (Chung et al. 2008), breakdown voltage (Saito et al. 2005), and off-state power (Saito et al. 2004). In metal oxide semiconductor high electron mobility transistors (MOSHEMTs) oxide layer is introduced under the gate in order to reduce the leakage current and degradation of drain current (Khan et al. 2000). The oxide dielectric layers like SiO_2 (Khan et al. 2000), HfO_2 (Liu et al. 2006), Al_2O_3 (Ganguly et al. 2011), Ga_2O_3 (Hong et al. 2007), Si_3N_4 (Shealy et al. 2003), MgO (Gila et al. 2006), Sc_2O_3 (Wu et al. 2009), or TiO_2 (Zhu et al. 2014a) can be used as insulators. However, a significant amount of reverse and forward leakage current densities of 10^{-1}A/cm^2 and 10 A/cm^2 respectively are found even in MOSHEMTs (Zhu et al. 2014a).

The conduction band energy profile of MOSHEMT along with the charge distribution is shown in Fig. 3. The charges at various edges remain similar to that present in HEMT structure except the additional oxide defect charges and gate voltage dependent oxide/barrier interface charges in MOSHEMT. The substrate layer provides support and proper thermal management, which can be Si (111) (Dumka et al. 2004), SiC (Tilak et al. 2001) or c-plane Sapphire (Chini et al. 2004). Researchers are in continuous effort to fabricate reliable HEMT devices on Si, as it is cost effective. Sapphire is widely used as a substrate in case of HEMTs because of its capability to sustain high temperature. But the overlying GaN buffer layer is not properly lattice matched. Hence, instead of growing the AlGaN barrier layer directly on Ga-faced c-plane GaN buffer layer, a wide band gap AlN called as nucleation layer is grown above the GaN layer as shown in Fig. 3 in order to increase the mobility and density of 2DEG (Zhu et al. 2014a). Above the barrier layer the insulator, may be oxide or nitride is grown by various methods such as sputtering (Maeda et al. 2005), atomic layer deposition (Ye et al. 2005), plasma enhanced chemical vapor deposition (Gregušova et al. 2009), or pulsed laser deposition (Sugiura et al. 2010) etc. In general, the growth techniques are insulator specific with contacts made through evaporation (Higashiwaki et al. 2006).

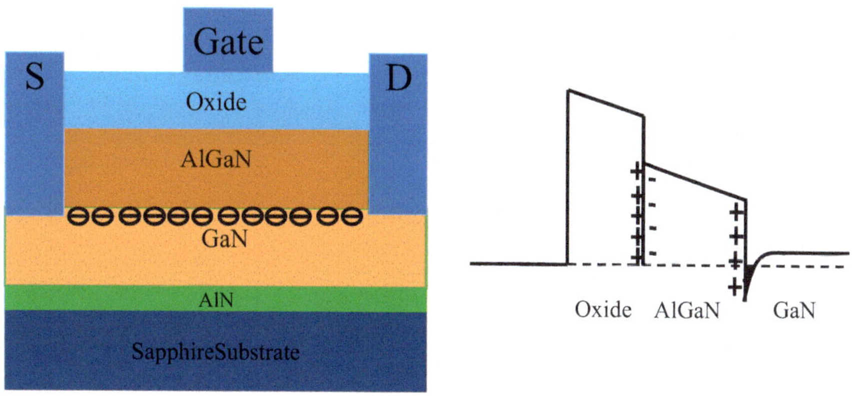

Fig. 3 AlGaN/GaN MOSHEMT Structure along with the conduction band diagram showing the formation 2DEG inside the triangular quantum well

4 GaN Based E-mode MOSHEMT

Generally, the GaN-based MOSHEMTs operate in E-mode due to the presence of inherent polarization dependent 2DEG layer inside the channel. However, for specific applications the demand of E-mode devices is also inevitable. Specifically in power electronics, E-mode devices can prevent failure due to short circuit at zero gate potential. As an example, the direct coupled FET logic is fabricated with integration of both E and D-mode HEMTs on a single wafer for high performance mixed signal applications. Moreover due to the lack of requirement of negative power supply, E-mode devices reduce the complexity of circuit and system design (Chen and Zhou 2011).

Similarly, in power electronic circuits such as SMPS, DC-DC and DC-AC converters, the role of E-mode MOSFET is also equally important. Due to immature normally-Off MOSHEMT technology, earlier Si-based MOSHEMTs were used in conjunction with D-mode GaN devices (Chowdhury and Mishra 2013). The schematic diagram of SMPS having one start-up circuit is shown in Fig. 4. The strat-up circuit employs two E-mode MOSHEMTs fabricated by Fluorine plasma treatment in order to deplete 2DEG (Tang et al. 2013).

Many methods have been employed so far to realize positive threshold voltage in both HEMTs and MOSHEMTs as evident from the literature. Kai Zhang et al. have fabricated E-mode HEMT with ultra-thin barrier layer having high Aluminium content using O_2 plasma implantation and obtained threshold voltage (V_T) of + 1 V (Zhang et al. 2015). Asamira Suzuki et al. have fabricated E-mode HFET with the Ge-doped GaN layer and obtained very low on resistance by selectively depositing NiO gate (Suzuki et al. 2014). Chang et al. had treated the gate area with O_2 plasma prior to make gate contact and fabricated E-mode AlN/GaN HEMTs (Chang et al. 2009). Huang et al. fabricated AlN/GaN MOSHEMT with V_T of 0.21 V by using ultra-thin AlN barrier with thickness of 1.5 nm and regrown source and drain contacts (Huang et al. 2014b). Another very useful method is to fabricate gate recessed MOSHEMT, in which the barrier layer below the gate foot region is etched away to reduce 2DEG content (Choi et al. 2014). However, thin barrier reduces the current driving capability. So in conjunction with the gate recess method some additional processes are

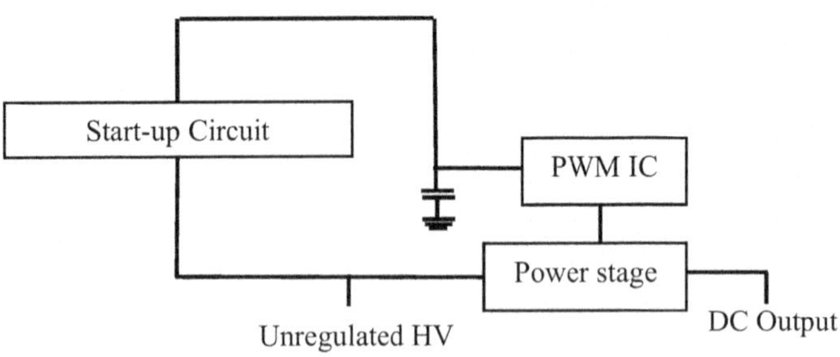

Fig. 4 Block diagram of SMPS

adopted such as, treatment with tetramethylammonium hydroxide after gate recess to compensate the damage due to plasma treatment, annealing after deposition of the oxide layer to reduce the gate leakage current, source and drain are re-grown to improve the resistance and V_T, controlling the stress to obtain high 2DEG and improve conductance, and use of the p-GaN as a back-barrier to reduce the buffer leakage current (Kim et al. 2013). Positive V_T can also be realized by depositing a P-type GaN layer below the gate electrode (Su et al. 2014). The P-GaN layer injects holes to the 2DEG to deplete it and thereby shifts the conduction band upward inside the quantum well. Another very reliable method is to consider sub-critical barrier thickness (Brown et al. 2014a, 2014b). Normally-Off AlGaN/GaN MOSHFETs on silicon substrate with TaN Gates and SiO_2 Tunnel Dielectric has also been demonstrated (Lee et al. 2010).

5 GaN Based MOSHEMTs—Modeling Perspective

There are several models present in literature for 2DEG density (n_s) both in AlGaAs/GaAs and AlGaN/GaN HEMTs. The models developed for n_s in AlGaAs/GaAs system are still valid for the AlGaN/GaN system with addition of polarization properties. Anderson et al. for the first time predicted the presence of electron inversion and accumulation phenomena at AlGaAs/GaAs heterojunction (Anderson 1960). Delagebeaudeuf et al. have obtained the expression for 2DEG density (n_s) and Electric field using energy states inside quantum well by solving Schrödinger's wave equation followed by Poisson's equation (Delagebeaudeuf and Linh 1982). In this model, it is assumed that the electrons reside at the heterojunction as the width of the quantum well is nearly zero. A special distribution of electrons in quantum well dependent self-consistent model is presented in Ref. (Ng et al. 1991). The n_s in the AlGaAs/GaAs system can also be predicted by the Montecarlo method using two-dimensional scattering rates (Tomizawa et al. 1985), two dimensional electric field profiles (Park et al. 1989) and quantum effects (Ravaioli and Ferry 1986). The n_s inside the triangular quantum well was calculated by considering two (Ravaioli and Ferry 1986), three (Tomizawa and Hashizume 1988) or five (Yokoyama and Hess 1986) lowest intra band energy levels in the conduction band of GaN. Another method is by solving Boltzmann transport and Poisson equation numerically which also considers the electron heating phenomenon through the hydro-dynamic model (Widiger et al. 1985).

Initially, the effect of polarization charges is neglected while modelling n_s in the AlGaN/GaN system (Stengel et al. 1996). However, both piezoelectric and spontaneous polarization charges significantly enhance 2DEG density in the AlGaN/GaN system over the AlGaAs/GaAs system nearly by a factor of 10 without any extra doping (Ambacher et al. 1999). Hence, an appropriate polarization dependent charge control model was developed by Rashmi et al. for n_s in both the strong and weak inversion mode (Rashmi et al. 2002). Similarly, there are some models revealing the consistency between n_s and fermi level E_F. In Ref. (Delagebeaudeuf and Linh

1982) the E_F is assumed as a fitting parameter and it is linearly related to n_s. On the contrary, the variation of E_F with n_s is represented by a second order analytical expression in Ref. (Kola et al. 1988; DasGupta and Dasgupta 1993). A charge control model for n_s is developed along with calculation of temperature dependency in Ref. (Li and Wang 2008). This approach has also been used by us in order to predict n_s in MOSHEMT. However, a unified compact model can be presented for 2DEG density in generic MOSHEMTs (Zhang et al. 2014).

In MOSHEMT, n_s can also be modeled by including the density of states (D_{it}) at the oxide/barrier interface with the existing polarization dependent model (Yu et al. 1997). In order to evaluate V_T, the n_s expression can be equated to zero and the corresponding gate voltage can be found. It is observed that there is a shift in C ~ V characteristics when D_{it} is varied (Miczek et al. 2008). So incorporating D_{it} in the existing V_T model can shift it towards positive which leads to normally-Off operation. Moreover, by reducing the charge density at the oxide/semiconductor interface and making it more negative leads to increase in positive V_T (Hung et al. 2014).

Eventually, the interface trap states get charged and discharged causing shift in C–V curve (Mizue et al. 2011). There are mainly three kinds of interface traps depending on the capture time. These are short traps, intermediate traps and long traps. However, these interface trap amounts can be calculated from C-V hysteresis. The long traps hold electrons for a longer time and mainly responsible for V_T shift (Johnson et al. 2013).

6 Gate Leakage Current Mechanisms–an Overview

There are several mechanisms which explain the leakage current under different field and temperature conditions separately for reverse and forward biases. The details of the mechanisms reported in different literatures are mentioned as follows.

7 Poole–Frenkel Emission (PFE)

PFE is a trap assisted tunneling mechanism mainly dependent on field and temperature as shown in Fig. 5. Oxygen vacancies present in oxide of MOSHEMT and nitride traps present in the barrier of HEMT are the main sites for electron trapping and help in tunneling. The current density due to this mechanism in HEMT is represented as.

$$J_{PFE} = q \cdot \mu_b \cdot N_C \cdot E_{ox} \cdot \exp\left[\frac{-q \cdot \left(\varphi_{t,b} - \sqrt{\left(q \cdot E_b / \pi \cdot \varepsilon_b\right)}\right)}{kT}\right] \quad (1)$$

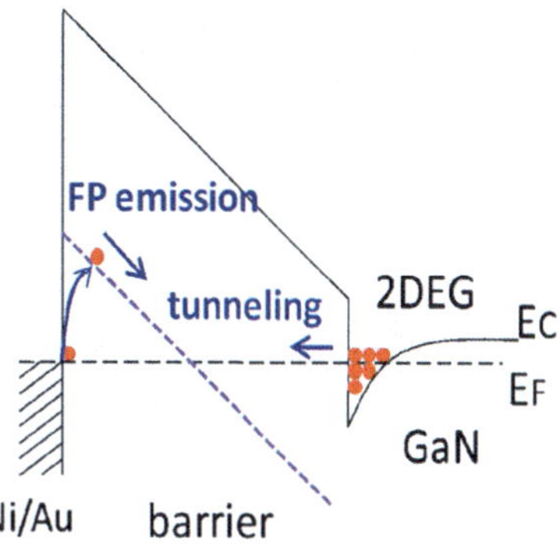

Fig. 5 PFE mechanisms in HEMT having Ni/Au as gate (Vollmann 1974)

Here, E_b is the electric field in the barrier layer, μ_b is the mobility of electron in the barrier, N_C is the density of states in the conduction band of the barrier and ε_b is the dielectric constant of the barrier (Vollmann 1974). However, it is a common practice that the product of q and N_C is treated as constant while extracting parameters from experimental data (Yan et al. 2010a).

8 Fowler–Nordheim Tunneling (FNT)

FNT determines the leakage current independent of temperature and under high electric field. It's a trap independent process and mainly increases with electric field. The expression for current density due to FNT is given by Zhang et al. (2006).

$$J_{FNT} = AE_b^2 \exp\left(-\frac{B}{E_b}\right) \tag{2}$$

where $A = \frac{q^2(m_e/m_n^*)}{8\pi h\varphi_b}$ and $B = \frac{8\pi\left(\sqrt{2m_n^*(q\varphi_b^3)}\right)}{3qh}$, ϕ_b is the effective barrier height at the metal to semiconductor contact, q, m_e and m_n^* are charge, free mass and effective mass of electron inside the conduction band respectively. The difference between direct tunneling and FNT is demonstrated in Fig. 6.

9 Thermionic Emission (TE)

This mechanism is dominant under forward gate bias. The gate leakage current density due to this mechanism is given by Turuvekere et al. (2013).

$$J_{TE} = J_0\left\{exp\left(\frac{qV}{\eta kT}\right) - 1\right\} \tag{3}$$

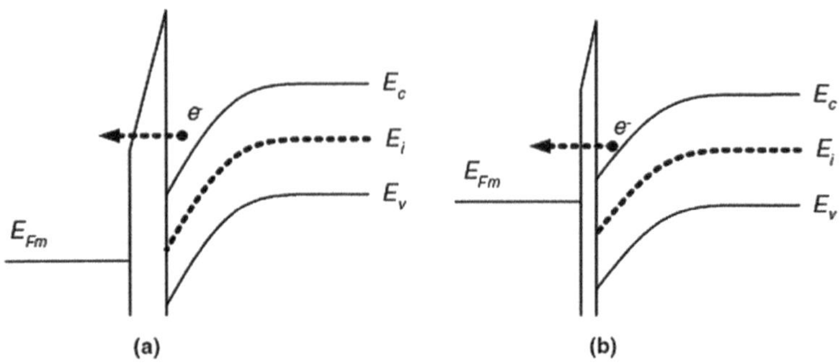

Fig. 6 **a** Tunneling through triangular barrier (Fowler–Nordheim tunneling) **b** Tunneling through trapezoidal barrier (direct tunneling) (Ranuarez et al. 2006)

$$J_0 = A^* T^2 exp\left(-\frac{q\varnothing_b}{kT}\right) \tag{4}$$

J_0 is the reverse saturation current density, A^* is the effective Richardson's constant and η is the ideality factor.

10 Trap Assisted Tunneling (TAT)

This mechanism is dominant at lower temperature when the electron does not get sufficient energy to be drifted towards the barrier conduction band so that the electron tunnels through triangular or trapezoidal shaped barrier depending on the electric field. The simplified expression for current density due to this mechanism is given by Turuvekere et al. (2013).

$$J_{TAT} = J_{02}\left\{\exp\left(\frac{q(V - V_0)}{\eta_2 kT}\right) - 1\right\} \tag{5}$$

where J_{02}, V_0, and η_2 are the fitting parameters.

11 Gateleakage Current Mechanisms in GaN Based HEMTs and MOSHEMTs

Miller et al. had emphasized on vertical tunneling and analytically modeled the gate leakage current in AlGaN/GaN HEMT (Miller et al. 2000). Karmalkar et al. presented that trap assisted tunneling is dominant at temperatures less than 500 K and direct tunneling is responsible for reverse leakage current at higher temperatures

(Karmalkar et al. 2003, 2006). TAT occurs through only a triangular barrier in the metal semiconductor (MS) structure, whereas both trapezoidal and triangular barrier when the insulator is inserted in between (Sathaiyaa and Karmalkar 2006). Poole–Frenkel emission model dominates gate leakage current in AlInN/AlN/GaN and AlGaN/GaN HEMT devices (Sathaiyaa and Karmalkar 2006; Arslan et al. 2009). Under reverse bias two gate leakage current mechanisms generally dominate in GaN-based HEMTs (Xia et al. 2013). One is Poole–Frenkel Emission (PFE) dominant at low electric field below 1.6 MV/cm and the other one is Fowler–Nordheim Tunneling (FNT) dominant at electric field more than 1.6 MV/cm. However, during forward bias thermionic emission mechanism dominates the gate leakage current (Turuvekere et al. 2013).

The gate leakage current in case of MOSHEMTs is well explained in Ref (Yan et al. 2010b; Liu et al. 2011) by extracting and fitting model parameters to the experimental results. There are several gate leakage current mechanisms that take place across an insulator such as, Schottky emission (SE), PFE, TAT, space charge limited current, carrier hopping and FNT (Liu et al. 2011). FNT dominates at high electric field and temperatures below 273 K. So for higher temperature ($T > 273$ K), PFE and TAT model dominates the leakage current. PFE mechanism is dominant particularly at $T > 388$ K. However during high electric field gate to channel series resistance replaces all other mechanisms and limits the gate leakage current. TAT is dominant at temperature below 388 K and responsible for forward gate leakage current in Al_2O_3/AlGaN/GaN MOSHEMT (Zhu et al. 2014b).

12 The Tri-Gate Architecture in GaN-Based E-mode HEMTs

12.1 Tri-Gate GaN-Based HEMT Design

Similar advancements as in Si MOSFETs can also be adopted in case of E-mode GaN-based HEMTs to be used in the RF and power electronics applications (Hisamoto et al. 1991). Si-based tri-gate MOSFETs, commonly known as FinFETs are transistors where the metal gate surrounds the channel from three sides (Hisamoto et al. 1991). Tri-gate MOSFETs have a number of benefits over top-gate MOSFETs in general (Doyle et al. 2003; Lee et al. 2008; Sun and Liu 2009). Tri-gate structures outperform top-gate structures in terms of electrostatics and effectively control short-channel effects. While the short channel effects are important during sub-threshold operation in dc state in digital MOSFETs, the RF applications need low drain current saturation and high drain conductance. Leading chip companies, including Intel, have already started mass producing tri-gate MOSFETs, which were initially designed for Si MOSFETs (Doyle et al. 2003; Karl et al. 2015, 2012; Ramey et al. 2013).

The tri-gate design is a recent substitute strategy to enhance the gate controllability of GaN-based HEMTs (Valizadeh and Altaibi 2011). This not only achieves

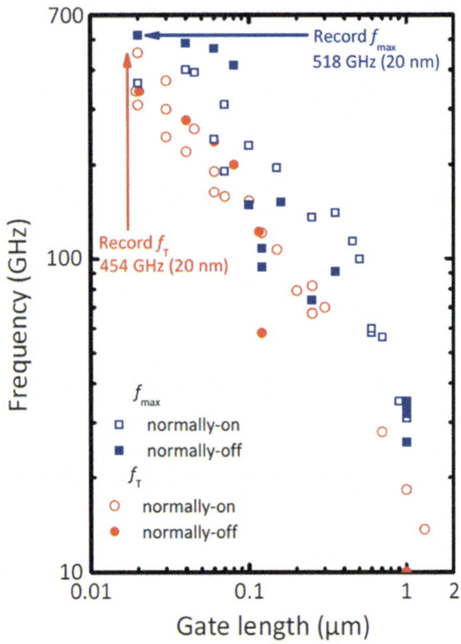

Fig. 7 The state-of-the art research on GaN HEMTs with normally-On and normally-Off operation demonstrating f_T and f_{max} as a function of gate length (Alsharef et al. 2013)

E-mode operation but also decreases short-channel effects suitable for GaN RF applications. This tri-gate HEMTs possess a higher V_T than typical planar transistors (Liu et al. 2012). There is a strong relationship between V_T of and the width of body and depth of etching (Alsharef et al. 2014; Alsharef et al. 2013). The tri-gate design has several advantages over typical planar designs, including enhanced electrostatics that resulted in greater SCE suppression (Ture et al. 2015; Subramanian et al. 2010). However, it can be observed from Fig. 7 that GaN RF tri-gate HEMTs continued to perform poorer than their single-gate planar counterparts until 2013. Many of the tri-gate advantages were not completely utilized due to lack of adequate research, particular benefits, technological constraints, optimized designs, and reliability difficulties. In addition, extensive study was needed to develop the design standards for high frequency targeted transistor structure.

12.2 *The E-mode AlGaN/GaN FinHEMT*

Even though Si-based MOSFETs are predominantly employed in digital circuits for RF applications, it is still highly beneficial to look into Si MOSFET advances. The first n-channel fin field-effect transistor (FinFET) was created in 1998 by a team at Taiwan Semiconductor Manufacturing Co., Ltd. (TSMC) under the direction of Chenming Hu (Hisamoto et al. 1998). In order to improve channel control, this construction has gates on three sides of the channel. The FinFET structure provides the following benefits as: altering the fin dimension may change the threshold voltage. It rises with narrowing fin width in n-channel MOSFETs. In addition to having a greater breakdown voltage than typical planar MOSFETs, those with superior electrostatic control of the active region also efficiently inhibit short-channel effects. This idea has achieved great success and is utilized extensively in processors developed by businesses like Intel. The FinFET concept was put up as an alternate approach to enhance GaN HEMTs gate controllability and eventually achieve E-mode HEMTs in 2012 (Liu et al. 2012). GaN-based FinFETs (FinHEMT) gain more from the ability to tailor the 2DEG they offer. People may manufacture devices with much better linearity (Choi et al. 2020), increased current injection efficiency (Xue et al. 2020), reduced gate voltage swings to attain the peak transconductance (g_m) (Lee et al. 2013), and improved thermal management (Ma et al. 2016) because of the greater gate controllability. It is obvious that the FinHEMT concept may be used to AlGaN channel HEMTs for comparable advantages. They will benefit more from the multi-gate design in terms of drain current density due to a bigger band gap and lower electron affinity. Figure 8 depicts schematic comparison of conventional AlGaN/GaN HEMT design to that of Tri-Gate AlGaN/GaN FinHEMT. In comparison to the traditional planar GaN HEMT design which has only one broad channel, multiple channels are realized in tri-gate design.

The tri-gate design is a practical architecture that has recently been employed in HEMTs and discovered to give a significant positive shift in V_T by improving both ON and OFF state performance. Dry etching procedures may be used to create several 2DEG channels and trenches on a regular basis. This structure makes use of three dimensional non planar design by depositing the gate material around AlGaN/GaN arrays from three sides as shown in Fig. 9.

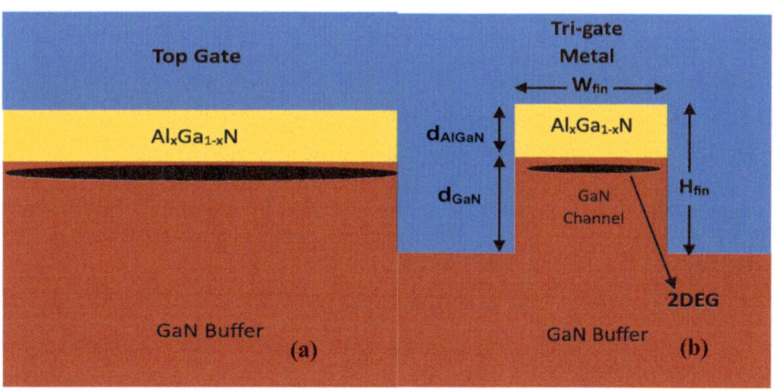

Fig. 8 **a** Conventional planer AlGaN/GaN HEMT **b** 2D view of Tri-GateAlGaN/GaN FinHEMT (Chakrabarty and Swain 2020)

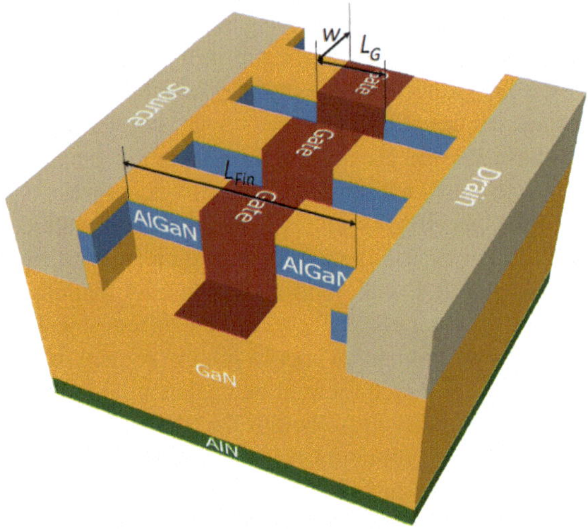

Fig. 9 3D view of AlGaN/GaN Tri-Gate FinHEMT (Alsharef et al. 2013)

13 Conclusion

A detailed review on GaN-based HEMTs and MOSHEMTs towards achieving E-mode operation is presented in this chapter. The study started from AlGaAs/GaAs based devices and ended at tri-gate AlGaN/GaN HEMTs (FinHEMTs). It can be concluded that GaN-based FinHEMTs with narrow AlGaN/GaN fins are promising devices to achieve E-mode operation due to lower 2DEG concentration at a particular Al mole composition within AlGaN. It is due to two factors such as, improved controlling power of gate from three sides of the channel and partial strain relaxation offered by narrow width of the AlGaN barrier layer. Moreover, improved electrostatics can also be obtained due to suppression of short channel effects which is due to further scaling down of the device and miniaturization.

References

Alsharef M, Granzner R, Schwierz F (2013) Theoretical investigation of Tri-gate AlGaN/GaN HEMTs. IEEE Trans Electron Dev 60:3335–3341

Alsharef M, Granzner R, Ture E, Quay R, Racko J, Breza J, Schwierz F (2014) Design of GaN tri-gate HEMTs. In: Advanced Semiconductor Devices Microsystems (ASDAM), 2014 10th International Conferenceon, pp 1–4

Ambacher O et al (1999) Two-dimensional electron gases induced by spontaneous and piezoelectric polarization charges in N and Ga-face AlGaN/GaN hetero structures. J Appl Phys 85:3222–3233

Anderson RL (1960) Germanium-gallium-arsenide heterojunctions. IBM J Res Dev 4:283–287

Arslan E, Bütün S, Ozbay E (2009) Leakage current by Frenkel-Poole emission in Ni/Au Schottky contacts on $Al_{0.83}In_{0.17}N$/AlN/GaNheterostructures. Appl Phys Lett 94:142106

Asif Khan M, Bhattarai A, Kuznia J, Olson D (1993) High electron mobility transistor based on a GaN-$Al_xGa_{1-x}N$ heterojunction. Appl Phys Lett 63(9):1214–1215

Brown R, Al-Khalidi A, Macfarlane D, Taking S, Ternent G, Thayne I, Wasige E (2014a) Novel high performance AlGaN/GaN based enhancement-mode metal-oxide semiconductor high electron mobility transistor. Phys Status Solidi C 11(3–4):844–847

Brown R et al (2014b) A sub-critical barrier thickness normally-off AlGaN/GaN MOS-HEMT. Electron Device Letters 35(9):906–908

Chabak KD et al (2011) High-Performance AlN/GaN HEMTs on sapphire substrate with an oxidized gate insulator. IEEE Electron Device Lett 32(11):1677–1679

Chakrabarty A, Swain R (2020) Modelling of fin width dependent threshold voltage in fin shaped nano channel AlGaN/GaN HEMT. Superlattices Microstructure. 141:106497

Chang CY, Pearton SJ, Lo CF, Ren F, Kravchenko II (2009) Development of enhancement mode AlN/GaN high electron mobilitytransistors. Appl Phys Lett 34:263505

Chen KJ, Zhou C (2011) Enhancement-mode AlGaN/GaNHEMT and MIS-HEMT technology. Phys Status Solidi A 208(2):434–438

Chini A, Buttari D, Coffie R, Heikman S, Keller S, Mishra U (2004) 12 W/mm power density AlGaN/GaN HEMTs on sapphire substrate. Electron Lett 40(1):73–74

Choi W, Chen R, Levy C, Tanaka A, Liu R, Balasubramanian V, Asbeck PM, Dayeh SA (2020) Intrinsically linear transistor for millimeter-wave lownoise amplifiers. Nano Lett 20(4):2812–2820

Choi W, Ryu H, Jeon N, Lee M, Cha Ho-Y, Seo KS (2014) Improvement of V_{th} Instability in Normally-Off GaN MIS-HEMTs Employing PEALD-SiN_xas an Interfacial Layer. IEEE Electron Device Letters 35:30–32

Chowdhury S, Mishra UK (2013) Lateral and Vertical Transistors Using theAlGaN/GaNHeterostructure. IEEE Trans Electron Devices 60(10):3060–3066

Chung JW, Roberts JC, Piner EL, Palacios T (2008) Effect of gate leakage in the subthreshold characteristics of AlGaN/GaN HEMTs. IEEE Electron Device Lett 29:1196–1198

Dabiran AM, Wowchak AM, Osinsky A, Xie J, Hertog B, Cui B, Look DC, Chow PP (2008) Very high channel conductivity in low-defect AlN/GaN high electron mobility transistor structures. Appl Phys Lett 93:82111–82113

DasGupta N, Dasgupta A (1993) An analytical expression for sheet carrier concentration vs gate voltage for HEMT modelling. Solid State Electron 36(2):201–203

Delagebeaudeuf D, Linh NT (1982) Metal-(n) AlGaAs-GaAs two-dimensional electron gas FET. IEEE Trans Electron Devices 29:955–960

Doyle BS, Datta S, Doczy M, Hareland S, Jin B, Kavalieros J, Linton T, Murthy A, Rios R, Chau R (2003) High performance fully-depleted tri-gateCMOS transistors. IEEE Electron Device Lett 24:263–265

Dumka D, Lee C, Tserng H, Saunier P, Kumar M (2004) AlGaN/GaN HEMTs on Si substrate with 7 W/mm output power density at 10 GHz. Electron Lett 40(16):1023–1024

Feng ZH et al (2010) 18-GHz 3.65-W/mm enhancement-mode AlGaN/GaN HFET using fluorine plasma Ion implantation. Electron Device Letter 31(12):1386–1388

Ganguly S, Verma J, Li G, Zimmermann T, Xing H, Jena D (2011) Presence and origin of interface charges at atomic-layer deposited Al_2O_3/III-nitride hetero junctions. Appl Phys Lett 99:1935041–1935044

Gila BP et al (2006) Novel dielectrics for gate oxides and surface passivation on GaN. Solid State Electron 50:1016–1023

Gregušova D et al (2009) Characterization of AlGaN/GaN MISHFETs on a Si substrate by static and high-frequency measurements. Semicond Sci Technol 24:075014

Higashiwaki M, Matsui T, Mimura T (2006) AlGaN/GaN MIS-HFETs With f_T of 163 GHz UsingCat-CVD SiN Gate-Insulating and Passivation Layers. IEEE Electron Device Lett 27:16–18

Hisamoto D, Kaga T, Takeda E (1991) Impact of the vertical SOI "DELTA" structure on planar device technology. IEEE Trans Electron Devices 38:1419–1424

Hisamoto D, Lee W-C, Kedzierski J, Anderson E, Takeuchi H, Asano K, King T-J, Bokor J, Hu C (1998) A folded-channel MOSFET for deep-sub-tenthmicron era, pp 1032–1034

Hong M, Chang Y, Lee Y, Chiu Y, Lin T, Wu S, Chiu H, Kwo J, Wang Y (2007) MBE grown high-k dielectrics $Ga_2O_3(Gd_2O_3)$ on GaN. J Cryst Growth 390:301–302
Huang S et al (2014a) High- f_{MAX} High Johnson's Figure-of-Merit 0.2-μm Gate AlGaN/GaN HEMTs on silicon substrate with AlN/SiN_x passivation. IEEE Electron Device Lett 35:315–317
Huang T, Zhu X, Lau KM (2014b) High- f_{MAX} High Johnson's Figure-of-Merit 0.2-μm Gate AlGaN/GaN HEMTs on Silicon Substrate With AlN/SiN_x Passivation. IEEE Electron Device Lett 35:315–317
Hubbard S, Pavlidis D, Valiaev V, Eisenbach A (2002) Metal-organic vapor phase epitaxy growth and characterization of AlN/GaN heterostructures. J Electron Mater 31:395–401
Hung TH, Park PS, Krishnamoorthy S, Nath DN, Rajan S (2014) Interface charge engineering for enhancement-mode GaN MISHEMTs. IEEE Electron Device Lett 35(3):312–314
Johnson DW et al (2013) Threshold voltage shift due to charge trapping in dielectric-Gated AlGaN/GaN high electron mobility transistors examined in Au-Free technology. IEEE Trans Electron Devices 60(10):3197–3203
Karl E., Wang Y., Ng Y., Guo Z., Hamzaoglu F., Bhattacharya U., Zhang K., Mistry K., Bohr M. (2012) A 4.6GHz 162Mb SRAM design in 22nm tri-gateCMOS technology with integrated active VMIN-enhancing assist circuitry. In: 2012 IEEE International Solid-State Circuits Conference, pp 230–232
Karl E, Guo Z, Conary JW, Miller JL, Ng Y, Nalam S, Kim D, Keane J, Bhattacharya U, Zhang K (2015) 17.1 A 0.6V 1.5GHz 84Mb SRAMdesign in 14nm FinFET CMOS technology. In: 2015 IEEE InternationalSolid-State Circuits Conference - (ISSCC) Digest of Technical Papers, pp. 1–3
Karmalkar S, Sathaiya DM, Shur MS (2003) Mechanism of the reverse gate leakage in AlGaN/GaN high electron mobility transistors. Appl Phys Lett 82:3976–3978
Karmalkar S, Satyan N, Sathaiya DM (2006) On the resolution of the mechanism for reverse gate leakage in AlGaN/GaN HEMTs. IEEE Electron Device Lett 27:87–89
Khan MA, Hu X, Simin G, Lunev A, Yang J, Gaska R, Shur MS (2000) AlGaN/GaN Metal oxide semiconductor heterostructure field effect transistor. IEEE Electron Device Lett 21:63–65
Kim DS et al (2013) Normally-off GaN MOSFETs on insulating substrate. Solid State Electron 90:79–85
Kola S, Golio JM, Maracas GN (1988) An analytical expression for Fermi level versus sheet carrier concentration for HEMT modeling. IEEE Electron Device Lett 9(3):136–138
Kuzmik J (2001) Power electronics on In AlN/(In)GaN: Prospect for a record performance. Electron Device Letters 22(11):510–512
Lee DS, Wang H, Hsu A, Azize M, Laboutin O, Cao Y, Johnson JW, Beam E, Ketterson A, Schuette ML et al (2013) Nanowire channel InAlN/GaNHEMTs with high linearity of gm and ft. IEEE Electron Device Lett 34(8):969–971
Lee C, Afzalian A, Yan R, Akhavan ND, Xiong W, Colinge J (2008) Drain breakdown voltage in MuGFETs: Infuence of physical parameters. IEEE Trans Electron Devices 55:3503–3506
Lee B, et al. (2010) Normally-Off AlGaN/GaN-on-Si MOSHFETs with TaN floating gates and ALDSiO_2 tunnel dielectrics. In: IEEE International Electron Devices Meeting (IEDM), pp 20.6.1–20.6.4
Lenka TR, Panda AK (2011a) Role of Nanoscale AlN and InN for the microwave characteristics of AlGaN/ (Al, In) N/GaN based HEMT. Semiconductors 45(9):1211–1218
Lenka TR, Panda AK (2011b) Characteristics study of 2DEG transport properties of AlGaN/GaN and AlGaAs/GaAs-based HEMT. Semiconductors 45(5):660–665
Li M, Wang Y (2008) 2-D analytical model for current–voltage characteristics and transconductance of AlGaN/GaN MODFETs. IEEE Trans Electron Devices 55(1):261–267
Liu C, Chor EF, Tan LS (2006) Investigations of HfO_2/AlGaN/GaN metal-oxide-semiconductor high electron mobility transistors. Appl Phys Lett 88:1735041–1735043
Liu ZH, Ng GI, Arulkumaran S, Maung YKT, Zhou H (2011) Temperature-dependent forward gate current transport in atomic-layer deposited Al_2O_3/AlGaN/GaN metal-insulator-semiconductor high electron mobility transistor. Appl Phys Lett 98:163501

Liu S, Cai Y, Gu G, Wang J, Zeng C, Shi W, Feng Z, Qin H, Cheng Z, Chen KJ, Zhang B (2012) Enhancement-mode operation of Nanochannel Array (NCA) AlGaN/GaN HEMTs. IEEE Electron Device Lett 33:354–356

Ma J, Santoruvo G, Tandon P, Matioli E (2016) Enhanced electrical performanceand heat dissipation in AlGaN/GaNSchottky barrier diodes using hybridtri-anode structure. IEEE Trans Electron Devices 63(9):3614–3619

Maeda N, Wang Ch, Enoki T (2005) High drain current density and reduced gate leakage current in channel-doped AlGaN/GaN heterostructure field-effect transistors withAl_2O_3/Si_3N_4 gate insulator. Appl Phys Lett 87:073504

Medjdoub F, Zegaoui M, Rolland N, Rolland PA (2011) Demonstration of low leakage current and high polarization in ultrathin AlN/GaN high electron mobility transistors grown on silicon substrate. Appl Phys Lett 98:2235021–2235023

Medjdoub F, Zegaoui M, Rolland N (2011) Beyond 100 GHz AlN/GaN HEMTs on silicon substrate. Electron Lett 47(24)

Medjdoub F, et al. (2006) Can InAlN/GaN be an alternative to high power/high temperature AlGaN/GaNdevices? IEEE Int Electron Devices Meet (IEDM), pp 1–4

Miczek M, Mizue C, Hashizume T, Adamowicz B (2008) Effects of interface states and temperature on the C-V behaviour of metal/insulator/AlGaN/GaNheterostructure capacitors. J Appl Phys 103:104510

Miller EJ, Dang XZ, Yu ET (2000) Gate leakage current mechanisms in AlGaN/GaNheterostructure field-effect transistors. J Appl Phys 88(12):5951–5958

Mimura T, Hiyamizu S, Fujii T, Nanbu K (1980) A new field-effect transistor with selectively doped GaAs/n-Al_xGa_{1-x}Asheterojunctions. Jpn J Appl Phys 19:L225–L227

Miyoshi M, Sakai M, Arulkumaran S, Ishikawa H, Egawa T, Tanaka M, Oda O (2004) Characterization of different-Al-content AlGaN/GaNheterostructures and high-electron-mobility transistors grown on 100-mm diameter sapphire substrates by metal organic vapor phase epitaxy. Jpn J Appl Phys 43(12):7939–7943

Mizue C, Hori Y, Miczek M, Hashizume T (2011) Capacitance–voltage characteristics of Al_2O_3/AlGaN/GaN structuresand state density distribution at Al_2O_3/AlGaN interface. Jpn J Appl Phys 50:021001

Ng S, Khoie R, Venkat R (1991) A two-dimensional self-consistent numerical model for high electron mobility transistor. IEEE Trans Electron Devices 38:852–861

Park D, Wang Y, Brennan KF (1989) Ensemble Monte Carlo Simulation of a 0.35-pm pseudomorphic HEMT. IEEE Electron Device Lett 10:107–110

Ramey S, Ashutosh A, Auth C, Clird J, Hattendorf M, Hicks J, James R, Rahman A, Sharma V, Amour AS, Wiegand C (2013) Intrinsictransistor reliability improvements from 22nm tri-gate technology. In: 2013 IEEE International Reliability Physics Symposium (IRPS), pp 4C.5.1–4C.5.5

Ranuarez JC, Deen MJ, Chen CH (2006) A review of gate tunneling current in MOS devices. Microelectron Reliab 46:1939–1956

Rashmi A, Kranti S, Haldar RS (2002) Gupta, "An accurate charge control model for spontaneous and piezoelectric polarization dependent two-dimensional electron gas sheet charge density of lattice-mismatched AlGaN/GaN HEMTs." Solid State Electron 46:621–630

Ravaioli U, Ferry D (1986) MODFET ensemble Monte Carlomodel including the quasi-two-dimensional electron gas. IEEE Trans Electron Device, ED-33:677–680

Saito W, Kuraguchi M, Takada Y, Tsuda K, Omura I, Ogura T (2004) High breakdown voltage UndopedAlGaN–GaN power HEMT on sapphire substrate and its demonstration for DC–DC converter application. IEEE Trans Electron Devices 51:1913–1917

Saito W, Kuraguchi M, Takada Y, Tsuda K, Omura I, Ogura T (2005) Influence of surface defect charge at AlGaN–GaN-HEMT upon Schottky gate leakage current and breakdown voltage. IEEE Trans Electron Devices 52:159–164

Sanabria C, Chakraborty A, Xu H, Rodwell MJ, Mishra UK, York RA (2006) The effect of gate leakage on the noise figure of AlGaN/GaN HEMTs. IEEE Electron Device Lett 9:19–21

Sathaiyaa DM, Karmalkar S (2006) Thermionic trap-assisted tunneling model and its application to leakage current in nitrided oxides and AlGaN/GaN high electron mobility transistors. J Appl Phys 99:093701

Shealy JR, Prunty TR, Chumbes EM, Ridley BK (2003) Growth and passivation of AlGaN/GaNheterostructures. J Cryst Growth 250:7–13

Shinohara K, et al. (2010) 220 GHz f_T and 400 GHz f_{MAX} in 40-nm GaN DH-HEMTs with re-grown ohmic. In: IEEE International Electron Devices Meeting (IEDM), 30.1.1–30.1.4

Stengel F, Noor Mohammad S, Morkoc H (1996) Theoretical investigation of electrical characteristics of AlGaN/GaN modulation doped field-effect transistors. J Appl Phys 80:3031–3042

Su LY, Lee F, Huang JJ (2014) Enhancement-mode GaN-based high-electron mobility transistors on the Si SubstrateWith a P-Type GaN Cap Layer. IEEE Trans Electron Device 61(2):460–465

Subramanian V, Mercha A, Parvais B, Dehan M, Groeseneken G, Sansen W, Decoutere S (2010) Identifying the bottlenecks to the RF performanceof FinFETs. In: 2010 23rd International Conference on VLSI Design, pp 111–116

Sugiura S, Hayashi Y, Kishimoto S, Mizutani T, Kuroda M, Ueda T, Tanaka T (2010) Fabrication of normally-off mode GaN and AlGaN/GaN MOSFETswith HfO_2 gate insulator. Solid State Electron 54:79–83

Sun X, Liu TK (2009) Scale-length assessment of the trigate bulk MOSFET design. IEEE Trans Electron Devices 56:2840–2842

Suzuki A, et al. (2014) Extremely low on-resistance Enhancement-mode GaN-based HFETusing Ge-doped regrowth technique. In: IEEE International Electron Devices Meeting (IEDM), pp 11.1.1–11.1.4

Tang Z, et al. (2013) Monolithically Integrated 600-V E/D-Mode SiN_x/AlGaN/GaN MIS-HEMTs and their applications in low-standby-power start-up circuit for switched-ModePower supplies. In: IEEE International Electron Devices Meeting (IEDM), pp 6.4.1–6.4.4

Tilak V et al (2001) Influence of barrier thickness on the high-power performance of AlGaN/GaN HEMTs. IEEE Electron Device Lett 22(11):504–506

Tomizawa K, Hashizume N (1988) Ensemble Monte Carlo simulation of an AlGaAs/GaAs heterostructure MIS-Like FET. IEEE Trans Electron Devices 35:849–856

Tomizawa M, Yoshii A, Yokoyama K (1985) Modeling for anAlGaAs /GaAs heterostructure device using Monte Carlo simulation. IEEE Electron Device Lett EDL-6:332–334

Ture E, Bruckner P, Raay FV, Quay R, Ambacher O, Alsharef M, Granzner R, Schwierz F (2015) Performance and parasitic analysis of submicron scaled tri-gate AlGaN/GaN HEMT design. In: Microwave Integrated Circuits Conference (EuMIC), 2015 10th European, pp 97–100

Turuvekere S, Karumuri N, Rahman AA, Bhattacharya A, DasGupta A, DasGupta N (2013) Gate leakage mechanisms in AlGaN/GaN and AlInN/GaN HEMTs: Comparison and modeling. IEEE Trans Electron Devices 60:3157–3165

Valizadeh P, AlOtaibi B (2011) Fin- and Island-Isolated AlGaN/GaNHFETs. IEEE Trans Electron Devices 58:1404–1407

Vollmann W (1974) Poole-Frenkel conduction in insulators of large impurity densities. Physica Status Solidi (a) 22:195–203

Widiger DJ, Kizilyalli IC, Hess K, Coleman JJ (1985) Two-dimensional transient simulation of an idealized high electron mobility transistor. IEEE Trans Electron Devices ED-32:1092–1102

Wu Y et al (2009) AlGaN/GaN MOSHEMTs with liquid-phase-deposited TiO_2 as gate dielectric. IEEE Trans Electron Devices 56:2911–2916

Xia L, Hanson A, Boles T, Jin D (2013) On reverse gate leakage current of GaN high electron mobility transistors on silicon substrate. Appl Phys Lett 102:113510

Xue H, Hussain K, Razzak T, Gaevski M, Sohel SH, Mollah S, Talesara V, Khan A, Rajan S, Lu W (2020) Al0.65Ga0.35N/Al0.4Ga0.6N micro-channel heterojunction field effect transistors with current density over 900 ma/mm. IEEE electron Device Letters 41(5):677–680

Yan D, Lu H, Cao D, Chen D, Zhang R, Zheng Y (2010a) On the reverse gate leakage current of AlGaN/GaN high electronmobility transistors. Appl Phys Lett 97:153503

Yan D, Lu H, Cao D, Chen D, Zhang R, Zheng Y (2010b) On the reverse gate leakage current of AlGaN/GaN high electron mobility transistors. Appl Phys Lett 97:153503

Ye PD et al (2005) GaN metal-oxide-semiconductor high-electron-mobility-transistor with atomic layer deposited Al_2O_3 as gate dielectric. Appl Phys Lett 86:063501

Yokoyama K, Hess K (1986) Monte Carlo study of electronictransport in $AI_{1-x}Ga_xAs$/GaAs single-well heterostructures. Physica ReviewE 33:5595–5606

Yu ET, Sullivan GJ, Asbeck PM, Wang CD, Qiao D, Lau SS (1997) Measurement of piezoelectrically induced charge in GaN/AlGaNheterostructure field-effect transistors. Appl Phys Lett 71:2794–2796

Zhang H, Miller EJ, Yu ET (2006) Analysis of leakage current mechanisms in Schottky contacts to GaN and $Al_{0.25}Ga_{0.75}N$/GaN grown by molecular-beam epitaxy. J Appl Phys 99:023703

Zhang J, Syamal B, Zhou X, Arulkumaran S, Ng GI (2014) A compact model for generic MIS-HEMTs based on the unified 2DEG density expression. IEEE Trans Electron Devices 61:314–323

Zhang K et al (2015) Enhancement-mode AlGaN/GaN HEMTs with thin and high Al composition-barrier layers using O_2 plasma implantation. Phys Status Solidi A 212(5):1081–1085

Zhu JJ, Ma XH, Hou B, Chen WW, Hao Y (2014b) Investigation of gate leakage mechanism in $Al_2O_3/Al_{0.55}Ga_{0.45}N$/GaN metal-oxide-semiconductor high-electron-mobility transistors. Appl Phys Lett 104:153510

Zhu JJ, Ma XH, Hou B, Chen WW, Hao Y (2014) Investigation of gate leakage mechanism in Al_2O_3/ $Al_{0.55}Ga_{0.45}N$/GaN metal-oxide-semiconductor high-electron-mobility transistors. Appl Phys Lett 104:153510–1–4

Zimmermann T, Cao Y, Li G, Snider G, Jena D, Xing H (2011) Subcritical barrier AlN/GaN E/D-mode HFETs and inverters. Physica Status Solidi (a) 208(7):1620–1622

GPSR Compliance

The European Union's (EU) General Product Safety Regulation (GPSR) is a set of rules that requires consumer products to be safe and our obligations to ensure this.

If you have any concerns about our products, you can contact us on ProductSafety@springernature.com

In case Publisher is established outside the EU, the EU authorized representative is:

Springer Nature Customer Service Center GmbH
Europaplatz 3
69115 Heidelberg, Germany

The manufacturer's authorised representative in the EU is Springer Nature Customer Service Centre GmbH, Europaplatz 3, 69115 Heidelberg, Germany. If you have any concerns regarding our products, please contact ProductSafety@springernature.com

Printed and bound by CPI Group (UK) Ltd, Croydon, CR0 4YY
30/09/2026
02225298-0005